New Techniques for the Detection of Nuclear and Radioactive Agents

NATO Science for Peace and Security Series

This Series presents the results of scientific meetings supported under the NATO Programme: Science for Peace and Security (SPS).

The NATO SPS Programme supports meetings in the following Key Priority areas: (1) Defence Against Terrorism; (2) Countering other Threats to Security and (3) NATO, Partner and Mediterranean Dialogue Country Priorities. The types of meeting supported are generally "Advanced Study Institutes" and "Advanced Research Workshops". The NATO SPS Series collects together the results of these meetings. The meetings are co-organized by scientists from NATO countries and scientists from NATO's "Partner" or "Mediterranean Dialogue" countries. The observations and recommendations made at the meetings, as well as the contents of the volumes in the Series, reflect those of participants and contributors only; they should not necessarily be regarded as reflecting NATO views or policy.

Advanced Study Institutes (ASI) are high-level tutorial courses intended to convey the latest developments in a subject to an advanced-level audience

Advanced Research Workshops (ARW) are expert meetings where an intense but informal exchange of views at the frontiers of a subject aims at identifying directions for future action

Following a transformation of the programme in 2006 the Series has been re-named and re-organised. Recent volumes on topics not related to security, which result from meetings supported under the programme earlier, may be found in the NATO Science Series.

The Series is published by IOS Press, Amsterdam, and Springer, Dordrecht, in conjunction with the NATO Public Diplomacy Division.

Sub-Series

A.	Chemistry and Biology	Springer
B.	Physics and Biophysics	Springer
C.	Environmental Security	Springer
D.	Information and Communication Security	IOS Press
E.	Human and Societal Dynamics	IOS Press

http://www.nato.int/science
http://www.springer.com
http://www.iospress.nl

Series B: Physics and Biophysics

New Techniques for the Detection of Nuclear and Radioactive Agents

edited by

Gul Asiye Aycik
Department of Chemistry
Mugla University
Mugla, Turkey

Springer

Published in cooperation with NATO Public Diplomacy Division

Proceedings of the NATO Advanced Training Course on
New Techniques for the Detection of Nuclear and Radioactive Agents
Mugla, Turkey
26–30 May 2008

Library of Congress Control Number: 2009920514

ISBN 978-1-4020-9599-3 (PB)
ISBN 978-1-4020-9598-6 (HB)
ISBN 978-1-4020-9600-6 (e-book)

Published by Springer,
P.O. Box 17, 3300 AA Dordrecht, The Netherlands.

www.springer.com

Printed on acid-free paper

CONTENTS

DIRECTORS

NATO ATC DIRECTOR: Gul Asiye AYCIK, Professor, Mugla University, Head of Chemistry Department, 48000 Mugla, **Turkey**, gulasiye@mu.edu.tr

NATO ATC CO DIRECTOR: Zeev B. ALFASSI, Professor, Ben Gurion University of the Negev, Nuclear Engineering Department, 84105 Beer Sheva, **Israel**, alfassi@bgu.ac.il

LIST OF SPECIALISTS

CROATIA:
Mária Ranogajec-KOMOR, Ph.D., Ruđer Bošković Institute, Bijenička, 54, 10000 Zagreb, **Croatia**, marika@irb.hr

GREECE:
Constantin PAPASTEFANOU, Professor, Aristotle University of Thessalonik Atomic and Nuclear Physics Laboratory, 54124 Thessaloniki, **Greece**, papastefanou@physics.auth.gr

ISRAEL:
Zeev B. ALFASSI, Professor, Ben Gurion University of the Negev, Nuclear Engineering Department, 84105 Beer Sheva, **Israel**, alfassi@bgu.ac.il

ITALY:
Flavia GROPPI, Professor, Milano University, Nuclear Physics Department, Health Protection Unit, Milano, **Italy**, flavia.groppi@mi.infn.it

ROMANIA:
Catalin STAN-SION, Professor, Institutul De Fizica Si Inginerie Nucleara "Horia Hulubei" Departamenul Fizica Nucleara Aplicata, **Romania**, stansion@ifin.nipne.ro

SLOVAK REPUBLIC:
Anton ŠVEC, Ph.D., Slovak Institute of Metrology, Center of Ionizing Radiations, Head of Radionuclide Metrology Group, Karloveska 63, 84255 Bratislava, **Slovak Republic**, svec@smu.gov.sk

SLOVENIA:
Borut SMODIŠ, Ph.D., Jožef Stefan Institute, Head, Radioecology Group of the Department of Environmental Sciences, **Slovenia**, Borut.Smodis@ijs.si

TURKEY:
Gul Asiye AYCIK, Professor, Mugla University, Head of Chemistry Department, 48000 Mugla, **Turkey**, gulasiye@mu.edu.tr

Yusuf GULAY, Turkish Atomic Energy Authority 9 km, 06530 Ankara, **Turkey**, yusuf.gulay@taek.gov.tr

USA:
Anthony GREENHOUSE, Ph.D., Senior Scientist Berkeley Laboratories, 788 Mickinley Avenue Oakland, CA. 94610-3833, **USA**, ag_oeha@pacbell.net

Mohammed K. ZAIDI, Idaho State University, College of Engineering, Pocatello, ID 83209-6080, **USA**, zaidmoha@isu.edu

LIST OF TRAINEES

ALBANIA:
Luan Qafmolla, Institute of Nuclear Physics (INP), **Albania**, l_qafmolla@hotmail.com

Shyqyri Arapi, Institute of Public Health (IPH), **Albania,** Sharapi2001@yahoo.com

AZERBAIJAN:
Azad Agalar Bayramov, Institute of Physics National Academy of Science of Azerbaijan G.Javide av.33, Baku AZ1143, **Azerbaijan**, bayramov_azad@mail.ru

Rauf Sardarly, Institute of Physics National Academy of Science of Azerbaijan, **Azerbaijan**, sardarli@yahoo.com

HUNGARY:
László Sági, Atomic energy Research Institute, **Hungary**, sagi@aeki.kfki.hu

KYRGYZ REPUBLIC:
Ainagul Jalilova, Institute of Biology and Pedology of National Academy of Sciences of the Kyrgyz Republic, **Kyrgyz Republic**, ainagulnas@mail.ru

Azamat Kalyevich Tynybekov, Kyrgyz Russian Slavonic University, **Kyrgyz Republic**, azamattynybekov@mail.ru

MOLDOVA:
Elena Mursa, National Agency for Regulation of Nuclear and Radiological Activities, **Moldova**, elmour_l@mail.ru

RUSSIAN FEDERATION:
Yury Sapozhnikov, Moscow State University, Chemistry Department, Radiohemistry Division, **Russian Federation**, yas@radio.chem.msu.ru

SLOVAK REPUBLIC:
Pavol Rajec, Comenius University, Faculty of Science Department of Nuclear Chemistry, **Slovak Republic**, rajec@fns.uniba.sk

TAJIKISTAN:
Anvarzhon Nasimzhonovich Akhmedov, Nuclear and Radiation Safety Agency, **Tajikistan**, anvari82@mail.ru

Shujoadin Nizomov, Oncology Research Centre, **Tajikistan**, shujo@mail.ru

THE FORMER YUGOSLAV REPUBLIC OF MACEDONIA:
Trajče Trajčev, Radiation Dosimetry Department – Public Health Protection Institute, **The Former Yugoslav Republic of Macedonia**, talepatale@gmail.com

TURKEY:
Ayse Nur Solmaz, Ankara University, Institute of Nuclear Sciences, Ankara, **Turkey**, Ayse.Solmaz@ankara.edu.tr

Aysun Ugur (Tanbay) Ege University, Institute of Nuclear Sciences, Izmir, **Turkey**, aysun.ugur@ege.edu.tr

Birkan Selcuk, Turkish Atomic Energy Authority, Saraykoy Nuclear Research and Training Center, Ankara, **Turkey**, birkan.selcuk@taek.gov.tr

Bulent Kirkan, Mugla University, Chemistry Department, Mugla, **Turkey**, b.kirkan@mu.edu.tr

Emin Yeltepe, Turkish Atomic Energy Authority, Saraykoy Nuclear Research and Training Center, Ankara, **Turkey**, emin.yeltepe@taek.gov.tr

Ezgi Eren, Mugla University, Chemistry Department, Mugla, **Turkey**, ezgi.eren@yahoo.com

Gamze Karayel, Mugla University, Chemistry Department, Mugla, **Turkey**, g.karayel@mu.edu.tr

Gokcen Topal, Celal Bayar University, Nuclear Medicine Department, Manisa, **Turkey**, yasemin.gultekin@hotmail.com

Hasan Dikmen, Turkish Atomic Energy Authority, Saraykoy Nuclear Research and Training Center, Ankara, **Turkey**, hasan.dikmen@taek.gov.tr

Meryem Seferinoglu, Turkish Atomic Energy Authority, Saraykoy Nuclear Research and Training Center, Ankara, **Turkey**, meryem.seferinoglu@taek.gov.tr

Nihat Tugluoglu, Turkish Atomic Energy Authority, Saraykoy Nuclear Research and Training Center, Ankara, **Turkey**, nihat.tugluoglu@taek.gov.tr

Pinar Esra Erden, Turkish Atomic Energy Authority, Saraykoy Nuclear Research and Training Center, Ankara, **Turkey**, esra.erden@taek.gov.tr

Sema Bilge Ocak, Turkish Atomic Energy Authority, Saraykoy Nuclear Research and Training Center, Ankara, **Turkey**, sema.ocak@taek.gov.tr

Senol Sert, Ege University, Institute of Nuclear Sciences, Ph.D. student, Izmir, **Turkey**, senol.sert@ege.edu.tr

Serdar Karadeniz, Turkish Atomic Energy Authority, Saraykoy Nuclear Research and Training Center, Ankara, **Turkey**, serdar.karadeniz@taek.gov.tr

Simay Yuksek, Turkish Atomic Energy Authority, Saraykoy Nuclear Research and Training Center, Ankara, **Turkey**, simay.yuksek@taek.gov.tr

Suleyman Inan, Ege University, Institute of Nuclear Sciences, M.Sc. student, Izmir, **Turkey**, suleyman.inan@ege.edu.tr

Turgay Karali, Ege University, Institute of Nuclear Sciences, Izmir, **Turkey**, turgay.karali@ege.edu.tr

Ugur Adnan Sevil, Turkish Atomic Energy Authority, Saraykoy Nuclear Research and Training Center, Ankara, **Turkey**, adnan.sevil@taek.gov.tr

Yasemin Parlak, Celal Bayar University, Nuclear Medicine Department, Manisa, **Turkey**, gokcentopal@hotmail.com

UKRAINE:

Arkadiy Batrak, Nuclear Regulatory Committee of Ukraine, **Ukraine**, batrak@hq.snrc.gov.ua

Namik Rashydov, Institute cell biology and genetic engineering NAS of Ukraine, **Ukraine**, nrashydov@yahoo.com

Oleg Udovyk, Ukrainian Academy of Sciences, **Ukraine**, Oleg_udovyk@hotmail.com

UZBEKISTAN:
Anvar D. Avezov, Institute of Nuclear Physics Republic of Uzbekistan, **Uzbekistan**, a.d.avezov@mail.ru

Regina Sattarova, Institute of Nuclear Physics Republic of Uzbekistan, **Uzbekistan**, regina.sattarova@stcu.int

Vitaliy Petrenko, Institute of Nuclear Physics Republic of Uzbekistan, **Uzbekistan**, petrenko@inp.uz

PREFACE

This book is based on the meeting of North Atlantic Treaty Organization (NATO) Advanced Training Course (ATC) that was devoted to *New Techniques for the Detection of Nuclear and Radioactive Agents*. The Course is being convened to review the global experience in monitoring and detecting (identification and characterization) of confiscated radioactive materials and agents. There is a growing need to evaluate and strengthen the current efforts being made in different countries. Such a basis is helpful in comparability of different laboratories generated radiological measurements.

NATO ATC was held in Mugla-Turkey on May 26–30, 2008 with purpose to bring into focus this important subject and provide a comprehensive overview to the advancement of environmental radionuclides. The book brings together contributions from the most eminent researchers in their field as specialists, and contributions from fundamental principles to materials, systems and applications in their countries by participants. A central theme of the book is focused on the new techniques based on radiation monitoring, measuring and analyzing radioactive-nuclear materials, agents and devices useful for environmental protection as quantitatively and qualitatively; preservation programmes and also nuclear material smuggling. Environmental problems caused by past or present military activities were also of great interest. These include spectroscopic techniques for low level radioactivity detection and sensing in environmental monitoring, safety and security, applications in practice.

Radionuclides, naturally occurred and/or artificially produced, in the environment are of great interest for a wide range of scientific and technological use from imaging in medical to tracking in oceanography.

More than 25 years after the Chernobyl accident and formation of new Republics from the Former Soviet Socialist Countries Block (SSCB), studies about the new techniques to detect the radionuclides became important and alternative methods in low level detection have been proposed, devised and developed. Besides that, new technologies for preventing illicit trafficking on the border of countries had also been developed.

To fashion new detection systems is necessary to improve the chances of detecting of radioactive agents. It is also a further need of nuclear smuggling. The problem of smuggling of fissile and radioactive materials is a worldwide one. Fissile materials are usable in covert nuclear proliferation problems. The fissile materials are also attractive for terrorists and can be used for terrorist purposes. Radioactive sources may be lost, through the bankruptcy of a holder of them, careless handling or other ways. These sources can be accidentally mixed in with scrap metal, leading to contaminated reclaimed metal. Contamination can also occur from other sources, such as the fallout from Chernobyl. Nations may be faced with many of problems relevant with radioactive-nuclear materials and devices.

The smuggling of fissile materials for proliferation purposes or for terrorism is a matter of increasing concern. It is of concern for countries that may be sources of the materials, for countries across which they transit, and especially for those countries that might be targets for terrorist activity.

The ongoing progress in related area including gamma spectroscopy, detection systems of alpha, beta, gamma and neutron sources as well as auxiliary methods and application areas in different countries have also been presented in this book.

I want to express my gratitude to the participants who had submitted their papers for publication. I extend my appreciation to co-director Professor Zeev B. Alfassi for his valuable contributions and to Mr. Mohammed Zaidi for his helps in coordinating this great event.

I am grateful to the North Atlantic Treaty Organization for their support of this Advanced Training Course on *New Techniques for the Detection of Nuclear and Radioactive Agents*, *Mugla, Turkey 2008*, which served as the original impetus for the publication of this book. I am thankful to Dr. Fausto Pedrazzini, NATO Science for Peace and Security Programme Directorate (Chemistry/Biology/Physics) for their support, in behalf of all participants.

I want to express my gratitude to the Mugla University for their material and moral support at every stage of the meeting.

I also convey my thanks to Springer Publishing Company for affording us the opportunity to publish this book.

Gul Asiye AYCIK
Mugla, Turkey, September 2008

OVERVIEW OF THE RADIOACTIVE AND NUCLEAR AGENTS IN THE ENVIRONMENT

GUL ASIYE AYCIK*
Mugla University, Chemistry Department 48000 Mugla, Turkey

Abstract. Radionuclides can be naturally occurred and artificially produced in the environment. Analyses and characterization of radionuclides can be achieved by using radiometric methods selected in according to the specification (i.e. energy, activity, chemical properties) of radionuclides. The radiometric methods are outlined as gamma spectrometric, alpha spectrometric, beta counting, track detector systems and in-site methods. Gamma spectrometric systems are traditional gamma spectrometry and underground gamma laboratories. Some of nuclide analysis methods including Inductively Coupled Plasma Mass spectrometry (ICPMS), Accelerator mass spectrometry (AMS), Thermal Ionization Mass Spectrometry (TIMS) and Neutron Activation Analysis (NAA) Methods are particularly useful for trace and ultra-trace analysis of environmental radionuclides.

Keywords: Environmental radionuclides, gamma spectrometry

1. Radionuclides in the Environment

Principally environmental radionuclides are originated by two sources as naturally occurring radionuclides and artificially sourced radionuclides. These include products from natural radioactive decay chains of uranium and thorium in terrain, cosmic radionuclides produced in the atmosphere from cosmic radiation as well as fallout from atmospheric weapons testing, nuclear fuel cycle activities and other low-level radiation sources such as hospitals or laboratories.

* To whom correspondence should be addressed. e-mail: gulasiye@mu.edu.tr

G.A. Aycik (ed.), *New Techniques for the Detection of Nuclear and Radioactive Agents*,

1.1. NATURALLY OCCURRING RADIONUCLIDES

A part of naturally occurring radionuclides, U-238, U-235 and Th-232, are of the primordial radionuclides found naturally on the earth's crust. These are the progenitor of a long series of radionuclides.

The other type of naturally occurring radionuclides is produced in the atmosphere, in terrain and in the water by influences from the cosmos. As a result of cosmic ray bombardment, natural radionuclide production occurs predominantly in the atmosphere with smaller contributions at the Earth's surface. These nuclides are called cosmic-ray-produced (CP) nuclides or cosmic-ray-induced radionuclides, because they are produced by nuclear reactions between high energy cosmic rays and nitrogen, oxygen and argon atoms in the air. These are nuclides such as Na-24, Mg-28, S-38, Cl-38, Cl-39, Ar-40 with short-lived as well as the nuclides with longer half-lives Be-7, H-3, C-14 and Na-22. Except in cases of an accident, cosmic radiation is mainly responsible for the very low level of atmospheric radionuclides. Fallout has been slowly deposited on the earth's surface and added to the environment.

1.2. ARTIFICIALLY PRODUCED RADIONUCLIDES

1.2.1. *Man-made radionuclides*

Man-made (anthropogenic) radionuclides are released into the environment as a result of nuclear activities, nuclear weapon testing, operation of nuclear power plants, production and reprocessing of nuclear fuel, disposal of radioactive waste and nuclear accidents at nuclear power plants. Nuclear activities are the main source of a number of long-lived radionuclides, occurring in the terrestrial environment. From the fission fragments and activation products generated by the spontaneous nuclear reactions of uranium (Oklo phenomenon) it becomes obvious that in principle, nearly all radionuclides might occur [1]. The development and proliferation of atomic weapons and atmospheric weapon tests have been the major source of radionuclides including certain actinides as U-236, Np-237, Pu-239,240 and fission products as Sr-90, Tc-99, I-129 and Cs-137. Additionally, Am-241 is a decay product of Pu-241.

1.2.2. *Technologically enhanced naturally occurring radionuclides*

A large amount of naturally occurring radionuclides (NORM) have been released to the environment by non-nuclear industries in the production of phosphate, petroleum, natural gas, heavy metals, rare earth elements and power plant burning coal. The newest detection of NORMs are mainly focused on the determination of total alpha, beta, total U, Th, Ra-226 and other radioisotopes by different techniques. The industrial regions sometimes can produce a significant amount of chemical polluting agents some of which are radioactive, especially Ra-226 discharged by the chemical fertilizer industries.

2. Analyses of Radionuclides

2.1. RADIOMETRIC METHODS

The factors influencing the algorithm of the analysis are such as the sample matrix, radionuclides to be determined, level of activity, ratios of radionuclides, the availability of sample material, the requested accuracy and

TABLE 1. List of determined natural radionuclides and applied analytical methods; D – directly, R – after radiochemical separation [2]

Nuclide	γ-Spectrometry	α-Spectrometry	LSC/β-counting	Emanometry	ICP-MS	XRF
U-238		R			D	D
Th-234	D					
Pa-234m	D					
U-234	D	R			D	
Th-230	D	R				
Ra-226	D, R			R		
Rn-222				D		
Pb-214	D					
Bi-214	D					
Pb-210	D, R		R			
Bi-210			R			
Po-210		R				
U-235	D	R			D	
Pa-231	D					
Ac-227		R				
Th-227	D	R				
Ra-223	D, R					
Rn-219	D					
Th-232		R			D	D
Ra-228	R					
Ac-228	D					
Th-228		R				
Ra-224	R					
Pb-212	D					
Tl-208	D					

detection limit, the available methods of measurement and economical aspects. According to these criteria radiometric methods such as γ-spectrometry, α-spectrometry β-Counting, LSC and in site measurements are used, Table 1.

2.1.1. *Gamma spectrometric systems*

The main advantages of γ-spectrometry are that it is able to provide absolute determination of isotopes directly, it does not require major pretreatment steps and chemical separation, it is nondestructive and fast, and it has best performance in radioactivity analysis. Best performance means; precision of the measurement, reliability of the measurements and traceability of laboratory measurements. However, the main disadvantage is often difficult corrections for self-attenuation in the sample matrix.

In gamma spectrometry method, samples were pretreated if it is necessary and conditioned to suitable geometries depending on their activities and the gamma spectrometry measurements were carried out using Ge-systems generally with the detectors of HPGe and HyPGe. Interactions of the produced charged particles, α, β and Compton electrons, with environmental matrices can give rise to enhanced levels of the continuous background.

The activity concentration is calculated as:

$$A = N/\varepsilon(E).\ f\gamma.m$$

where: A is the activity concentration at the time of the measurement of the sample [Bq/kg], N is the background corrected counting rate in the selected photopeak from the radionuclide of interest [cps], $\varepsilon(E)$ is the detection efficiency corrected for self-attenuation and coincidence summing for the (E) energy from the radionuclide of interest, $f\gamma$ is the gamma yield for the selected energy from the radionuclide of interest and m is the mass of the sample [kg].

In a radioactivity measurement the standard deviation is given by the square root of the counting rates of the sample and the background, but this is only the statistical uncertainty of counting. The measurement uncertainty must also include other sources of uncertainty such as the efficiency corrected for self-attenuation and coincidence summing, the gamma yield, the mass of the sample, tracer activity concentration, chemical recovery. The evaluation of uncertainty requires the analyst to look closely at all stages of the method and at all possible sources of uncertainty. The combined standard uncertainty is calculated through the following equation [3].

$$U_A = [\Sigma_i (\partial A/\partial x_i)^2 U^2 (x_i)]^{1/2}$$

where: ($\partial A/\partial xi$) are the sensitivity coefficients, x_i are represented by all counting rates, efficiency, gamma yield, mass and chemical recovery and Ux_i are the corresponding standard uncertainties. The decision limit, the detection limit and the minimum detectable activity (MDA) of the spectrometric system have also to be determined for sample geometry.

2.1.1.1. Efficiency Determination in Gamma Spectrometry

The absolute determination of activities of radionuclides using gamma spectrometry Ge detectors requires reliable and accurate determination of the detector's photo peak efficiencies. Because all gamma-ray emission-rate measurements made with solid-state detectors are essentially relative, the sample is either compared directly with a standard source of the same gamma emitter or indirectly with standard sources of other gamma emitters used to produce an efficiency calibration curve for a particular source-detector set-up. The experimental determination of the efficiency of Germanium detectors is the most accurate approach. The best way is to do an efficiency calibration for all the sample matrices to be measured. This often requires extensive delicate laboratory work by source preparations and measuring time. Accurate experimental calibration of Ge detectors in the range 1–100 keV is complicated because of highly absorption by matter. For efficiency calibration of gamma detectors, liquid standards are more usable because of their less self-attenuation and because of being easily homogenized and conditioned to different detector geometries.

In gamma spectrometric measurements the efficiency is based on many parameters of detector, sample and counting geometry. A simple semi empirical approximation approach can be suggested for analyses of samples in cylindrical geometry but unknown composition. The efficiency $\varepsilon(E)$ as a function of energy E split into two terms:

$$\varepsilon (E) = \varepsilon_{geom}(E) \,.\, \varepsilon_{\mu}(E)$$

$\varepsilon_{geom}(E)$ should include detector properties altogether, sample size (diameter, height) and counting geometry. $\varepsilon_{\mu}(E)$ describes the influence of sample material which is characterized by density δ and mass absorption coefficient μ' [cm^2/g]. The product of both is the linear absorption coefficient $\mu = \mu'\delta$ [cm^{-1}]. The task is to obtain $\varepsilon_{geom}(E)$ for a definite detector and a given cylinder geometry. Once found, $\varepsilon_{geom}(E)$ is valid for all samples of that geometry. The second task is to find $\varepsilon_{\mu}(E)$ for an unknown sample material. Any new sample material requires setting up its own function $\varepsilon_{\mu}(E)$. The efficiency function $\varepsilon(E)$ for a certain sample is obtained by multiplication of $\varepsilon_\mu(E)$ with $\varepsilon_{geom}(E)$ for a sufficient number of points [4]. Some calibration samples are necessary with the same geometrical parameters but of different materials, containing known activities which homogeneously distributed. Their densities are measured, their mass absorption coefficients may be unknown. These calibration samples are positioned in the counting geometry, for instance directly on the detector. The efficiency function $\varepsilon(E)$ for each sample is gained by measuring the gamma spectra and evaluating all usable gamma energy peaks. From this $\varepsilon(E)$ the common valid $\varepsilon_{geom}(E)$ is deduced.

Another method for complete evaluation of the photo peak efficiency of germanium detectors is to use Monte Carlo simulations. Semi-empirical Monte Carlo approaches provide a more accurate (within 5%) alternative for the simulation of the performance of Ge detectors at energies below 40 keV [5]. For Monte Carlo simulations uncertainty sources have to be known. These are; matrix of the sample, homogeneity of the sample, amount of the sample, density of the sample and counting geometry. Good statistical agreement is obtained between the experimental and simulated efficiencies after charge collection correction for both well type Ge-detectors and for photon energies above 40 keV [6].

Sources of Background and Its Reduction

In gamma measurements especially in the low count rate systems, the parameter of detection (LD) and the determination limits (LOD) are used to estimate the capabilities of the measurement process. These parameters are determined predominantly by the error structure arising from the backgrounds. The background on the gamma spectrometer originates from:

- The intrinsic radioactivity of the detector and its assembling
- Surrounding environment and
- The cosmic radiation

The radiations from the first two origins can be reduced by rigorous selection of the low-activity materials. Then the cosmic-ray-induced background become predominant, but it can be suppressed if the spectrometer is installed in an underground laboratory [7]. Other alternatives are the active shield method with plastic scintillating plates and coincidence electronics techniques.

Background reduction is one way to improve the sensitivity of gamma ray measurements. Gamma spectrometric measurements of low-level radionuclides are always done in close source detector geometry. This geometry causes coincidence summing effects for measurements of multi-photon emitting nuclides. The measurements are also influenced by the absorption of photons in the materials which have to be analyzed. These effects must be taken into account by correction factors with respect to an energy-specific efficiency calibration of the detector system for a given geometry and a given composition of the calibration source.

Self-absorption

As a gamma photon passes through any material (including the sample in which it is contained), it undergoes interactions that result in it losing energy. This loss of energy will result in fewer gamma-rays reaching the detector with their full energy and, thus, reduce the count rate observed under the photopeak. This phenomenon is often referred to as sample self absorption or sample attenuation. The degree of self absorption will depend on a number of factors, including sample density, elemental composition and sample size and gamma energy.

In gamma spectrometric measurements, especially for energies below 150–200 keV, self-absorption corrections are essential. Both sample density and sample composition affect the strength of the corrections required. The self-absorption correction method based on Monte Carlo simulations involving different matrices and sample-detector geometries is the precise method used for samples where the composition is properly known. The sample density and composition are input parameters to these simulations. Special attention have to be given to high Z elements which can, even in small concentrations, increase substantially the absorption power of any matrix [5].

Other methods used mainly for samples which the composition is not well known, are outlined as:

- To calculate self-absorption correction factors using the direct transmission method [8] a point source is placed on top of each sample container and counted until adequate counting statistics are achieved for the photopeaks of interest. The source is then placed in the same position on an identical empty container and the spectrum collects as before. By calculating the background-corrected photopeak count rates for both these measurements with and without the sample, efficiency correction factors are obtained for each photopeak energy and sample type.
- The method based on the preparation of the sets of gamma absorption curves by using samples with different compositions and densities and spiked with known amounts of radioactive standards is used to determine the self absorption effect. The photon attenuation is then calculated as the difference between the added and the measured activity. Since various samples are used, some scattering in the data can be observed. Photon transmittance functions are fitted to the data taking as variables the sample mass and the photon energy.

- Another convenient method is to prepare sets of gamma absorption curves. This approach involves making a series of direct transmission measurements for samples of varying densities and compositions but similar type. Sets of matching samples, both spiked and unspiked, were prepared and density correction factors determined using the direct transmission method and the spiked sample approach.
- Another experimental method commonly used is to add standardized activity solutions to the sample of interest. From the measurements without and with these solutions the self-attenuation can be directly calculated.
- However, while most laboratories measuring environmental samples by gamma-ray spectrometry can prepare the samples in essentially identical shape and size, they must contend with a large variety of sample densities and compositions.

Table 2 shows the mass attenuation coefficients for several mixtures and compounds. It has to be pointed out that the higher the photon energy is the lesser the mass attenuation coefficient comparing to each other and the attenuation for higher photon emission only a function of the density whereas for lower photon energy the attenuation varies strongly with the composition of the material. Therefore every calculation of self-attenuation in a sample for photon energies, up to 200–300 keV, requires experimental data about the elemental composition of the sample.

TABLE 2. Mass attenuation coefficients Np for several mixtures and elements as a function of the photon energy, E [9]

E (keV)	μ/ρ (cm^2/g)			
	H_2O (ρ = 1 g cm^{-3})	SiO_2 (ρ = 1.5 g cm^{-3})	Ge (ρ = 5.33 g cm^{-3})	Pb (ρ = 11.34 g cm^{-3})
50	0.2262	0.3165	3.314	8.041
100	0.1707	0.1682	0.5525	5.550
300	0.1187	0.1076	0.113	0.4026
500	0.09687	0.08738	0.08212	0.1613
1,000	0.07070	0.06366	0.05728	0.07103
1,500	0.05755	0.05185	0.04658	0.05222

Coincidence Summing

Gamma spectrometric measurements of low-level radioactivity are always done in close source detector geometry. The largest solid angles, typically of the order of 98% to 99% from 4π, are realized by measurements in a well-type detector. A consequence of the large solid angle is the presence of important coincidence summing effects for multi-photon emitting nuclides. In the well geometry, coincidence-summing effects are high and make the construction of the full energy peak efficiency curve a difficult task with a usual calibration standard, especially in the high energy range. Besides the principal emission lines at 605 and 796 keV (with emission probabilities of 97.6% and 85.4%, respectively), the sum peak at 1,401 keV can be clearly seen in the spectrum with a comparable peak height Note that there is no Cs-134 emission line with this energy and that coincidence-summing is entirely responsible for this peak.

For samples measured by using high efficiency detectors, high detection efficiency makes the calibration procedure difficult. These are:

- The magnitude of the true coincidence summing effects is directly proportional to the detection efficiency. This is a negligible effect for small volume Ge crystals but the major process for high volume well type Ge detectors.
- Affecting the detection efficiency is the self-attenuation effect in the sample. This is particularly important for low-energy photons.

Like the efficiency, coincidence-summing effects are greatly enhanced. Ignoring these effects can lead to a typical error of a factor of 2 in the determination of Co-60 activity in well-type geometry. The occurrence of coincidence-summing effects in the well-type geometry has led to the difficulties specific to that geometry. The photopeak efficiency calibration curve construction and the calculation of coincidence-summing corrections. To solve these difficulties, an effective method to help in calibrating and in computing correction factors seems to be the Monte Carlo method. Many reports on Monte Carlo calculation of the photopeak efficiency or coincidence-summing corrections have shown that the accuracy with regard to experimental results can usually reach 2–3% and 5–10% in the case of complex geometries and at low energy. By using Monte Carlo method it can be evaluated for the determination of the calibration photo-peak efficiency curve of a well-type HPGe detector and of coincidence-summing corrections [10].

From the metrological point of view, more efforts seem to be necessary, in particular in order to improve the coincidence summing corrections for the 'high-efficiency measurement geometries'. The estimation of the measurement uncertainties should be performed consistently in accordance with an international standard in order to achieve a better comparability of the results [11].

Underground Gamma Laboratories

The underground laboratories are generally situated at a depth of 35 m we (water equivalent). These laboratories are equipped with ventilation and air conditioning maintaining overpressure and stable humidity and temperature levels all year round. This laboratory is equipped with a special airflow system with filters that prevent the introduction and accumulation of Rn-222 from the building materials and it is protected against external electromagnetic waves by a Faraday shield. The background measured with a 30% well-type 35 m underground installed detector is reduced by a factor of about 50 in comparison with above ground installations. In comparison with results from 225 m depth underground laboratory, 35 m depth laboratory's background is higher by a factor of about 10. A large amount of the remaining background in the gamma spectrometry is still attributable to the neutrons from cosmic rays and the surrounded rocks [7]. Required and measured detection limits for water samples in 2 L of Marinelli beakers as reference volume without any chemical separation are shown in Table 3.

TABLE 3. Required and measured detection limits for water samples in 2 L of Marinelli beakers [7]

	Detection limits (mBq/L)		
	Required depth	Above ground 0 m	Under ground 110 m
Ra-226	40	310	52
Pb-210	40	920	150
Ra-224	20	31	5
Ra-228	20	55	10

2.1.2. *Alpha spectrometry systems*

The alpha spectrometry systems are used to measure natural (polonium, uranium, thorium) and anthropogenic (plutonium, americium, neptunium) alpha emitters. The elements are first separated and purified chemically before being either electroplated or precipitated in order to be in the proper geometry needed for alpha particles counting. The samples are counted under vacuum and, because the levels are low, good background conditions are requested as well as reliable calibration parameters.

The main advantage of alpha spectrometry systems is the excellent low limit of detection. The main disadvantages are being destructive, the need of radiochemical separation, high time consuming and many chemical treatment procedures to prepare alpha particle source.

2.1.3. *Beta counting systems*

The main advantage is the relatively low limit of detection similarly to alpha spectrometry systems. The main disadvantages are being destructive, the need of radiochemical separation and beta-particle source preparation. Liquid Scintillation Counter (LSC) is mainly used for determination of low energy beta emitters, H-3 and C-14. LSC has the big advantage over most other methods of a 4π geometry and therefore high counting efficiencies for both beta and alpha emitters. Sample preparation is in most cases easy and there is no self absorption to correct for. Many extraction methods can be used to give a solution ready to count. Alpha-emitters can also be measured by LSC, but routine applications were scarce. Alpha-emitters have a counting efficiency in LSC of approximately 100%. This is the big advantage over solid state spectrometry. The big disadvantage is the relatively poor energy resolution [12].

2.2. NUCLIDE ANALYSIS METHODS

Mass spectrometric (MS) and spectroscopic techniques, including resonance-ionization MS (RIMS), accelerator MS (AMS), inductively coupled plasma-MS (ICP-MS), Capillary electrophoresis-inductively coupled plasma mass spectrometry (CE-ICP-MS), thermal ionization mass spectrometry (TIMS), time-resolved laser-induced fluorescence spectroscopy (TRLIF), Electro spray ionization mass spectroscopy (ESI-MS) and inductively coupled plasma-atomic emission spectrometry (ICP-AES) have all been used for the quantitative determination of long-lived radionuclides.

Accelerator Mass Spectrometry: Accelerator mass spectrometry (AMS) is an analytical technique the detection of long-lived radionuclides which cannot be practically analyzed with decay counting or conventional mass spectrometry. Long-lived radionuclides, such as C-14 (5.73 ka), Be-10 (1.6 Ma), Al-26 (720 ka), Cl-36 (301 ka), Ca-41 (104 ka), and I-129 (16 Ma) are commonly analyzed with an isotopic sensitivity sensitivities down to as low as 10^{-15}. Other low-level radionuclides occurring in nature at ultra-trace levels or produced by nuclear technology, including U-236 and Pu-244, can also be analyzed by AMS. Unfortunately, the use of AMS is limited by the expensive accelerator technology required [13].

Time-Resolved Laser-Induced Spectroscopy: TRLIF has been used to investigate speciation of the fluorescent actinides (mostly limited to U(VI), Am(III) and Cm(III) and lanthanides in simple aqueous systems [14].

Nuclear Magnetic Resonance, Raman and UV–Vis Spectroscopy: Raman spectroscopy has been used to characterize radionuclides based on vibrational spectral observations. However this technique has traditionally had limited application to environmental radionuclide analysis due to a requirement for relatively high analyte concentration (mM–M) and matrix interferences It has been used for simple observations of uranyl complexation with in different solvent media. Recent studies include the use of C-13 NMR spectroscopy to characterize the stable uranyl(V)-carbonate complex [15] and H-1 NMR has been used to provide structural information about the protonation of a Th-desmethyldesferrithiocin complex [16].

X-ray Absorbance Spectroscopy: The development of synchrotron X-ray sources for research in the 1990s led to a series of qualitative analyses making use of X-ray absorbance spectroscopy, especially in uranium-speciation analysis [17] .

Electro Spray Ionization Mass Spectroscopy: ESI-MS was recently applied to the determination of radionuclides, where the applicability of the method initially compared favorably especially with actinide-speciation [18].

Instrumental Neutron Activation Analysis: Among nuclear analytical techniques, neutron activation analysis (NAA) is particularly useful for trace and ultra-trace analysis of samples. In trace element work associated with pollution, instrumental NAA is a powerful technique for multi-element surveys, in particular when combined with other spectroscopic techniques. Analyses and characterization of radionuclides can be achieved by using radiometric and nuclide analysis methods selected in according to the specification (energy, activity, chemical properties, etc.) of radionuclides.

References

1. Tykva R (2004) Sources of environmental radionuclides and recent results in analyses of bioaccumulation, a review, Nukleonika 49:3–7
2. Bothe M, Quality Assurance in the Analysis of Natural Radionuclides-Measures and Results (1996) Methods and Applications of Low-Level Radioactivity Measurements, Proceedings of Workshop, Rossendorf/Dresden, Editor: J. Fietz
3. Taylor BN, Kuyatt CE (1994) Guidelines for Evaluating and Expressing the Uncertainty of NIST Measurement Results, NIST Technical Note 1297
4. Bothe M, Approximate Determination of Efficiency for Activity Measurements of Cylindrical Samples (1996) Methods and Applications of Low-Level Radioactivity Measurements, Proceedings of Workshop Rossendorf/Dresden, Editor: J. Fietz
5. Hernández F, El-Daoushy F (2002) Semi-empirical method for self-absorption correction of photons with energies as low as 10 keV in environmental samples, Nucl Instrum Meth Phys Res A 484:625–641
6. Hernández F, El-Daoushy F (2003) Accounting for incomplete charge collection in Monte Carlo simulations of the efficiency of well type Ge detectors, Nucl Instrum Meth Phys Res A 498:340–351
7. Niese S, Köhler M, Low-level counting techniques in the underground laboratory "Felsenkeller" in Dresden (1996) Methods and Applications of Low-Level Radioactivity Measurements, Proceedings of Workshop Rossendorf/Dresden, Editor: J. Fietz
8. Cutshall NH, Larsen IL, Olsen R (1983) Direct analysis of Pb-210 in sediment samples: self-absorption corrections, Nucl Instrum Meth 206:309–312

9. McMahon CA, Fegan MF, Wong J, Long SC, Ryan TP, Colgan PA (2004) Determination of self-absorption corrections for gamma analysis of environmental samples: comparing gammaabsorption curves and spiked matrix-matched samples, Appl Radiat Isotopes 60:571–577
10. Laborie JM, Le Petit G, Abt D, Girard M (2000) Monte Carlo calculation of the efficiency calibration curve and coincidence-summing corrections in low-level gamma-ray spectrometry using well-type HPGe detectors, Appl Radiat Isotopes, 53:57–62
11. Wershofen H et al. (2008) An inter-laboratory comparison of low-level measurements in ground-level aerosol monitoring, Appl Radiat Isotopes 66:737–741
12. Schönhofer F, Low level measurements with liquid scintillation spectrometry development and application (1996) Methods and Applications of Low-Level Radioactivity Measurements, Proceedings of Workshop Rossendorf/Dresden, Editor: J. Fietz
13. Tuniz C, Norton G (2008) Accelerator mass spectrometry: new trends and applications, Nucl Instrum Meth Phys Res B 266:1837–1845
14. Geipel G (2006) Some aspects of actinide speciation by laser-induced spectroscopy, Coordin Chem Rev 250:844–854
15. Mizuoka K, Grenthe I, Ikeda Y (2005) Structural and kinetic studies on uranyl(V) carbonate complex using 13C NMR spectroscopy, Inorg chem 44:4472–4474
16. Jiang J, Renshaw JC, Sarsfield MJ, Livens FR, Collison D, Charnock JM, Eccles H (2003) Solution chemistry of uranyl ion with iminodiacetate and oxydiacetate: a combined NMREXAFS and potentiometry calorimetry study, Inorg Chem 42:1233–1240
17. Tokunaga TK, Real-time x-ray absorption spectroscopy of uranium, iron, and manganese in contaminated sediments during bioreduction (2008) Lawrence Berkeley National Laboratory (University of California) LBNL-718E
18. Pasilis S, Somogyi A, Herrmann K, Pemberton JE (2006) Ions generated from uranyl nitrate solutions by electrospray ionization (ESI) and detected with Fourier transform ion-cyclotron resonance (FT-ICR) mass spectrometry, J Am Soc Mass Spectrom 17:230–240

THE LOCALIZATION OF GAMMA EMITTING POINT SOURCE IN A LARGE MEDIUM BY MULTI DETECTORS MEASUREMENT

ZEEV B. ALFASSI*
Department of Nuclear Engineering, Ben Gurion University, Beer Sheva, 84105, Israel

Abstract. This paper describes a method for localization of a small radioactive source in a large medium. The method is based on calibration of various points in the medium with several detectors and using the count-rates of the various detectors as components of a vector. The unknown source vector is compared to all the vectors of the calibration (library) by vector analysis.

Keywords: Localization, a radioactive source, multi-detector measurement

1. Introduction

The search for a small radioactive source and its location and identification in a relatively large bulky sample like a human being, a box or a suitcase was studied by gamma spectrometry using the full-energy peak of the γ photons [1–8]. It was found that by the use of two detectors a radioactive source positioned on the line connecting the two detectors can be located [1–4]. If the line on which the source is located is known there are only two unknowns, the position on the line and the source activity; thus two measurements (detectors) are sufficient. In the general case there are four unknowns, the three coordinates in space and the activity of the source. Hence four detectors are the minimal number to obtain all the four unknowns. An extension of this study to four detectors shows that in case of approximately known shape and contents (lung phantom) the gamma emitting source cannot be localized by writing four equations for the four unknowns, especially when the distance of the source-detectors is small.

* To whom correspondence should be addressed. e-mail: alfassi@bgu.ac.il

This is due to the fact that for small detector-source distances there are not simple equation for the count-rate versus the source-detector distance. For large detector distances the equation using the virtual point detector concept (see the appropriate chapter in this book) is more accurate and hence the equations can be used more accurately. Thus this method can be used for large containers but not for measuring emitters of low-energy photons in the human body as it is required for example in Whole Body Counter (WBC) or in the more specific case of the lung counter. In this case another method is necessary [5–8], however this method can be used only for specific geometry samples since the method is based on calibration of each sample with small radioactive source positioned in different places in the sample (or a phantom of the sample).

2. Theoretical Background

The subject of source location identification by multi detector measurement is a special case of multi-parameter identification [9, 10]. This problem was discussed in detail for unknown molecule identification from their electron impact (EI) mass spectra using a catalogue (library) of the EI mass spectra of many molecules. Rasmussen and Isenhour [11] studied several criteria to find the best fit. The basic test is the sum of the absolute intensity differences $\sum_{j=1}^{n} \left| I_{L,j},-I_{u,j} \right|$, were the subscripts L and u stand for library and unknown, respectively. Another method is the geometric (Euclidean) distance: $\sqrt{\sum_{j=1}^{n} (I_{L,j}-I_{u,j})^2}$. Stein and Scott [12] gave a clearer presentation of the search methods. They suggested looking on the ion intensities as the components of a vector and normalizing the vectors to unit length. Each individual normalized vector can be described as a single point on a sphere with unit radius in a hyperspace of n dimensions, where n is the number of components of the vector. If two vectors are identical in all the values of their components, they will be a perfect "match" and will be the same point in the hyperspace. However, because instrumental variability and instability and because of the statistical nature of the measurements, very rarely the unknown point will coincide with a point of the library of standards. The similarity of two normalized vectors can be seen as the inverse of their distance. Rasmussen and Isenhour [11] suggested two criteria for Matching Factor (MF):

Geometric distance:
$$MF_g = \left[1 + \frac{\sum (u_i - s_i)^2}{\sum u_i}\right]^{-1} \tag{1}$$

Absolute difference:
$$MF_d = \left[1 + \frac{\sum |u_i - s_i|}{\sum u_i}\right]^{-1} \tag{2}$$

where u_i *and* s_i are the $i-th$ components of the unknown sample and of the standard from the library, respectively.

Stein and Scott [12] following a technical report of Finnigan Corporation [13] suggested a different approach to test for similarity of the vectors. This method is based on the calculation of the cosine of the angle between the two vectors, through the use of their scalar product:

$$\underline{x} \cdot \underline{y} = |\underline{x}| \cdot |\underline{y}| \cdot \cos\theta \tag{3}$$

Where $|\underline{x}|$ and $|\underline{y}|$ are the lengths of the two vectors and the angle between them is θ. Both the scalar product and the lengths can be calculated from the components of the vectors:

$$\underline{x} \cdot \underline{y} = \sum x_i y_i \quad ; \quad |\underline{x}| = \sqrt{\Sigma x_i^2} \text{ and } |\underline{y}| = \sqrt{\Sigma y_i^2} \tag{4}$$

Where the summation is done over all the components of the two vectors.

Hence:
$$\cos\theta = \frac{\Sigma x_i \cdot y_i}{\sqrt{\Sigma x_1^2 \cdot \Sigma y_i^2}} \tag{5}$$

Thus the third matching factor was defined by Stein and Scott [12] as:

$$MF_\theta = \cos^2\theta \tag{6}$$

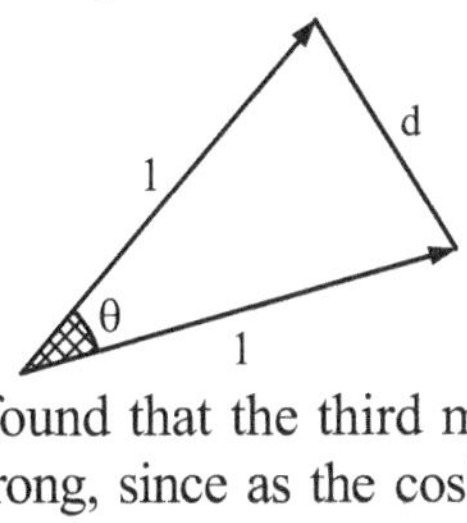

Stein and Scott [12] found that the third matching factor is the best one. However this must be wrong, since as the cosine theorem relates the distance between the heads of the two vectors – d and the cosine of the angle between the two vectors for normalized vectors:

$$d^2 = 1^2 + 1^2 - 2 \cdot 1 \cdot 1 \cdot \cos\theta \;\Rightarrow\; d^2 = 2(1 - \cos\theta) \;\Rightarrow\; \cos\theta = \; 1 - 0.5d^2 \quad (7)$$

The distance and cosθ must lead to the dame library vector for the unknown. The error of Stein and Scott [12] was that Equation (1) normalizes wrongly the vectors [14]. Each vector, either the unknown or each of the library vectors, must be of unit length by dividing by its own length. In Equation (1) Stein and Scott [12] normalizes both the unknown and the library (standards) by the length of the unknown.

A commercial case of four detectors is the Lung Counter, which is used to measure radioactive contamination in the lung. In normal routine measurements the activity of the contamination is calculated by assuming that the lung is contaminated uniformly, but it was shown that if the contamination is point wise [15] this assumption can lead to large error. This chapter shows that our method can almost eliminate this error.

3. The Experimental System and the Method

The measurements were performed by the NRC-Negev Lung Counter system, which consists of four Semi-Planar HPGe detectors, manufactured by Eurisys Mesures – France. The active area of each detector is 3,800 mm^2, with a diameter of 70 mm and a thickness of 25 mm. Each detector is equipped with a 1.2 mm thick carbon window, and is mounted in a cryostat of type SBF-00PA6, which can contain 6 L of liquid nitrogen. The liquid nitrogen is supplied automatically from a central container by a computer controlled system. The Lung Counter system is placed in an "old" low background steel shielding, built of 25 cm thick blocks, lined with 1 mm cadmium, 1 mm lead, 2 mm copper and 5 mm polypropylene. The air supply to the room passes through an absolute filter system.

The detectors are calibrated using a realistic phantom, designed by the Lawrence Livermore National Laboratory [13] and manufactured by Humanoid, The phantom can adapt either loaded lungs containing calibrated radioactive material, which is dispersed homogeneously in the lungs volume or inert (unloaded) lungs, made of the same material, but having cylindrical holes. In one of the holes a point source of natural uranium oxide was placed. The point source was 1.85 kBq natural uranium oxide U_3O_8 (99.3% ^{238}U and 0.7% ^{235}U) in a polyethylene capsule (diameter 3 mm and length 5 mm). All other holes are filled with cylinders made of the same tissue equivalent plastic as the lungs, so that the absorbing properties of the lungs remain practically unchanged and fully simulates the human lungs. The unloaded lungs have 56 points for placing the sources at different locations distributed all over the lungs. Twenty-eight points are located on the upper surface of the lungs and 28 on the lower surface. The four detectors assembly placed over the phantom is shown in Figure 1.

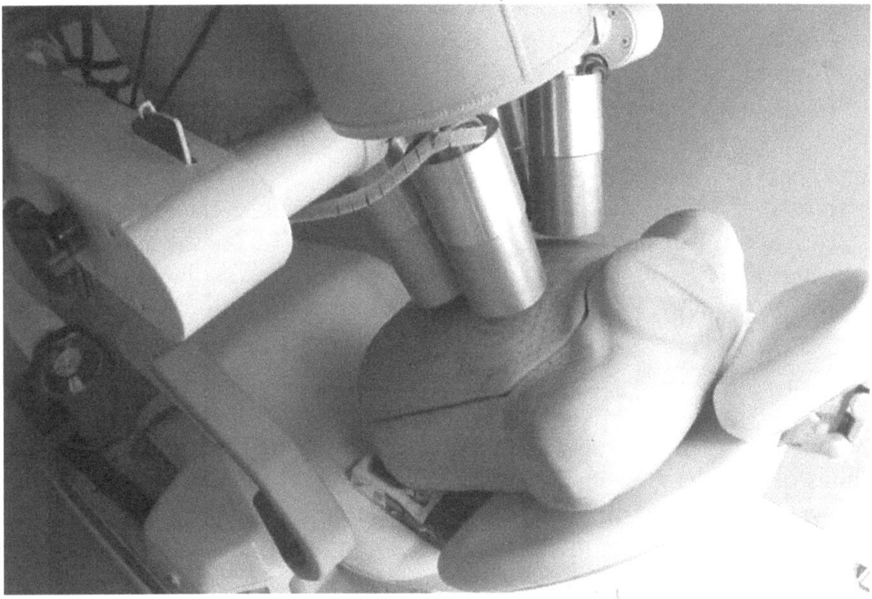

Figure 1. The four lung counterplaced over the phantom

The information from each detector was analyzed separately by a multi-channel analyzer, which is integrated in a computer system. At the end of the acquisition, all spectra from the detectors were analyzed by the computer. For quantitative analysis of the content of natural uranium at the different points, the areas of the gamma peaks were used. 224 measurements were done, four measurements of 3 h for each of the 56 points. Average counts of the four results for each of the detectors, for the 56 source positions, were used in order to produce the library data base.

Based on the minimum distance (or the maximum of cos θ), a "predicted point" was determined, and compared to the actual point location. If the predicted point activity was evaluated based on the system efficiency for the "predicted point" and was compared to the known source activity. The difference between these activities was defined as the error for the specific evaluation.

4. Results

Two gamma lines were used for the measurements of the activity (amount) of the uranium contamination, the 92 keV photons are due to a daughter of ^{238}U, while the 186 keV line is due to ^{235}U. The 186 keV line leads to 81%

of hits and an average error of activity of 10.8%, while the 92 keV line yields 78% hits and an average error of activity of 18.6% The lower accuracy of the 92 keV line is due to higher absorption of the photons in the phantom leading to larger statistical error of the count rates. One problem of this method is that we are using each measurement both as the test case and in the calculation of the vectors of the library, by averaging the four repeated measurements. If we use for the library the average of only two measurements, while using the other two measurements as the test case the percentage of hits is smaller, mainly due to the larger statistical error of the vectors of the library. In this case it was found that the average percentage of hits (from the six possible combinations of two from four) is 69% (186 keV) and 64% (92 keV).

4.1. HOMOGENOUS VERSUS HETEROGENEOUS DISTRIBUTION

Application of this method to a homogeneously distributed source in the lungs of the phantom results in the best "fit" for one of the points in the center. The error in the activity is quite large in this case, almost 40%. Thus, it seems that this method can work only when it is known that there is only a small region of radioactive contamination. However, this problem can be solved by adding another vector to the set of the standards. This vector is the response of the four detectors to homogeneously distributed radionuclide, which can be defined as a "virtual point". Another possibility for a vector representing the homogenously distributed lungs is to take the arithmetic mean of the 56 calibrating points for each one of the four detectors. With this extended set of standards, the "guessed point" for homogenously lungs was always found by this virtual vector and the error in the calculated activity was below 2%. Thus, one can actually distinguish between a homogeneous radioactive distribution and a point contamination. Even if only one lung is contaminated the method leads to this "virtual point" with error of activity of only 5–6%.

4.2. IMPROVEMENTS OF THE METHOD

The method can be improved by two ways. The first way is to use weighted count-rates instead of using the directly measured count-rates as the vectors components. Thus, for example Stein [16] found that the identification in EI-MS gets better if, instead of raw ion intensities, the values (ion intensity $\oplus$ ion mass)$^{1/2}$ are taken as the components of the vectors; that increases the weights of low-intensity peaks produced by ions of big masses. The second way to increase the accuracy of this method is by increasing the dimension of the vectors. The increase in the vector dimensionality can be done by either using more detectors or by using the count-rates of more than one line. In the second case both gamma lines should be due to the same nuclide [6]. In case of two gamma lines due to two isotopes of the same element a double dimension vector can be used but in this case, the count-rates of each isotope should be normalized separately [8].

4.3. RESULTS DUE TO IMPROVEMENTS

Table 1 shows the various matching factors due to different weighing factors used in this study. Table 2 shows the percentage of "hits" and the average error in the activity calculated for the 224 measurements done in this study. The smallest errors in the activity are due to weighing factors 11 and 14.

TABLE 1. The various matching factors

1	2	3	4	5
$\cos\theta=\frac{(\sum L_i\cdot U_i)^2}{\sum L_i^2\cdot\sum U_i^2}$	$\lvert r\rvert=\frac{\sum(U_i-\bar{U})(L_i-\bar{L})}{\sqrt{\sum(U_i-\bar{U})^2\sum(L_i-\bar{L})^2}}$	$\sum_{i=1}^{n} L_i\cdot\lvert U_i-L_i\rvert$	$\sum_{i=1}^{n} L_i\cdot(U_i-L_i)^2$	$\sum_{i=1}^{n}\left\lvert\frac{U_i}{L_i}-1\right\rvert$
6	**7**	**8**	**9**	**10**
$\sum_{i=1}^{n}\left(\frac{U_i}{L_i}-1\right)^2$	$\sum_{i=1}^{n}\frac{\lvert U_i-L_i\rvert}{U_i}$	$\sum_{i=1}^{n}\frac{\lvert U_i-L_i\rvert}{\sqrt{U_i\cdot L_i}}$	$\sqrt{\sum_{i=1}^{n}(L_i-U_i)^2}$	$\sum_{i=1}^{n}\lvert L_i-U_i\rvert$
11	**12**	**13**	**14**	**15**
$\sum_{i=1}^{n}\frac{(U_i-L_i)^2}{(U_i+L_i)}$	$\sum_{i=1}^{n}\left[\frac{U_i-L_i}{U_i+L_i}\right]^2$	$\sum_{i=1}^{n}\frac{(U_i-L_i)^2}{U_i\cdot L_i}$	$\sum_{i=1}^{n}\frac{(U_i-L_i)^2}{\sqrt{U_i\cdot L_i}}$	$\sum_{i=1}^{n}\frac{\lvert U_i-L_i\rvert}{U_i+L_i}$

In both of them the distance between the heads of the vectors is divided by the average (either arithmetic or a geometric one) of the count rates of the library and the unknown. Since the criterion is the minimum distance, it gives more weight to the high count rates which have lower statistical errors. Table 2 shows also that when using eight dimension vectors constructed from both the count-rates of the two lines (92 and 186 keV) the errors are considerably smaller. Table 3 gives the results due to increased number of detectors. In this case several detectors were in front of the phantom and assigned by A, B, C and D. Detector A is located opposite the upper part of the right lung. Detector B is located opposite the upper part of the left lung. Detector C is located opposite the lower part of the left lung, and detector D is located opposite the lower part of the right lung. The detectors in the back have similar notation but with the prefix

B; thus detector BA is in the back of the lungs opposite the upper part of the right lung. In order to see the effect of increasing the number of detectors we choose the case that the library is composed of only 112 measurements (two repeating measurements), while the test cases were the other 112 measurements. It can clearly be seen, mainly in the case of using only the 186 keV results, that increasing the number of detectors increases the percentage of hits and reduces the average error of the calculated activity. This effect is not only due to the increase in the number of detectors but also due to the fact that some of the detectors are in the back.

TABLE 2. The results obtained when either the 186 or the 92 keV peaks where used separately and when both peaks where used simultaneously, using the library of 224 measurements

	92 keV		186 keV		92 + 186 keV	
Critreion number	Numberof "hits" from 224 measurements (%)	Average error in activity calculation (%)	Number of "hits" from 224 measurements (%)	Average error in activity calculation (%)	Number of "hits" from 224 measurements (%)	Average error in activity calculation (%)
1	78	17.6	81	10.9	90	5.5
2	67	23.8	71	16.8	73	9.5
3	67	19.2	72	20.4	74	13.0
4	70	21.0	77	16.1	79	10.7
5	79	14.6	81	11.4	88	6.0
6	81	14.0	80	11.0	89	5.5
7	82	11.3	81	11.4	89	5.5
8	82	11.3	80	12.3	90	5.4
9	78	17.6	81	10.9	90	5.5
10	80	14.5	82	10.4	88	6.3
11	82	11.2	84	9.8	94	4.1
12	82	13.9	81	10.9	90	5.7
13	82	13.9	81	10.9	90	5.7
14	82	11.3	82	9.5	94	3.8
15	83	11.3	80	12.3	90	5.4

5. Possible Other Uses of the Method of Vector Analysis

This method of vector analysis can be used for every multi-parameter analysis. Some examples where it can be applied are:

1. Archeologists know that the source of antique items is not always where it was found, as many items in the ancient days were imported from other areas. In some cases, the origin of clay potteries can be determined by its stylistic appearance. However, in many cases this stylistic approach does not always lead to a definite answer as hybrids of different stylistic

TABLE 3. The average percentage of "hits" and the average error in the activity calculation with 5, 6, 7 and 8 detectors for 56 points (112 measurements, average of six combinations)

Number of detectors	Detectors	92 + 186 keV		186 keV	
		Number of "hits"(%)	Error in activity(%)	Number of "hits" (%)	Error in activity (%)
5	A, B, C, D, BD	95.8 ± 1.6	2.1	92.5 ± 2.3	2.6
5	A, B, C, D, BB	97.2 ± 1.2	2.1	93.7 ± 1.5	2.9
5	A, B, C, BA, BB	95.7 ± 1.6	2.4	92.8 ± 3.5	3.1
5	A, B, D, BA, BC	99.0 ± 0.6	1.7	98.0 ± 0.6	1.9
5	A, C, D, BA, BB	95.8 ± 1.8	2.8	95.8 ± 1.6	2.7
5	A, C, D, BC, BD	98.5 ± 0.5	1.8	97.5 ± 1.4	1.8
5	B, C, D, BA, BB	96.0 ± 0.9	2.3	96.2 ± 1.2	2.0
5	B, C, D, BB, BD	97.8 ± 1.0	1.9	96.5 ± 1.6	1.9
5	B, C, D, BC, BD	98.8 ± 0.8	1.6	96.5 ± 1.2	1.9
6	A, B, D, BA, BB, BC	99.5 ± 0.8	1.5	98.8 ± 0.8	1.5
6	A, B, D, BA, BB, BD	99.2 ± 0.8	1.7	97.8 ± 1.0	1.8
6	B, C, D, BA, BB, BD	98.0 ± 0.6	1.8	96.5 ± 1.0	1.8
6	A, B, C, D, BA, BB	97.3 ± 1.0	1.9	97.8 ± 1.2	1.8
6	A, B, C, D, BC, BD	99.2 ± 0.4	1.4	99.2 ± 0.8	1.4
7	A, B, C, D, BA, BB, BC	99.3 ± 0.5	1.5	98.7 ± 0.5	1.5
7	A, B, C, D, BB, BC, BD	99.7 ± 0.5	1.3	99.2 ± 0.8	1.3
7	A, B, C, D, BA, BB, BD	99.3 ± 0.8	1.6	98.3 ± 1.2	1.6
7	A, B, C, D, BA, BC, BD	99.2 ± 0.8	1.4	99.0 ± 0.6	1.4
8	A, B, C, D, BA, BB, BC, BD	99.5 ±0.5	1.3	99.0 ± 0.6	1.3

features were found, as for example Mycenaean decorations on bowls of typical Cypriote shape. Thus, it is clear that independent methods of determining pottery proveniences are required. The most fruitful approach was found to be to measure the elemental composition of the pottery. Since the major elements are the same for all clays from different sources, the origin of antique potteries can be identified only from the composition of its trace elements. Only specific concentrations of several trace elements can lead to unequivocal identification of the source of the pottery [17–20]. Let us say that we are able to measure the concentration of six trace elements whose concentrations vary from one location to another (in many experiments more than ten trace elements are determined).

In order to identify the source of the clay we take a sample of earth or of known local old pottery from several, let us say ten possible locations, and determine in each of them the concentration of those six trace elements. The issue of identification is to determine to which of these ten sets of six numbers of the standard ceramics, the unknown combination of six numbers of the unknown pottery fits best. Similar trace elements determination for the identification of the origin was done for coins [21], glass artifacts [22], obsidian tools [23] and other archaeological items.

2. A new system to analyze the concentration of various gases and vapors is the "electronic nose". This term refers to a system that mimics human action by combining the responses of an array of chemical sensors, which have differing sensitivities towards various compounds or classes of compounds. It is used as a reliable, quick-acting, low-cost detector for various gases and volatile compounds [24–27]. Several hundreds papers appear in the literature in the last 10 years about the development of various "electronic noses" and their applications. It is used for example for the determination of spoilage of various food products such as processed milk, vacuum-packed meat, seafood, beer, tea, coffee, apple aroma, etc. Various combinations of sensors are used in these gas and vapor detectors. For example, one kind of "electronic nose" consists of different kinds of semi conducting metal oxide sensing films that are deposited on silicon elements, which can be heated ("microhotplate"). A change in the physical properties, e.g. electrical conductance, at a given temperature or at specific time after the beginning of the process, can result from the chemical interactions of the adsorbed gas molecules with the metal oxide surface. Each detector, in the array of the detectors, measures the change of some physical property. The data set formed by the measured physical properties is the source of analytical information. Each of the sensor elements may be made from different semiconductor materials, different microstructures, or have different catalytic metal

additives deposited on it, resulting in unique interactions between analyte and individual sensor elements. Thus, for example the "electronic nose" designed at NIST [28] forms an 80-component response database (four sensing films at 20 operating temperatures) measuring the ratios of the response onset and recovery time constants of the different sensor materials at different temperatures. Measuring these 80 values for each standard combination of gases (types and concentrations) and for the unknown, allows us to find to which of the standards the unknown fits the best. The sensitivity of the analysis will depend on the number of the standards used in the calibration.

References

1. Presler O, Pelled O, German U, Leichter Y, Alfassi ZB (2002) Determination of a source in a box with two detectors. I. Non-absorbing media. Instr Meth Phys Res 491A:314–325
2. Presler O, German U, Alfassi ZB (2004) Location-independent determination of the activity of a point source in absorbing media. Appl Radiat Isotopes 60:213–216
3. Presler O, German U, Golan H, Alfassi ZB (2004) Determination of a source in a box with two detectors: the general case. Instr Meth Phys Res A 527:632–647
4. Presler O, German U, Alfassi ZB (2005) Radioactive point source localization in a bulky volume. Instr Meth Phys Res A 547:628–637
5. Pelled O, Tsroya S, German U, Haquin G, Alfassi ZB (2004) Locating a "hot spot" in the lungs when using an array of four HPGe detectors. Appl Radiat Isotopes 61:107–111
6. Pelled O, Tsroya S, German U, Abraham A, Alfassi ZB (2005) Improved localization of a "hot spot" in the lungs for an array of four HPGe detectors – the simultaneous use of two gamma energies. Instr Meth Phys Res A 551:553–562
7. Alfassi ZB, Bonardi ML, German U, Groppi F, Pelled O (2007) Improved measurement of the activity of a radioactive point source inside a bulky medium by the use of several detectors – the special case of lung counter. Appl Radiat Isotopes 65:234–238
8. Pelled O, Tsroya S, German U, Abraham A, Alfassi ZB (2008) Localization of a "hot spot" of uranium in the lungs by an array of four HPGe detectors. The effect of variation in the isotopic composition. Instr Meth Phys Res A 584:406–411
9. Alfassi ZB (2005) The vector analysis of multi measurements identification. J Radioanal Nucl Chem 266:245–250
10. Alfassi ZB (2005) Identification of analyte by multi-measurement analysis. In: Statistical Treatment of Analytical Data, Z.B. Alfassi Z. Boger, Y. Ronen (Eds), Blackwell, Oxford
11. Rasmussen GT, Isenhour TL (1979) The evaluation of mass spectral search algorithms. J Chem Inf Comput Sci 19:179–186
12. Stein SE, Scott DR (1994) Optimization and testing of mass spectral library search algorithms for compound identification. J Am Soc Mass Spectrom 5:859–866
13. Sokolow S, Karnofsky J, Gustafson P (1978) The Finnigan Library Search Program, Finnigan Application Report 2. San Jose, CA
14. Alfassi ZB (2004) On the normalization of a mass spectrum for comparison of two spectra. J Am Soc Mass Spectrom 15:385–387

15. Pelled O, German U, Pollak G, Alfassi ZB (2006) Estimation of errors due to inhomogeneous distribution of radionuclides in lungs. Instr Meth Phys Res A 564:491–495
16. Stein SE (1999) An integrated method for spectrum extraction and compound identification from gas chromatography/mass spectrometry data. J Am Soc Mass Spectrom 10:770–781
17. Wilson AL (1978) Elemental analysis of pottery in study of its provenance – review. J Archeol Sci 5:219–236
18. Perlman I (1981) Applications to archaelogy. In: Nondestructive Activation Analysis, S. Amiel (Ed.), Elsevier, Amsterdam
19. Kuleff I, Djingova R (1990) Activation Analysis in Archaeology in Activation Analysis, Z.B. Alfassi (Ed.), Vol. 2, CRC press, Boca Raton, FL
20. Zhang ZQ, Cheng HS, Xia HN, Jiang JC, Gao JC, Yang FJ (2002) Trace elements measurement by PIXE in the appraisal of the ancient potteries. Nucl Instrum Meth Phys Res B:Beam Interactions with Materials and Atoms B 190:497–500
21. Constantinescu B, Sasianu A, Bugoi R (2003) Adulterations in first century BC: the case of Greek silver drachmae analyzed by x-ray methods Spectrochim Acta B 58:777–781
22. Jokubonis C, Wobrauschek P, Zamini S, Karkowski M, Trnka G, Stadler P (2003) Results of quantitative analysis of Celtic glass artefacts by energy dispersive x-ray fluorescence spectrometry. Spectrochim Acta B 58:627–633
23. Gratuze B (1999) Obsidian characterization by laser ablation ICP-MS and its application to prehistoric trade in the Mediterranean and the Near East: sources and distribution of obsidian within the Aegean and Anatolia. J Archaeol Sci 26:869–881
24. Snopok BA, Kruglenko IV (2002) Multisensor systems for chemical analysis: state-of-the-art in electronic nose technology and new trends in machine olfaction. Thin Solid Films 418:21–41
25. Martin YG, Oliveros MCC, Pavon JLP, Pinto CG, Cordero BM (2001) Electronic nose based on metal oxide semiconductor sensors and pattern recognition techniques: characterisation of vegetable oils. Anal Chim Acta 449:69–80
26. Dutta R, Kashwan KR, Bhuyan M, Hines EL, Gardner JW (2003) Electronic nose based tea quality standardization. Neural Networks 16:847–853
27. Delpha C, Siadat M, Lumbreras M (2001) Identification of forane R134a in an air-conditioned atmosphere with a TGS sensor array. Sensors Actuators B Chem 50:1370–1374
28. Boger Z, Meier DC, Cavicchi RE, Semancik S (2003) Rapid identification of chemical warfare agents by artificial neural network pruning of temperature-programmed microsensor databases. Sensor Lett 1:86–92

ENVIRONMENTAL RADIONUCLIDES MEASURED BY AMS

CATALIN STAN-SION*, MIHAELA ENACHESCU AND MARIUS DOGARU
National Institute for Physics and Nuclear Engineering, Bucharest, Romania

Abstract. The Accelerator Mass Spectrometry analyzing method is presented with applications for detection of existing nuclear pollution, with the aim of protecting the environment against it and also of accurately mapping current and past nuclear contamination. After a brief description of an AMS facility and how it produces its analysis with the highest known sensitivity (10^{-15}), the main isotopes used to monitor nuclear pollution are each described and presented in interaction with the respective monitoring process. This paper concludes with the applications for detecting and preventing the nuclear pollution of the environment.

Keywords: Accelerator mass spectrometry, ^{129}I, ^{36}Cl, ^{239}Pu, ^{240}Pu, ^{3}H

1. Introduction

Nowadays, far more than ever in our history, besides all positive impact, the explosive rapid development of the technology and science has generated negative consequences on the environment and on our health. Among them, some of the worst ones are the chemical and the nuclear pollution. They are responsible for sever perturbations produced anywhere from within small entities like the genomes up to the large scale climatic changes. Unfortunately, the top scientific interest in this field stayed focused mainly on the general prevention of nuclear accidents and on the nuclear fuel and waste safety transportation issues. Meanwhile, only a small interest was paid to the rising global nuclear contamination consisting on ongoing nuclear releases which are yet steaming from the nuclear reactors and from the nuclear reprocessing power plants straight into the atmosphere or into the planetary ocean.

* To whom correspondence should be addressed. e-mail: stansion@nipne.ro

G.A. Aycik (ed.), *New Techniques for the Detection of Nuclear and Radioactive Agents*,

In June 2008, the NATO Science for Peace and Security Programme has organized at Mugla, Turkey, an advanced Training Course about New Techniques for the Detection of Nuclear and Radioactive Agents. The basic research of this paper was presented there in the frame of the Training Course. It deals with the new and modern experimental method called the Accelerator Mass Spectroscopy (AMS), which is the most advanced, top of the world technology for elemental analysis. Totally opposite but complementary to the classical techniques, the AMS method is not detecting the radioactive agents by their emitted radiation (α, β, γ), but it selects and counts them individually, one by one, each radioactive atom. The AMS method has the highest analyzing sensitivity known today, which is 10^{-15} (ratio: isotope/element). This sensitivity is equivalent with the real possibility to select and register one single type of atom from a million of billions of other types of atoms.

The AMS facility is very complex, gathering many modern methods from accelerator, atomic and nuclear physics like accelerator and focusing elements, electromagnetic analyzers, and particle detection systems etc.. A brief description of an AMS facility and of the way it produces its high sensitivity analysis will be presented in Chapter 2.

The main radioisotopes used to monitor nuclear pollution are ^{129}I, ^{36}Cl, ^{239}Pu, ^{240}Pu and ^{3}H. Section 3 presents applications with these isotopes, which are used for monitoring and investigating of their concentration in the environment, for controlling the level of nuclear pollution in the ecosystem and in the vicinity of the nuclear plants (water, air, earth), for the supervision of the nuclear power plants activities, for the reconstruction of nuclear pollutions in the past and for controlling the integrity of nuclear installations, of water quality and overall of life. Section 4 presents the conclusions of this research.

2. The Accelerator Mass Spectrometry Method

Accelerator Mass Spectrometry (AMS) is a powerful method to measure very small concentrations of radionuclides in different materials. Long lived radionuclides are detected by AMS free of molecular interferences and detector background at isotopic ratios as low as 10^{-15}. As already presented in the introduction, this top of the world experimental method selects and counts individually the atoms with the highest sensitivity known today.

How is this possible, what are the main apparatus and the physics principles implied that govern this outstanding method? These are questions we will try to answer and make AMS a clear and friendly method to be used by any scientific research team that needs this analysis of high isotopic sensitivity. In Figure 1 we present the general lay out of our AMS facility in Bucharest [1, 2].

The central component of the facility is a tandem type accelerator. The terminal potential can be up to 15 MV high. In the last years, the small and compact AMS facilities used tandems of only 3.5 MV with very good results. The dedicated components of the AMS analysing facility, called "filters" of atomic selection will be described in the following.

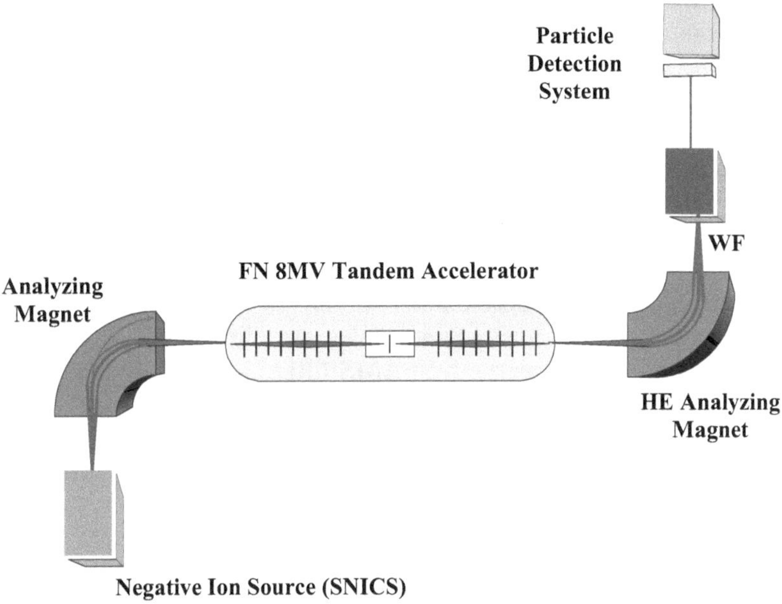

Figure 1. Schematic lay-out of the AMS analysing facility at NIPNE Bucharest, Romania

The first filter (F1) is the high current ion sputter source itself, Figure 2.

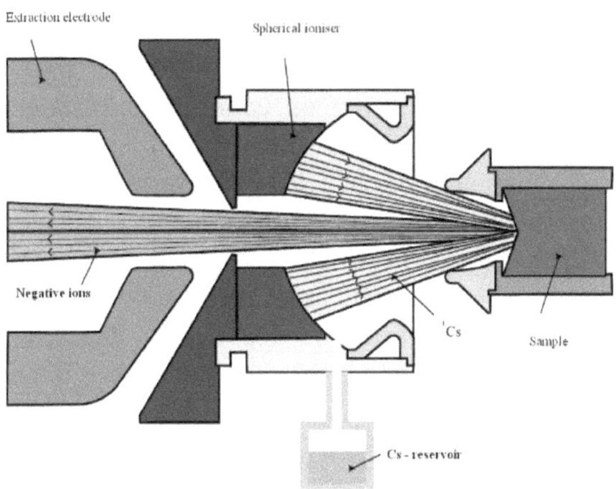

Figure 2. The sputter negative ion source and the trajectories of positive (spattering) and negative extracted ions

The samples to be analysed for their small isotopic concentration are introduced into the sputter ion source. Inside this ion source, the sputter process is produced by the bombardment of the sample with an accelerated $^{133}Cs^{+}$ ion beam. This bombardment is produced by heated ^{133}Cs vapours that are mounting from the Cs reservoir into the sputter camber of the ion source. There the Cs vapours come in contact with the surface of a high heated spherical ioniser (made of Wolfram and heated at 1,200°C). At this contact, the Cs atoms loose one electron and become ionised as Cs^{+}. These positive ions are then accelerated in the potential difference of about 10 kV between the spherical ioniser and the cathode (which is the sample holder containing the sample material). Negative ions, positive ions and neutrals are than formed by the collision of the Cs^{+} with the sample material. The highest productivity is for the production of neutrals and positive ions. In the best case, a maximum of 2–3% from the total sputter yield are the negative ions. However, only these negative ions are extracted from the ion source by the Pierce extraction electrode (the only ions possible to accelerate in a tandem accelerator). In this way, most of the possible interferences with the positive molecules are eliminated. Therefore, the sputter source itself acts as **the first AMS filter**.

The extracted negative ions, produced by sputtering form the sample material, leave the ion source with energy of about 30 keV and will be then mass separated by a first analysing magnet, Figure 1. **This is the second filter (F2)** of the AMS analysing system.

At the exit of the AMS injector deck, the selected negative ion beam is pre-accelerated in an acceleration tube with an energy gain of 50 keV. Then, the beam is injected into the tandem accelerator with a total incoming energy of 80 keV (30 + 50 keV). This energy is sufficient to adapt the ion source emittance to the acceptance of the tandem accelerator. Entering the tandem accelerator, the atomic and remaining molecular species will be accelerated and will pass the stripper foil placed in the tandem terminal. This stripping process causes the molecular ions to dissociate and to be destroyed. Consequently, the tandem stripper acts as a filter. This is **the third AMS filter (F3)**. At the exit of the tandem accelerator, the high energy particles are selected and analyzed by the large 90° analyzing magnet (120 MEP) of the tandem accelerator. It represents **the fourth filter (F4) of the AMS system.**

Unfortunately, the magnetic analyzers are not able to select ions having all the same (p/q) ratio (magnetic degerancy). This can be seen from the relation characterizing the ion optic transport through magnetic systems:

$$(BR)^2 = 2\ ME/q^2 = (p/q)^2 = \text{constant} \quad (1)$$

where B is the magnetic field, R is the magnet bending radius, E the ion energy, M the atomic mass and q the ion charge state.

One way to remove this p/q ambiguity is to use a velocity Wien filter (WF) [3]. This is a device that has orthogonal electric and magnetic fields and that separates ions according to their velocity. In this field geometry, one can write the equilibrium of forces:

$$qvB = qE$$

The WF selects according to the settings of B and E only the ions having a precise velocity (v = E/B) and solves the ambiguity of the p/q ratio. This system represents **the fifth AMS filter (F5)**.

The final part of the AMS facility contains the ion detector. It is **the last and most powerful filter of the analysing system (F6)**. Different types of detectors are used for light, medium heavy and heavy nuclei.

In the following, let's now discuss about how one has to proceed in order to perform an AMS analysis.

The AMS experimental system has to select and detect with a very high sensitivity (10^{-15}) ions of a certain kind. Such ions will have extremely low currents. They arrive at the detector with frequencies not more than 1 event/min or so. The detector situated at the end of the AMS system, is the only one capable of measuring this events and no other diagnose systems can be used to measure the ion beam intensity. The transport of an extremely low intensity ion beam (few atoms/second), on its way from the sputter ion source all the way (about 40 m) to the detector (see Figure 3.1) has to be optimised. AMS solves this problem by using a so called "pilot beam" formed by ions of mass close to the ions that have to be analysed, but with measurable current beam intensities (approximately 100 nA). Different AMS ions measured for environmental studies with their pilot beams are given in the Table 1 below.

TABLE 1. Some AMS ions with their corresponding pilot beams

AMS ion	^{36}Cl	^{129}I	^{3}H	^{239}Pu
Pilot beam ion	^{35}Cl	^{127}I	^{12}C	^{240}Pu

One has to mix the carrier (pilot) material with the sample material to be analysed. Consequently, each sample material contains the two ion species: the ions from the carrier, producing the macroscopic pilot beam and the AMS ions (those to be measured for their concentration) forming the microscopic ion beam. Form the total ion beam current produced by the ion source, one first selects (with the AMS analysing magnet on the injector deck) the current produced from the carrier material (the pilot beam). This beam is used to optimise the transport through the entire system, to tune the values of each filter and the values of the tandem accelerator. However, due to its large current intensity, the pilot beam will never enter the detector. This is due to the detector that has an extremely high sensitivity and dose not support more than 200 kHz event rate. Once the ion optics for the transport of the pilot beam is optimised, one starts the analysis by tuning the AMS facility to select the microscopic beam (ions to be analysed). With this aim, two changes have to be performed:

1. Change the B field of the AMS magnet according to the new calculated value with relation (1):

$$B_2 = B_1\ (m_2/m_1)^{1/2}$$

2. Change the terminal voltage of tandem accelerator to the new value U_{T2}, in order that the microscopic beam should follow exactly the way defined by the pilot beam. U_{T2} can be easily calculated with relation (1):

$$2\ ME/q^2 = 2M\ U_T\ (1 + q) = \text{constant}, \quad U_{T2} = (M_1/M_2)\ U_{T1}$$

It is assumed that the same charge state was selected for the microscopic ion beam and for the ions of the pilot beam. In this way, AMS counts the individual atoms produced from the analyzed sample into the ion source. As a result, calibration and absolute concentration values are obtained by the use of the standard samples.

3. AMS Method Applications to Monitor the Nuclear Pollution in the Environment

3.1. ENVIRONMENTAL RADIONUCLIDES MEASURED BY AMS

The high analyzing sensitivity of the AMS makes possible to established specific applications for environmental research based on radionuclides specific for such investigations. A list of the major radionuclides already used in the AMS research of the environment is presented below, in Table 2.

Next, we will emphasis the radionuclides used by the AMS to determine and to control the nuclear environmental pollution. These are ^{36}Cl, ^{129}I, ^{3}H, ^{239}Pu and ^{240}Pu and will be discussed below.

TABLE 2. AMS ions with their applications

Izotope	Applications
^{3}H	Diagnose of fusion experiments, biomedicine
^{10}Be	Astrophysics, geology, oceanography
^{14}C	Archeology, medicine
^{26}Al	Medicine, geology
^{36}Cl	Atmospheric physics, tracer of the stratosphere–troposphere–biosphere exchanges, hydrology (determination of flow paths and groundwater age, ecology, radioactive monitoring of nuclear reactors)
^{41}Ca	Determination of fossil age, biomedicine
^{55}Fe	Biosphere studies
^{60}Fe	Supernovas investigation
^{59}Ni, ^{63}Ni	Safeguard and monitoring of nuclear activities, astrophysics
^{129}I	Monitoring of nuclear activities, oceanography
^{244}Pu	Astrophysics (Supernovas), nucleo-synthesis

3.1.1. *^{36}Cl*

Chlorine 36 is a radionuclide with a half life $T_{1/2} = 3.01\ 10^5$ years. It is a good signature of nuclear pollution produced by nuclear power plants (NPP). To have a good estimate of the produced nuclear pollution one has to take into account all the other sources of ^{36}C production at a certain location.

In the atmosphere, ^{36}C is produced by the spallation reactions $^{40}Ar(n, p4n)^{36}Cl$ and $^{40}Ar(p, 2p3n)^{36}Cl$. [4, 5]. The experimental determined values for the cross sections of these reactions are used to calculate the mean global atmospheric production rate of ^{36}Cl to be 24 atoms $m^{-2}\ s^{-1}$ [4].

In order to describe the earth latitudinal dependence of the ^{36}Cl fallout, a two-dimensional atmospheric transport model (ATM) was constructed. In this model the earth's atmosphere is horizontally divided in 18 latitude belts of 10° each and vertically in one troposphere and four stratospheric layers. Except for a few basic rate constants, most exchange rate constants are determined by the assumed steady state conditions. The basic rate constants are derived from simulations of temporal and latitudinal dependencies of the ^{90}Sr and ^{14}C bomb pulses (open nuclear explosions and experiments).

Figure 3 shows the calculated ^{36}Cl deposition fluxes in comparison with the measured ones. Most of the measured ^{36}Cl fluxes are larger than the simulations for atmospheric cosmogonist production. The difference is larger for the northern than for the southern hemisphere. After correction for local precipitation rates, the difference is largest for low northern latitudes and can be up to about one order of magnitude. The highest values at latitude 30° are due to a large NPP close to Shanghai, showing clearly the enhanced nuclear pollution at that location.

There are also additional sources of atmospheric ^{36}Cl concentrations that can contribute to the tropospheric ^{36}Cl deposition. These sources are: the cosmogenic *in-situ* production of ^{36}Cl in the lithosphere, the Chernobyl fall-out and the ^{36}Cl production by nuclear weapon tests.

First, in the rock material on the Earth's surface, ^{36}Cl is mainly produced by neutron-induced spallation reactions on potassium and calcium. For vanishing erosion rates the surface saturation concentration ratios were obtained to be 3.6 10^{-15} for $^{36}Cl/Ca$ and 4.8 10^{-15} for $^{36}Cl/K$. For a mean global erosion rate of 60 µm/year [9] the surface saturation concentrations are about two orders of magnitude lower [10]. Due to eroded land, with lithospheric abundances of 3.65% Ca and 2.58% K, the ^{36}Cl flux would be 0.3 m^{-2} s^{-1}. This estimate shows that the erosion of in-situ produced ^{36}Cl cannot explain the observed differences in the ^{36}Cl fluxes.

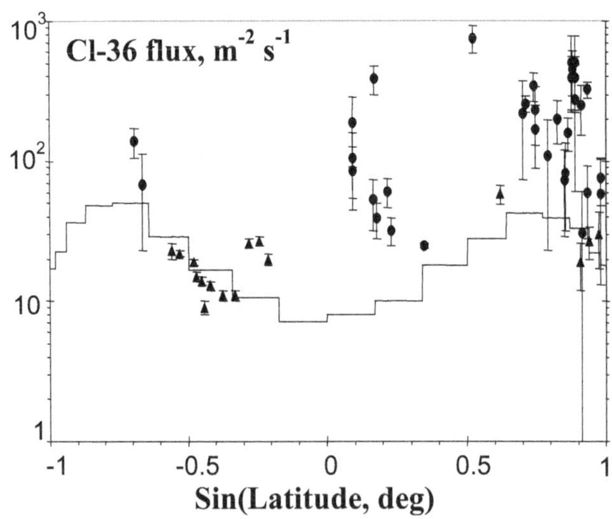

Figure 3. Measured ^{36}Cl deposition fluxes in comparison with calculated values (histogram) and other experimental results [6–8] as a function of latitude

Secondly, from a global point of view, the Chernobyl accident in 1986 was a local event, since the radioactivity did not reach the earth's stratosphere for a global distribution and since it was washed out from the troposphere within weeks. Therefore, the accident cannot explain the increased ^{36}Cl fall-out in Africa. In the central Europe, area concentrations of 6,400 Bq of ^{137}Cs and of 1.6 10^{11} ^{36}Cl atoms/L were measured in the rainwater collected on April 30, 1986. Using the total ^{137}Cs release from this accident [11], the total ^{36}Cl release can be estimated to be about 2 10^{24} ^{36}Cl atoms, corresponding to 6 years of global natural atmospheric production. In ice cores of Dye 3 (Greenland, 65° N) the ^{36}Cl bomb-pulse was measured to have a fluence (time integrated flux) of 2.4 10^{12} atoms m^{-2} [12].

Finally, if part of the bomb ^{36}Cl is stored in the biosphere and re-enters the troposphere after a mean residence time of several tens of years, the tropospheric fallout will be increased. The storage of bomb ^{36}Cl into the biosphere was first proposed by Milton et al. [13]. The organic compound to be considered re-entering the troposphere is methyl chloride (CH_3Cl). It has to be emphasized that the tropospheric OH concentrations are highest in the tropical regions, higher in the northern than in the southern hemisphere and higher on the land than on the sea [14]. Simulation calculations were performed with a modified atmospheric box model. It was observed a higher deposition in the northern hemisphere compared to the southern hemisphere. That is due to the larger fallout of bomb ^{36}Cl (61%) and to the higher OH concentrations, i.e. enhanced destruction of $CH_3{}^{36}Cl$ with subsequent deposition, in the northern hemisphere. The conclusion is that the recirculation of the ^{36}Cl bomb in the biosphere is the mechanism to account for the observed discrepancies between measured and calculated cosmogenic fall-out.

Finally, the calculated curve of ^{36}Cl deposition fluxes can be used at any location on the earth to determine by AMS an increase of the nuclear pollution in the environment.

3.1.2. *^{129}I*

^{129}I ($T_{1/2}$ = 15.7 My) is a fission product. Starting with the 1960s, the ^{129}I was emitted by nuclear installations either in the liquid or in the gas form and became the highest source of nuclear pollution for the environment. Furthermore, between 1966 and 1994 the biggest Western Europe nuclear reprocessing plants from La Hague (France) and Sellafield (Great Britain) contributed with an input of 906 kg of ^{129}I, discarded in the English Channel and an input of 534 kg of ^{129}I, discarded in the Ireland Sea.

The only stable isotope of iodine is the ^{127}I. The natural production of the ^{129}I is based on the spallation reaction of cosmic rays on atmospheric Xe with a low production yield. As a consequence, the natural background for the measurements is practically negligible. Due to its high volatility, when entering into the atmosphere, the ^{129}I is transported over large distances. The major problem in the application of the ^{129}I has always been the combination of a low natural abundance with the long average life, which results in a very low activity. Given the required limit of detection, which is usually dictated by the mass of available sample (some milligrams of iodine), detection is only possible to achieve by using the AMS technique.

In the light of the ultra-sensitive measurements of the ^{129}I, the monitoring possibilities of the AMS were evaluated by the Vienna Agency of Atomic Energy (IAEA). This was done as a follow up to the inter-comparison exercise from July 1997 that was done in cooperation of the US Department of Energy and the International Safeguards Division. As a result, IAEA has identified and recommended the measurement of the ^{129}I as a potential signature of the reactor or reprocessing operations, or for the detection of undeclared nuclear activities [15].

3.1.3. ^{3}H

Tritium ($T_{1/2}$ = 12.35 years), was widely dispersed during the atmospheric testing of nuclear weapons in the mid-1950s and in the early 1960s. The quantity of tritium existing in the atmosphere produced by the weapons testing peaked in 1963 and it has been decreasing ever since. Tritium may be released as steam or it may leak into the underlying soil and ground water. Such releases are usually small and are required to meet federal environmental standards. Because of its dangerous and extremely noxious nature, there is a special interest for tritium research for health and environment protection. In the environment, tritium comes with a variety of chemical forms because it can substitute an element in any molecule that has hydrogen atoms.

Tritium, as HTO, produced by heavy water reactors, has a high mobility in the hydrologic cycle. It does not have external radiotoxicity for life beings, but once inside the body it will attach to the life cells. Since tritium is almost always found as water, it goes directly into soft tissues and organs. The associated dose to these tissues is generally uniform and dependent on the tissues water content. Recent studies have shown the correlation between this kind of infestation and the appearance of carcinomas. Today, sources of tritium include biological research laboratories, commercial nuclear reactors, research reactors, nuclear reprocessing plants and weapons production plants. There is a high interest to perform precise measurements of the T concentrations in the water samples from rivers which are used as a heat exchange medium for the nuclear power plants. It significantly helps controlling for the possible water infestation.

The AMS analyzing method has a supplementary capacity to determine the concentration of elements by performing a depth profiling. This type of analysis is especially interesting for tritium, which has great penetration mobility through diverse materials. Therefore, the concentration profiling at the surface and at high depth, is often playing a deciding role for obtaining the correct data. Due to their high sensitivity, the AMS measurements can determine the Tritium leaks from the unsealed parts of the nuclear installations, from the tritium separation facilities and also the respective spread in the neighboring clean spaces [16, 17]. In this way, the identification of any unsealed or defective parts/procedures becomes possible.

3.1.4. *^{240}Pu and ^{239}Pu*

Plutonium concentration in sediments and in water, together with the $^{240}Pu/^{239}Pu$ ratio, can be used to identify the sources of plutonium and to monitor its dispersal away from nuclear installations. As a result of the atmospheric weapon testing and of the reprocessing operations of the spent nuclear fuel, together with the ^{236}U, these isotopes are widely dispersed in the environment. Because the long half lives ($t_{½}$ ^{240}Pu = 6,550 years, $t_{½}$ ^{239}Pu = 24,000 years, $t_{½}$ $^{236}U = 2.3 \times 10^7$ years), these isotopes will persist in the environment for periods of time much longer than our lifetime and will be a very negative heritage for our successors. As will be explained below, the ratio $^{239}Pu/^{240}Pu$ provides a sensitive separation between the pollution produced by Weapon Nuclear Reactors or by the Electric Nuclear Power Plants [18].

3.2. THE AMS STRATEGY FOR DETERMINING THE NUCLEAR POLLUTION OF THE ENVIRONMENT

Using the above described radioisotopes, AMS can be applied for measuring, monitoring and investigation of the nuclear pollution in the environment, for controlling the level of radionuclides in the eco-systems and in the vicinity of the nuclear plants (water, air, earth), for the supervision of the nuclear power plants activities, for controlling the integrity of nuclear installations, of water quality and of life. AMS is also able to reconstruct the history of pollution many years back.

With this goal, an AMS research uses the accumulation of radioisotopes in the natural archives like glaciers, lakes and centennial tree rings. The outspreading or nuclear pollution is calculated according to the Global Transport Model (GTM) [19]. This computer code was developed starting from the ATM, as applied for the ^{36}Cl fall-out and described in Section 3.1. The initial model was upgraded and extended to permit a detailed description of the troposphere propagation and of the biosphere exchanges of specific elements.

Generally, an environmental pollution study by AMS implies the following steps: (i) the identification of the pollution source (location, dimension, etc.), (ii) the calculation of the expected quantities of expelled nuclear material from the source, (iii) AMS measurements of the nuclear pollution in the natural archives, (iv) the use of the GTM for internal calibrations and calculation of the pollution at different locations, (v) followed by the comparison with experimental data. In this way, one can obtain the value of pollution at any location on earth. Let us consider few examples.

In Europe, Figure 4, the nuclear pollution comes form the nuclear reprocessing plants and from the plutonium reactors. The largest facilities are at Sellafield (GB), La Hague and Marcoule (F).

As already presented above, by the Sellafield and La Hague facilities, large discards of ^{129}I were measured in the English Channel and in the Ireland Sea. On one hand, in this way, a large pollution of the Atlantic Ocean was measured. On the other hand, a high pollution of the earth's

Figure 4. Europe map showing the three largest nuclear reprocessing power plants

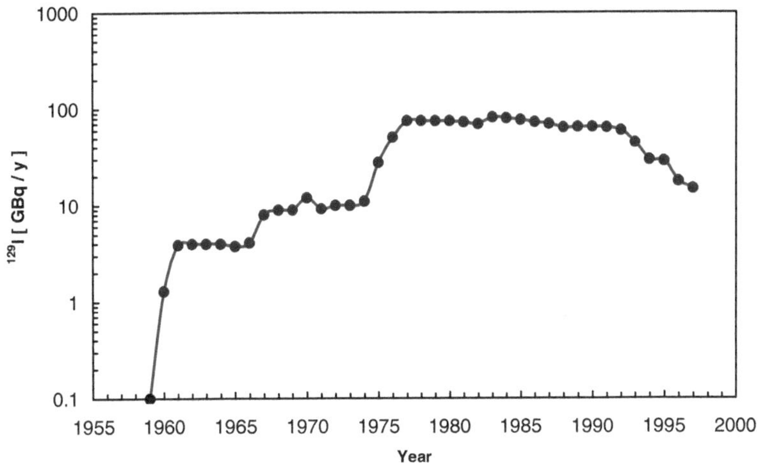

Figure 5. ^{129}I radioactivity expelled from Marcoule reprocessing power plant into the atmosphere

atmosphere was also measured. The Marcoule P1, plutonium extraction facility, expelled between 1959–1993 an important amount of ^{129}I in the atmosphere, Figure 5. The overall produced pollution from these three sources is visible even far away from them, in the natural archive of the Fisher Horn Glacier in Switzerland.

Figure 6 represents the Fisher Horn Glacier where the AMS measurements on core drillings revealed the time evolution of ^{129}I nuclear pollution at high altitudes in Central Europe. A steady increase of the concentration was registered until 1990, Figure 7, when the activity of these nuclear faculties was either diminished or stopped. However, it is to be emphasised that the Iodine concentration decreases dramatically with height. Therefore, one has to consider that only a small amount (ca. 40%) of the iodine can reach the height of the glacier. Obviously, no direct comparison with iodine fall-out measured from soil or rainwater can be performed.

Furthermore, a nuclear pollution input in the Fisher Horn comes also from the nuclear weapon experiments. The fission products ^{129}I and ^{137}Cs are produced from ^{238}U bombarded by fast neutrons, with yields of 5.57% and 1.58%, respectively [20]. After about 30 years from such events, the deposition ratio ^{137}Cs/^{129}I is estimated to be 1.6. These values are closed to the values measured in [21]. The overall conclusion, obtained is that the deposition at Fisher Horn until 1963 was generally dominated by the fall-out of nuclear weapon experiments.

Figure 6. Photo of the Fischer Horn Glacier

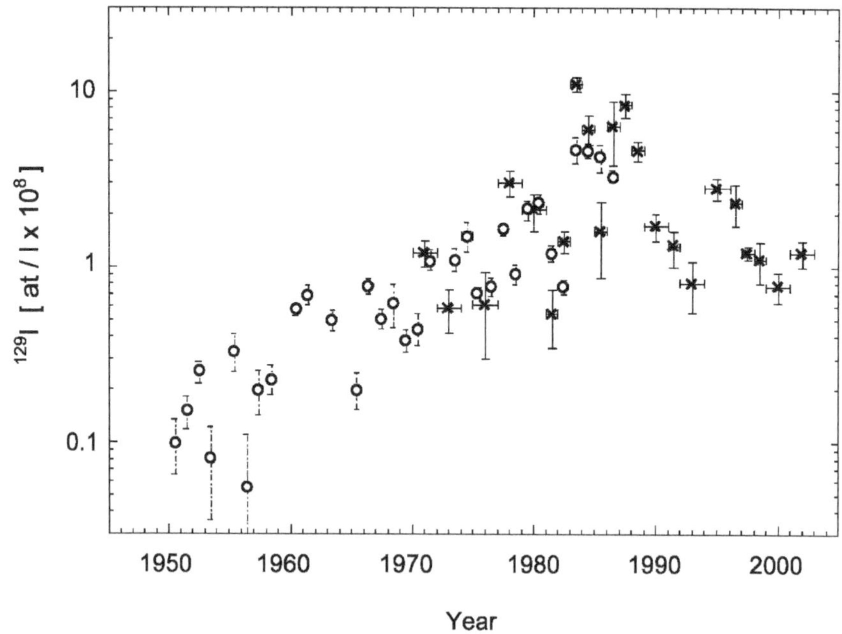

Figure 7. Timely ^{129}I deposition (fall out) on the Fisher Horn Glacier [22]

Similarly, as a consequence of such experiments, enhanced concentrations of ^{129}I were observed and measured in big lakes, with long flushing times, located in Central Europe. In Bodensee, Chimesee, Stanberger See and Ammersee the averaged concentration value is 2 10^8 atmos/L. This value is three orders of magnitude over the natural background, before the start of any nuclear activities. The AMS measured pollution values in the lakes archives are of high importance for the calibration of GTM. This important database will be enhanced with the new AMS measurements from the centennial tree rings from the trees situated in the vicinity of nuclear installations.

Similar to Europe, the former nuclear installations from USSR and Asia have produced important nuclear pollution to the environment. The most important Nuclear Power Stations, from Tschelniabisk (Majak), Tomsk and Krasnoyarsk, can be seen in the map below, Figure 8. All these large power plants are situated in a reasonable distance from important natural archives: the glacier from Belucha in Altai mountains, the Baikal lake and the Issyk-Kul lake.

Figure 8. The map of the former USSR with the main nuclear installations and with natural archives for the AMS research, like the Belucha glacier and the Baikal sea

In order to reveal the history of nuclear pollution in Asia, the Belucha glacier offers very good archives. In 2001, a core drilling of 140 m length in the glacier was performed by an international expedition [23]. For transportation, the drill core was frozen. The data of the stratification was obtained by using the Tritium peak from 1963 which was produced by the nuclear weapon tests and the radioactivity of ^{210}Pb. Therefore, the first 86 m of the core drill was able to reveal the pollution history from 1941 to 2001. As shown in Figure 9 the AMS measurements showed a fast increase of ^{129}I fall-out starting in 1950, followed by a constant value between 1958 and 1976. In 1977 a sudden increase of the fall-out is to be noticed. It is correlated with an unreported nuclear event that happened in Majak. Based on AMS and on the GTM, the measured data from the Belucha glacier also determined the separate contribution to the environmental pollution of the mentioned region: 44% by Majak, 33% by Tomsk, 15% by Krasnojarsk-26 and 8% contribution from Europe.

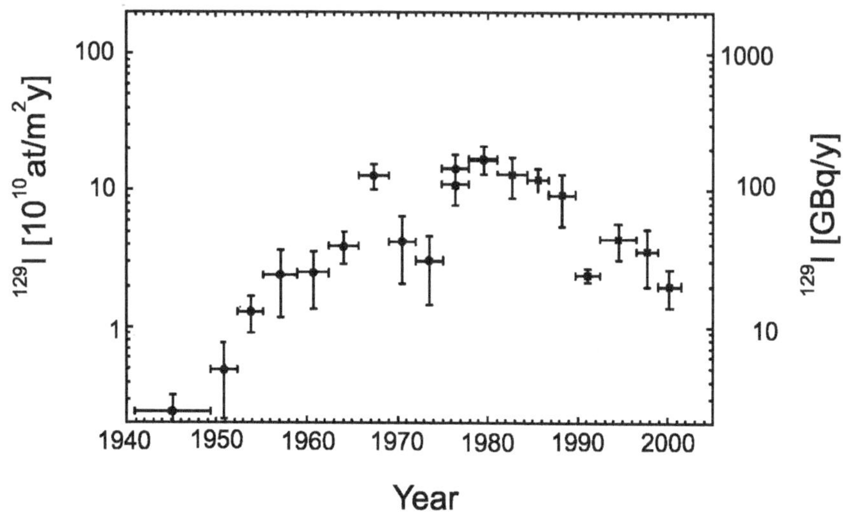

Figure 9. ^{129}I fall out measured at the Belucha glacier. The equivalent radioactivity deposition is shown on the *right* hand side

An important feature is that the AMS analysis can distinguish between the nuclear pollution produced by the Electric Nuclear Power Plants or by the Weapon Nuclear Reactors. The difference is given by the value of the ratio ^{240}Pu/^{239}Pu. For weapons production, the so called "low burn-up" reactor fuel is used. ^{239}Pu is produced from ^{238}U, after (n,γ), followed by a double β-decay:

$$^{238}\mathrm{U}\ (\mathrm{n},\gamma)\ ^{239}\mathrm{U} + 2\beta \ \rightarrow \ ^{239}\mathrm{Pu}$$

Depending on the burning time, ^{240}Pu is also produced in the reactor. For weapon use, ^{240}Pu should be less than 6% of the ^{239}Pu content and this is attained after about 120 days of fuel burning in the reactor. The burning time dependence of the production yields of the two radionuclei is given in Figure 10.

Thus, the low burn-up limit is considered up to ratio $\Delta\eta/\eta = 0.24$. The maximum ^{240}Pu/^{239}Pu ratio of 6% should not be exceeded. Significant

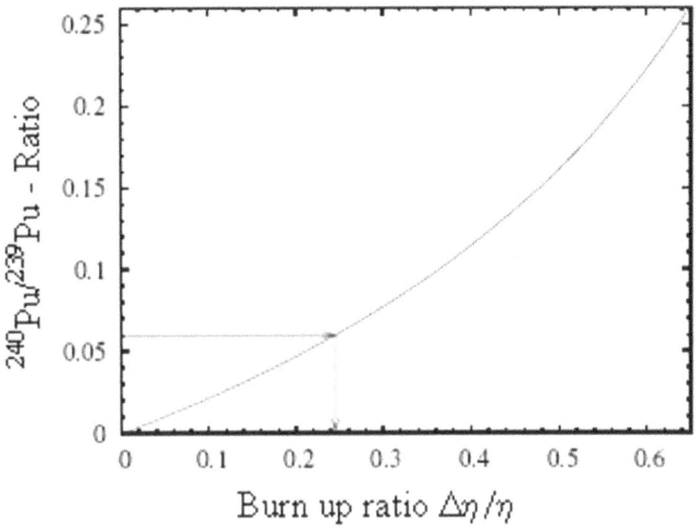

Figure 10. Production yield ratio of ^{240}Pu and ^{239}Pu. The critical value of 6% corresponds to a burn up ratio of 0.24

quantities of ^{240}Pu in an explosive device would make it hazardous to the bomb makers, as well as unreliable and unpredictable. Thus, one can easily determine the source of pollution.

A straight forward calculation gives the ratio of atomic yields for $^{129}I/^{131}I$ in the low burn-up limit of 120 days, Figure 11. The ratio value $^{129}I/^{131}I = 2.6$ is expressed by the number of atoms, produced at the moment of the end of the burn up in the reactor. At that moment, about 8% of the fission products are still as ^{129}Te.

After that moment, a fuel "cooling" time should follow. It is not precisely known how long this cooling time was. Many times, it was depending on the political interest for producing as fast as possible many bombs. Therefore, the cooling time was varying during the years of functioning of the nuclear installations. From unclassified data it results an average cooling time used for the nuclear fuel of such reactors in USSR of about 36 days. After this time, the ratio of nuclides increases to $^{129}I/^{131}I = 50$.

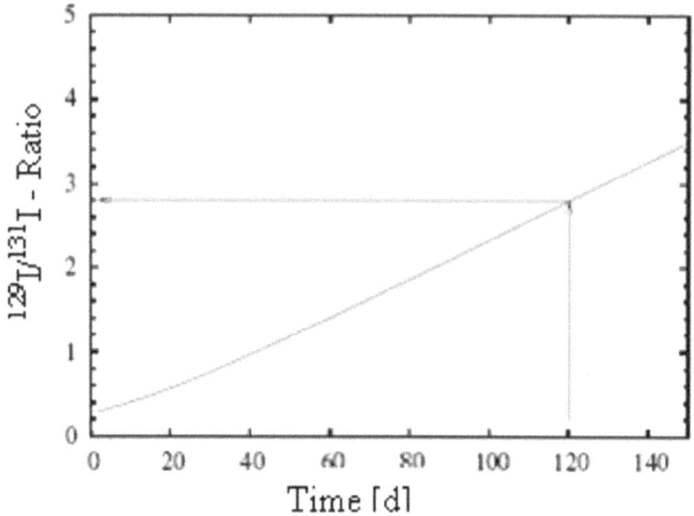

Figure 11. The ratio value $^{129}I/^{131}I$. After 120 days of burning time the ratio becomes 2.6

This atomic ratio value exists also at the beginning of reprocessing operations and might be the same at the moment when the radionuclei have been released or escaped into the environment. Using this calculation, the GTM computer code together with the AMS measurements from the Belucha archives, both correlated with measurements from the European archives, one can obtain the produced radioactivity pollution values at the location of Majak that are 12.5 GBq of ^{129}I and 179 PBq of ^{131}I.

These evaluations are larger than the previous evaluation from but more realistic. Similarly, for Majak, separate contributions to the measured pollution come from the weapon reactors and from the RT-1 civil nuclear power plant. In the former, up to 6% burn-up fuel was used where as in the later, over 60% burn-up fuel. Obliviously, the different contribution to the plutonium inventory and $^{239}Pu/^{240}Pu$ are to be taken into account for estimates of the total production.

The Chernobyl accident produced large nuclear pollution over the entire earth's environment. Different from the weapon nuclear reactors, the nuclear power plants, devoted for production of electric energy or research will function in the high burn-up fuel regime. The Chernobyl reactor, initially loaded with 192 t of ^{238}U at η_0 = 2% At the moment of the accident the reactor had a burn-up ratio $\Delta\eta/\eta_0$ = 67.6%. The produced ^{239}Pu was 907 kg. To the inventory of ^{129}I contributed the fission of ^{235}U and ^{239}Pu. The radioactivity released in the atmosphere was 67.4 GBq of

^{129}I and 3,010 PBq of ^{131}I. This values correspond to an atomic ratio ^{129}I/^{131}I = 16.

In an independent research Mironov et al. [21] measured the pollution in Belarus. They used radioactivity measurements for determination of ^{137}Cs and the nuclear activation analysis method (NAA) for determination of ^{131}I. For ^{129}I measurements the AMS method was applied. In this way, the ratios of ^{131}I/^{137}Cs and ^{129}I/^{137}Cs in Soils from 68 locations in Belarus were measured. They derived values at the time of the Chernobyl accident were ^{131}I/^{137}Cs = 10 Bq/Bq and ^{129}I/^{131}Cs = 15.2 g/g.

This measured ratios support the hypothesis of relatively little fractionating of iodine and cesium during the migration and deposition of the radioactive cloud. The measured fractional factor of ^{129}I/^{137}Cs = 1.16(54) suggest [21] that ^{137}Cs can be used to give correct estimates of the ^{131}I releases from Chernobyl.

4. Conclusions

AMS brings a complementary and useful contribution to the investigation, reconstruction and surveillance of nuclear pollution of the environment. It selects and counts individually the atoms, determining concentrations with the highest sensitivity known today, which is 10^{-15} (ratio: isotope/element). Together with a Global Transport Model for the radionuclides in the earth's atmosphere it provides reliable information of the outspreading of nuclear pollution of the environment.

AMS with ^{36}Cl is an ultra sensitive method for the detection of nuclear releases into the atmosphere.

^{129}I is a very useful radioisotope for the reconstruction of the nuclear releases from the past and for the monitoring of the nuclear activities in the present. IAEA has identified and recommended the AMS measurement of the ^{129}I as a potential signature of the reactor or reprocessing operations, or for the detection of any pollution produced by nuclear activities. AMS measurements of the ratio ^{239}Pu/^{240}Pu provide a sensitive separation between the pollution produced by Weapon Nuclear Reactors or by the Electric Nuclear Power Plants. It contributes with very important information concerning the discovery of undeclared nuclear activities at any geographic location on our planet.

^{3}H is an important isotope to be constantly monitored in the vicinity of heavy water reactors and of the nuclear installations handling this isotope, in order to preserve human health and overall, in order to preserve life.

AMS measurements demonstrated that radioactivity measurements of ^{137}Cs provide reasonably good information of nuclear pollution from the past and can be used to obtain estimates for the ^{131}I contamination. Finally, the AMS measurements, sometimes correlated with other classic methods of radioactivity detection methods brings an important contribution to the overall protection of the environment against nuclear pollution.

References

1. Radulescu M, Dima R, Ivascu M, Ivascu I, Marinescu L, Plostinaru D, Stan-Sion C (1998) The accelerator mass spectrometry at the Institute of Nuclear Physics and Engineering. Bucharest Romania J Phys 43:121–130
2. Stan-Sion C, Ivascu M, Plostinaru D, Catana D, Marinescu L, Radulescu M, Nolte E (2000) Nuclear instruments & methods in physics research section b-beam interactions with materials and atoms. 172:29–33 Nucl Instrum Meth Phys Res B 172:957–977
3. Catana D, Rohrer L, Stan-Sion C, Enachescu M, Plostinaru D, Vata I (2001) Design and construction of a WIEN velocity filter for AMS facilities. J Phys 46:595–602
4. Huggle D, Blinov A, Stan-Sion C, Korschinek G, Scheffel C, Massonet S, Zerle L, Beer J, Parrat Y, Gaeggeler H, Hajdas W, Nolte E (1996) Production of cosmogenic ^{36}Cl on atmospheric argon. Plane Space Sci 44:147–151
5. Stan-Sion C, Huggle D, Nolte E, Blinov A, Dumitru M (1996) AMS measurements of the production cross sections of ^{36}Cl with protons up to 1 GeV. Nucl Instrum Meth B 117:26–30
6. Lukasczyk C (1994) ^{36}Cl im groenlandeis. Doctors thesis, ETH-Zurich, Swiss
7. Hainsworth LJ, Mignerey AC, Helz GR, Sharma P, Kubik PW (1994) Modern chlorine-36 deposition in southern Maryland USA. Nucl Instrum Meth B 92:345–349
8. Keywood MD, Fifield LK, Chivas AR, Cresswell RW (1997) Fallout of chlorine-36 to the Earth's surface in the Southern hemisphere. J Geophys Res 103:8281–8286
9. You CF, Lee T, Brown L, Shen JJ, Chen JC (1988) ^{10}Be study of rapid erosion in Taiwan. Geochim Cosmochim Acta 52:2687–2691
10. Heisinger B, Niedermayer M, Hartmann FJ, Korschinek G, Nolte E, Morteani G, Neumaier S, Petitjean C, Kubik P, Synal A (1997) Ivy-Ochs S in-situ production of radionuclides at great depths. Nucl Instrum Meth B 123:341–346
11. OECD/NEA (1996) Chernobyl Ten Years on Radiological and Health Impact Paris, France
12. Synal HA, Beer J, Bonani HJ, Suter M, Woelfli W (1990) Atmospheric transport of bomb produced ^{36}Cl. Nucl Instrum Meth B 52:483–487
13. Milton JCD, Andrews HR, Chant LA, Cornett RJJ, Davies WG, Greiner BF, Imahori Y, Koslowsky VT, McKay JW, Milton GM (1994) ^{36}Cl in the Laurentian Great Lakes basin. Nucl Instrum Meth B 92:440–444
14. Andreae MO, Crutzen PJ (1996) Atmospheric aerosols: biogeochemical sources and role in atmospheric chemistry. Science 276:1052–1058
15. UCRL-ID-128212 in Rep LLNL July (1997)
16. Enachescu M, Lazarev V, Stan-Sion C (2006) Unfolding procedure for AMS depth profiling. J Phys D Appl Phys 39:2876–2880
17. Stan-Sion C, Roth J, Krieger K, Enachescu M, Ertl K, Lazarev V, Reithmeier H, Nolte E (2007) AMS-sensitive tool used as nuclear safeguard and to diagnose fusion experiments. Nucl Instrum Meth Phys Res B 259:694–701
18. Lind OC, Oughton DH, Salbu B, Skipperud L, Sickel MA, Brow JE, Fiefield LK, Tims SG (2006) Transport of low $^{240}Pu/^{239}Pu$ atom ratio plutonium-species in the Ob and Yenisey Rivers to the Kara Sea. Earth Planet Sci Lett 251:33–43

19. Lazarev V (2003) The cosmogenic and anthropogenic ^{36}Cl in the environment. Ph.D. thesis, Technical University Muenchen, Germany
20. UN Sources and effects of ionizing radiation UNSCEAR 2000 (2000) Report to the General Assembly annex C: exposure from man-made sources of radiation UN, New York
21. Mironov V, Kudrjashov V, Yiou F, Raisbeck GM (2002) Use of ^{129}I and ^{137}Cs in soils for the estimation of ^{131}I deposition in Belarus as a result of the Chernobyl accident. J Environ Radioact 59:293–307
22. Reithmeier H (2005) ^{129}I in Umweltproben als Tracer für die atmosphärischen ^{131}I-Freisetzungen in Majak. Ph.D. thesis, Technical University Muenchen, Germany
23. Oliver S, Bajo S, Fifield LK, Gaeggeler HW, Papina T, Santschi PH, Schrotter U, Schwikowski M, Wacker L (2004) Plutonium from global fall out recorded in an ice core from the Belucha Glacier Siberian Altai. Environ Sci Technol 38:6507–6512

COMBINATION OF RADIOCHEMICAL AND ACTIVATION TECHNIQUES FOR THE DETECTION OF RADIONUCLIDES

BORUT SMODIŠ*[1] AND JUDMILA BENEDIK[1,2]
[1]*Jožef Stefan Institute, Jamova cesta 39, SI-1000 Ljubljana, Slovenia*
[2]*European Commission, Joint Research Centre, Institute for Reference Materials and Measurements, Retieseweg 111, B-2440 Geel, Belgium*

Abstract. Determination of radionuclides may be performed in principle either by direct activity measurement, usually termed radiometric analysis, or by mass measurement. A technique for the mass determination available to many nuclear research institutes is neutron activation analysis (NAA). Basic approaches for the NAA determination of radionuclides (both independent and in combination with α spectrometric measurement) are presented and examples are given for the determination of ^{99}Tc, ^{129}I, ^{135}Cs, uranium and thorium isotopes, ^{231}Pa and for ^{237}Np.

Keywords: NAA, ^{99}Tc, ^{129}I, ^{135}Cs, uranium isotopes, thorium isotopes, ^{231}Pa, ^{237}Np

1. Introduction

There are many naturally occurring radionuclides in the environment, including the isotopes of uranium and thorium decay chains, ^{40}K, and those produced from the cosmic ray reactions. Besides them, a large number of radionuclides have been produced and released to the environment as consequence of human activities such as operation of nuclear power plants, research reactors and nuclear fuel reprocessing plants, nuclear weapons testing, as well as radionuclides applied in industry and hospitals. For the radiation protection purpose, the specific activities of these radionuclides should be monitored in relevant environmental, waste and biological samples. The radionuclides released can also be frequently used as environmental tracers for following various transport processes. In Table 1

* To whom correspondence should be addressed. e-mail: Borut.smodis@ijs.si

radionuclides, having half-lives longer than 10 years, which are often required to be measured in environmental, waste and biological samples, are shown.

TABLE 1. Selected radionuclides of interest in environmental, waste and biological samples [1]

Nuclide	Generation	Half-life	Decay mode
^{3}H	^{2}H (n, γ) ^{3}H; ^{3}He (n, p) ^{3}H; ^{6}Li (n, α) ^{3}H	12.3 years	β^-
^{14}C	^{14}N (n, p) ^{14}C; ^{13}C (n, γ) ^{14}C; ^{17}O (n, α) ^{14}C	5.73 ky	β^-
^{36}Cl	^{35}Cl(n, γ) ^{36}Cl; ^{40}Ar(p, nα) ^{36}Cl; ^{36}Ar(n, p) ^{36}Cl; ^{39}K(2n, 2p) ^{36}Cl; ^{40}Ca(n, 2n3p) ^{36}Cl; ^{40}Ca(μ⁻, α) ^{36}Cl; ^{39}K(n, α) ^{36}Cl	301 ky	β^-
^{41}Ca	^{40}Ca (n, γ) ^{41}Ca	103 ky	EC
^{59}Ni	^{58}Ni (n, γ) ^{59}Ni	75 ky	EC + β^+
^{63}Ni	^{62}Ni (n, γ) ^{63}Ni; ^{63}Cu (n, p) ^{63}Ni	100 years	β^-
^{79}Se	^{78}Se (n, γ) ^{79}Se; ^{235}U (n, f) ^{79}Se	480 ky	β^-
^{90}Sr	^{235}U (n, f) ^{90}Sr	28.6 years	β^-
^{99}Tc	^{235}U (n, f) ^{99}Tc; ^{98}Mo (n, γ) ^{99}Mo (β) ^{99}Tc	211 ky	β^-
^{129}I	^{129}Xe (n, p) ^{129}I; ^{235}U (n, f) ^{129}I; ^{127}I (2n, γ) ^{129}I	15.7 My	β^-
^{135}Cs	^{235}U (n, f) ^{135}Cs	2.3 My	β^-
^{137}Cs	^{235}U (n, f) ^{137}Cs	30.2 years	β^-
^{210}Pb	Naturally occurring; ^{238}U decay	22.3 years	β^-
^{226}Ra	Naturally occurring; ^{238}U decay	1.60 ky	α
^{228}Ra	Naturally occurring; ^{232}Th decay	5.75 years	β^-
^{229}Th	Naturally occurring; ^{233}U decay	7.88 ky	α
^{230}Th	Naturally occurring; ^{238}U decay	75.4 ky	α
^{232}Th	Naturally occurring	14.05 Gy	α
^{233}U	^{232}Th (n, γ) ^{233}U	159.2 ky	α
^{234}U	Naturally occurring; ^{238}U decay	245.5 ky	α
^{235}U	Naturally occurring	0.704 Gy	α
^{236}U	^{235}U (n, γ) ^{236}U	23.4 My	α
^{238}U	Naturally occurring	4.47 Gy	α
^{237}Np	^{238}U(n, 2n) ^{237}U → ^{237}Np; ^{235}U(n, γ) ^{236}Np(n, γ) ^{237}U → ^{237}Np	2.14 My	α
^{238}Pu	^{235}U (n, γ) ^{236}U (n, γ) ^{237}U (β⁻) ^{237}Np (n, γ) ^{238}Np (β⁻) ^{238}Pu; ^{238}U (n, 2n) ^{237}U (β⁻) ^{237}Np (n, γ) ^{238}Np (β⁻) ^{238}Pu	87.7 years	α
^{239}Pu	^{238}U (n, γ) ^{239}U (β⁻) ^{239}Np (β⁻) ^{239}Pu	24.1 ky	α
^{240}Pu	^{238}U (n, γ) ^{239}U (β⁻) ^{239}Np (β⁻) ^{239}Pu (n, γ) ^{240}Pu	6.56 ky	α
^{241}Pu	^{238}U (n, γ) ^{239}U (β⁻) ^{239}Np (β⁻) ^{239}Pu (n, γ) ^{240}Pu (n, γ) ^{241}Pu	14.35 years	β^-
^{242}Pu	^{238}U(n, γ)^{239}U(β⁻)^{239}Np(β⁻) ^{239}Pu(n, γ)^{240}Pu(n, γ)^{241}Pu(n, γ)^{242}Pu	375 ky	α
^{244}Pu	Nucleosynthesis	80.0 My	α
^{241}Am	^{238}U(n, γ)^{239}U(β)^{239}Np(β)^{239}Pu(n, γ)^{240}Pu(n, γ)^{241}Pu(β)^{241}Am	432.2 years	α

In general, determination of radionuclides may be performed either by direct activity measurement, usually termed radiometric analysis, or by mass measurement. However, the required sensitivity usually limits the choice of mass measurement to spectroscopic techniques, mass spectrometry and neutron activation analysis (NAA). NAA is a form of isotopic mass measurement depending on a nuclear transformation reaction followed by a radiometric (usually gamma spectrometric) measurement of the newly induced (by neutron capture) radionuclide or its daughter products. When applying radiometric methods, the radionuclides of interest are determined by their characteristic radiation. The decay rate (number of decays per unit of time) of the radionuclide of interest is measured and the number of atoms present is calculated based on the statistical property of the decay of the radionuclide using its specific half-life. From the basic physical equations is evident that the shorter the half-life of the radionuclide is, the higher is its specific radioactivity. This means that the radiometric methods are generally more sensitive for short-lived radionuclides whilst mass measurement methods are relatively more advantageous for long-lived radionuclides.

Many institutes involved in environmental radioactivity measurements have also access to neutron reactors. Therefore, NAA applied to radionuclides can play a useful role in providing supplementary measurement results thus improving general quality assurance of analytical data.

2. Radiometric Methods

Radionuclides are unstable and de-excite to stable state by radioactive decay with a specific rate (half-life). There are several types of processes involved in the de-excitation of radionuclides, such as α decay, β decay, electron capture, internal conversion, γ-ray emission, and spontaneous fission. Alpha emitters are usually measured by α-spectrometry, beta emitters by beta counter or a liquid scintillation counter (LSC), and the radionuclides with emission of gamma rays by γ-spectrometry. A radionuclide may have more than one decay process; in this case it can be measured by more radiometric methods. For example, ^{129}I is a β-emitter, but also emitting γ-rays with energy of 39.6 keV. Consequently, it can be measured by both beta counting and γ-spectrometry.

3. NAA for Radionuclide Determination

As already mentioned above, NAA is more favorable for low specific activity, i.e. longer lived nuclides. NAA, however, only becomes worth considering when the nuclear characteristics are highly favorable, i.e., the target nuclide has a large capture cross-section for formation of a product nuclide of relatively short half-life with good measurement properties for gamma spectrometric measurement. In the most favorable cases non-destructive, so-called instrumental NAA (INAA) can be employed. This is usually the case for determination of ^{238}U via $^{238}U\ (n, \gamma)\ ^{239}U\ (\beta^-)\ ^{239}Np$ and ^{232}Th via $^{232}Th\ (n, \gamma)\ ^{233}Th\ (\beta^-)\ ^{233}Pa$, in many materials at natural levels. Often radiochemical separation of the induced radionuclide needs to be done after irradiation (radiochemical NAA, RNAA) to improve the signal-to-noise ratio and the sensitivity. This measurement has an important advantage over normal radiometry of the original nuclide in that added carrier could be used to optimize and control chemical recovery, and crucially, the procedure is not subject to blank corrections. Sometimes, however, an element of interest should be separated or concentrated before the irradiation in order to improve the detection limit or even to allow for its determination.

Byrne and co-workers introduced the so-called advantage factor (AF) to quantify the advantages of NAA with respect to radiometric determination of the original radionuclide [2–4]. The advantage factors for some radionuclides are shown in Table 2 [4].

TABLE 2. Values of advantage factor AF for NAA of some long-lived radionuclides [4]

Nuclear reaction involved	AF
$^{238}U\ (n, \gamma)\ ^{239}U$	7.0×10^6
$^{238}U\ (n, \gamma)\ ^{239}U\ (\beta^-)\ ^{239}Np$	8.0×10^5
$^{232}Th\ (n, \gamma)\ ^{233}Th\ (\beta^-)\ ^{233}Pa$	4.0×10^5
$^{230}Th\ (n, \gamma)\ ^{231}Th$	27
$^{237}Np\ (n, \gamma)\ ^{238}Np$	640
$^{231}Pa\ (n, \gamma)\ ^{232}Pa$	106

As shown in Table 2, extremely high values of AF are found for NAA of ^{238}U and ^{232}Th, and lower but still favorable values for ^{237}Np, ^{231}Pa and ^{230}Th.

4. Examples

4.1. TECHNETIUM-99

^{99}Tc levels in the environment are dominated by the releases from the nuclear fuel cycle, mostly through discharges from reprocessing plants and nuclear bomb tests. Its metastable isomer ^{99m}Tc is widely used in nuclear medicine. Due to its high mobility it is also interesting as an environmental tracer. ^{99}Tc is a pure beta emitter with maximum beta particle energy of 294 keV. Radiometric methods using beta counting by gas proportional counter or LSC are therefore the main techniques for its determination. They require a thorough chemical separation from the matrix and other radionuclides because of the difficulties of spectrometric isotope identification for beta emitters. However, ^{99}Tc can also be determined by pre-separation RNAA using the 16-s ^{100}Tc, based on the reaction ^{99}Tc (n, γ) ^{100}Tc. A detection limit of 2.5 mBq (4.0 pg) has been reported [5].

4.2. IODINE-129

^{129}I is formed by uranium fission and cosmic ray reaction with Xe. Its levels in the environment are dominated by discharges from nuclear reprocessing facilities. ^{129}I decays by emitting beta particles with a maximum energy of 154.4 keV; it also emits gamma rays of 39.6 keV and X-rays (29–30 keV). It can therefore be measured by γ-spectrometry and β-counting using LSC. The direct gamma spectrometric measurement with relatively high detection limit can be applied, but chemical separation of iodine from the matrix and interfering radionuclides can improve the detection limit to around 20 mBq. Using LSC and measuring ^{129}I separated from the matrix and other radionuclides results in a slightly better detection limit of about 10 mBq [6]. RNAA determination (involving pre-concentration and post-separation steps) via the reaction ^{129}I (n, γ) ^{130}I ($T_{1/2}$ = 12.36 h) is more sensitive, with the detection limit of 1 μBq [7]. However, in this case one should take into account several interfering nuclear reactions in order to optimize the analytical process: ^{235}U(n, f) ^{129}I(n, γ) ^{130}I; ^{133}Cs(n, α) ^{130}I; ^{128}Te(n, γ) ^{129m}Te(β^-) ^{129}I(n, γ) ^{130}I.

4.3. CESIUM-135

^{135}Cs, a fission product of ^{238}U, is a pure β^- emitter with maximum beta particle energy of 269 keV. Therefore, it can be detected by using a beta counter. However, the presence of ^{137}Cs which is also a beta emitting fission product and usually present in much higher concentrations, makes the detection of ^{135}Cs by beta counting impossible. ^{135}Cs can be also measured by gamma spectrometry by counting its 268.2 keV γ-rays. Due to its low specific activity, radiometric determination is not very sensitive. ^{135}Cs can be determined by RNAA using the 13.16-day ^{136}Cs, based on the reaction ^{135}Cs (n, γ) ^{136}Cs. Due to the very low concentration of ^{135}Cs in the environmental samples, a pre-irradiation concentration and post-irradiation separation has to be carried out for improvement of the detection limit. A detection limit of 0.1 mBq (1 pg) of ^{135}Cs has been reported for a sample with a ratio of ^{133}Cs:^{135}Cs:^{137}Cs of 1:1:1 [8].

4.4. URANIUM AND THORIUM ISOTOPES

Among the naturally occurring uranium and thorium isotopes, there are some very long-lived (^{238}U, ^{235}U, ^{232}Th), some with intermediate half-lives (^{234}U, ^{230}Th) and some with a short half-life (^{234}Th, ^{231}Th, ^{228}Th). As already mentioned in the Chapter 3, $^{238}U$and ^{232}Th can be determined by NAA, by measuring its activation products $^{239}U/^{239}Np$ and ^{233}Pa, respectively.

A so-called LICSIR (Long Irradiation, Cooling, Short Irradiation, and Radiochemistry) technique for the simultaneous RNAA of ^{238}U and ^{232}Th was developed, with the detection limits at pg levels for both radionuclides [9]. In the first, long irradiation ^{233}Pa (27.0 days) is induced by neutron capture on ^{232}Th and then the sample is cooled for several weeks. A second short irradiation to induce ^{239}U (23.5 m) is followed by a rapid sequential radiochemical separation by solvent extraction of ^{239}U and ^{233}Pa. Chemical yields of ^{239}U and ^{233}Pa are measured for each sample aliquot using added ^{235}U and ^{231}Pa tracers from the γ-spectra of the separated fractions.

NAA offers also an interesting possibility of applying internal standard method in alpha spectrometric determination of uranium (^{234}U, ^{235}U, ^{238}U) and thorium (^{228}Th, ^{230}Th, ^{232}Th) radioisotopes using INAA [10]. When applying this approach in e.g., environmental samples, first ^{238}U and ^{232}Th are accurately determined by INAA, by measuring the induced ^{239}Np and ^{233}Pa activities, respectively. From the known mass concentrations of ^{238}U and ^{232}Th, their activity concentrations are derived, and these values are then used as internal standards in the alpha spectrometric analysis. This is performed on separate sample aliquots by the usual dissolution, separation, thin source preparation and α-spectrometric procedures. However, since the activity concentrations of ^{238}U and ^{232}Th are already known, only the relative peak heights of the U and Th radioisotopes in the respective alpha spectra are needed to obtain the absolute specific activities of ^{234}U and ^{235}U, and of ^{228}Th and ^{230}Th. Thus the advantages of this procedure are that neither the chemical yield (recovery) of the radiochemical separations nor the counting efficiency of the α detectors need be known. Hence the use of expensive, calibrated, external radioisotopic tracers for uranium and thorium is eliminated.

Somehow similar approach named ESRR (Endogenous Standard, Radioisotopic Ratio) method in RNAA [11] can be applied for the determination of ^{231}Pa (^{235}U decay chain, $t_{1/2}$ = 32.8 ky) and ^{237}Np. In this approach the endogenous standard is an element or radionuclide (other than the determinant) whose content is already known, or can be determined independently with good accuracy. If a comparator with a known ratio of the determinant and endogenous standard are co-irradiated with the sample, the determinant content is derived in terms of the endogenous standard content and the activity ratios of the two induced nuclides in the sample and comparator. Thus, knowledge of the sample mass and the radiochemical yield is eliminated, and uncertainties due to measurement conditions are substantially reduced. Two examples are shown in Table 3.

TABLE 3. Examples of determinant – endogenous standard pairs used in the ESRR method

Determ.	End. std.	Induced nuclear reactions		Isotopic pair
^{231}Pa	Th	^{231}Pa (n, γ) ^{232}Pa	^{232}Th (n, γ) ^{233}Th (β^-) ^{233}Pa	^{232}Pa-^{233}Pa
^{237}Np	U	^{237}Np (n, γ)^{238}Np	^{238}U (n, γ) ^{239}U (β^-) ^{239}Np	^{238}Np-^{239}Np

4.5. NEPTUNIUM-237

^{237}Np has been produced in nuclear bomb testing and in nuclear reactors followed by releases from spent fuel reprocessing. It is also produced as a consequence of the decay of ^{241}Am. It is highly mobile in the environment and consequently one of the most hazardous radioisotopes in spent nuclear fuel. ^{237}Np is an alpha emitter so α-spectrometry has therefore traditionally been used for its measurement. However, ^{237}Np can also be determined by pre-separation RNAA using the 2.2-days ^{238}Np, based on the reaction ^{237}Np (n, γ) ^{238}Np. A detection limit for ^{237}Np as low as 0.01 mBq (0.5 fg) in environmental and biological samples has been reported [12].

5. Conclusion

It has been shown on a number of examples that neutron activation analysis, independent or in combination with alpha spectrometry, is an important method for the determination of long-lived radionuclides in environmental and biological samples. Important parameters that need to be considered before applying this method include neutron cross sections for the reactions involved, half-lives of the induced radionuclides, neutron irradiation conditions, interfering nuclear reactions, cooling and counting times and potential presence of spectral interferences. However, the procedures involved in the NAA-determinations are not always simple; in most cases, a pre-irradiation or post-irradiation separation step or both of them should be carried out in order to concentrate the determinant and/or achieve adequate decontamination from other activation/impurity radionuclides.

Acknowledgement: The Slovenian Research Agency (Contract No. P2-0075) is greatly acknowledged for its financial support.

References

1. Magill J, Pfennig G, Galy J (2006) Karlsruher Nuklidkarte, 7th Edition 2006, EC–DG JRC–ITE, Karlsruhe, Germany
2. Byrne AR (1986) Determination of ^{237}Np in Cumbrian (UK) sediments by neutron activation analysis: preliminary results. J Environ Radioact 4:133–144
3. Byrne AR (1993) Review of neutron activation analysis in the standardization and study of reference materials, including its application to radionuclide reference materials. Fresenius J Anal Chem 345:144–151
4. Byrne AR, Benedik L (1999) Applications of neutron activation analysis in determination of natural and man-made radionuclides including Pa-231. Czech J Phys 49/S1:263–270
5. Foti S, Delucchi E, Akamian V (1972) Determination of picogram amounts of technetium in environmental samples by neutron activation analysis. Anal Chim Acta 60:269–276
6. Suarez JA, Espartero AG, Rodriguez M (1996) Radiochemical analysis of ^{129}I in radioactive waste streams. Nucl Instrum Meth Phys Res A 272:275–279
7. Hou XL, Dahlgaard H, Rietz B, Jacobsen U, Nielsen SP, Aarkrog A (1999) Determination of ^{129}I in sea water and some environmental materials by neutron activation analysis. Analyst 124:1109–1114
8. Chao JH, Tseng CL (1996) Determination of ^{135}Cs in sodium from an in-pile loop by activation analysis. Nucl Instrum Meth Phys Res A 272:275–279
9. Benedik L, Byrne AR (1995) Simultaneous determination of trace uranium and thorium by radiochemical neutron activation analysis. J Radioanal Nucl Chem 189:325–331
10. Byrne AR, Benedik L (1997) An internal standard method in α spectrometric determination of uranium and thorium radioisotopes using instrumental neutron activation analysis. Anal Chem 69:996–999
11. Byrne AR, Dermelj M (1997) An endogenous standard, radioisotopic ratio method in NAA. J Radioanal Nucl Chem 223:55–60
12. Germian P Pinte G (1990) Neptunium-237 in the marine environment determination in animal and plant species in the English Channel: biological indicators and trophic relationships. J Radioanal Nucl Chem 138:49–61

THE CONCEPT OF VIRTUAL POINT DETECTOR FOR VOLUMINOUS GAMMA DETECTORS

ZEEV B. ALFASSI*
Department of Nuclear engineering, Ben-Gurion University of the Negev, Beer-Sheva, 84105, Israel

Abstract. The replacement of voluminous γ detectors by virtual point detectors allows simple interpolation or extrapolation of the efficiency of the detector with the source-detector distance. The distance between the virtual point detector and the detector's cap was found to depend for cylindrical detectors on both the radius and the height of the detector. This concept can be used also for disk sources and explains the variation of the efficiency of cylindrical sources with the height of the cylinder. This effect was used to calculate the optimal dimensions for a cylindrical source with constant volume.

Keywords: Point detector, disk sources, cylindrical sources, optimal size of cylinder

1. Introduction

In order to measure the full energy of a gamma photon it should lose all its energy in the detector. Due to the large penetration of high energy photons the detector must be of a large volume. Thus a common scintillator for gamma measurement is a 3" × 3" NaI(Tl) and HPGe detectors of more than 100 cm^3 are used in many studies. The large volume of the gamma detector makes it difficult to extrapolate (or interpolate) the efficiency of the detector for different distances or different shapes. This problem was found to be partially solved by the concept of the virtual point detector. This concept replaces the voluminous detector by a point detector, which allows the calculation of distance variation or the use of different geometries [1] using the reciprocal distance law together with the exponential absorption law [2]. However it should be stressed that this point detector has no physical meaning and it is just a mathematical assumption to simplify the

* To whom correspondence should be addressed. e-mail: alfassi@bgu.ac.il

calculations. The distance of the point detector from the detector cap depends both on the detector shape, the source geometry and the energy of the γ photons.

2. Formulation and Results

The distance of the virtual point detector from the detector cap is assigned by h_0. Thus, for a non-absorbing media, equating the emission rates when the source is positioned at two measuring distances from the detector's cap (x_0 and x), and assigning the corresponding counting rates by $C(x_0)$ and C(x) (assuming isotropic emission of the γ photons) yields:

$$C(x)(x+h_0)^2 = C(x_0)(x_0+h_0)^2 \qquad (1)$$

Rearranging Equation (1) yields:

$$\sqrt{\frac{C(x_0)}{C(x)}} - 1 = \frac{1}{h_0 + x_0}(x - x_0) \qquad (2)$$

Or, for $x_0 = 0$, i.e. when the reference point is on the detector cap:

$$\sqrt{\frac{C_0}{C(x)}} - 1 = \frac{1}{h_0} \cdot x \qquad (3)$$

If the virtual point concept is valid, then $C(x)^{-1/2}$ should be linearly dependent on x for all values of x. Notea [3] and Debertin and Helmer [4] proved the existence of this correlation for Ge detectors, when the source is positioned on the central axis of the detector. Presler et al. [5] proved that this concept is valid also for off-center efficiency.

Notea [3] named h_0 as "the effective interaction depth", however a preferred name is the virtual point detector distance, since this is not a physical point which has a physical meaning, but rather a point which has only a mathematical meaning for the use of equations for interpolation and extrapolation of efficiency versus distance.

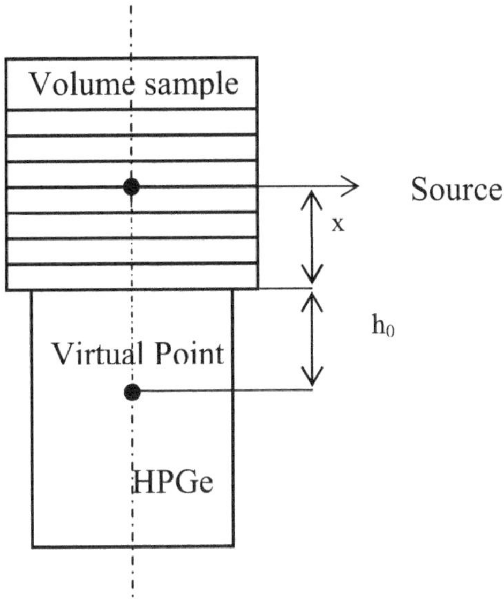

Figure 1. The schematic geometry of a HPGe detector and a point source within a cylindrical absorber

In the case where the volume between the detector cap and the source is an absorbing medium, Figure 1, with an absorption coefficient μ, the count rates for a point source on the symmetry axis (Equation 1) on the absorbing material and on the detector's cap are related by:

$$C(x) \cdot (x + h_0)^2 = C_0 \cdot h_0^{\,2} \cdot e^{-\mu \cdot x} \tag{4}$$

This leads to:

$$\sqrt{\frac{C_0}{C(x)} \cdot e^{-\varpi \cdot x}} - 1 = \frac{1}{h_0} \cdot x \tag{5}$$

Figure 2 shows an example of the results plotted according to Equation (5) for various absorbers. Table 1 gives the various h_0 which were obtained from the analysis of the experimental results according to Equation (5). It can be seen that h_0 varies with the energy of the photons and the absorbing material. It can be seen that in general the h_0 increases with the energy of the photons and except in the case of lead it increases with the linear absorption coefficient, μ [6].

TABLE 1. The calculated h_a values (from the experimental data) for different absorbers and energies

Absorber media	Photons energy (keV)	h_a (cm)
Air	661.65	1.48 ± 0.01
	1,173.2	1.66 ± 0.01
	1,332.5	1.57 ± 0.02
Aluminum	661.65	1.76 ± 0.05
	1,173.2	2.11 ± 0.05
	1,332.5	2.13 ± 0.07
Iron	661.65	2.31 ± 0.02
	1,173.2	2.61 ± 0.04
	1,332.5	2.70 ± 0.04
Copper	661.65	2.39 ± 0.02
	1,173.2	2.63 ± 0.04
	1,332.5	2.72 ± 0.05
Lead	661.65	2.24 ± 0.02
	1,173.2	2.61 ± 0.05
	1,332.5	2.76 ± 0.02

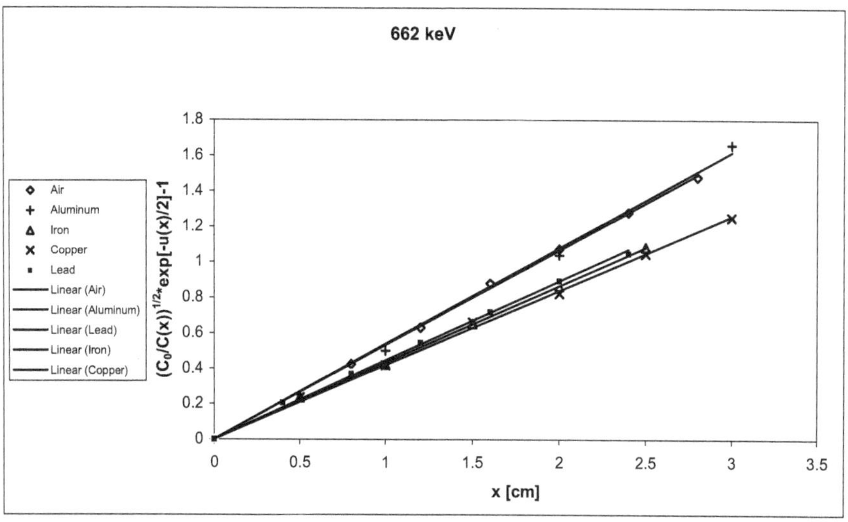

Figure 2. The fitted linear curves to the experimental results for various absorbing media, for a ^{137}Cs source. h_a is determined from the slope of the fitted lines

Presler et al. [7] found that the concept of virtual point detector can be applied not only to Ge detectors but also to scintillation detectors, such as NaI(Tl) and BGO. It was found that similarly to Ge detectors this assumption is not very accurate close to the detector cap [8]. However, for distances larger than 4 cm Equation (2) is quite accurate, as can be seen in Figure 3. For the smaller distances a quadratic equation is very accurate.

Debertin and Helmer [4] stated that the virtual point detector lies inside the detector and its distance from the cap is increasing with the photon energy till an asymptotic value of half the thickness of the detector. However, Alfassi et al. [9] found that for planar and semi-planar Ge detectors the virtual point detector lies outside of the detector. This is logical since the distance from the center of the cap of the detector to its periphery is larger than its thickness. Since it is almost impossible to obtain many detectors with different radii and thicknesses, the effect of the radius, R_d, and the thickness of the detector, H_d, of cylinder-shaped detectors were studied by Monte-Carlo simulations. The calculations were done using the MCNP code, for transport of γ photons, for 49 detectors (seven different radii and seven different heights) for seven energies from 0.186 to 1.836 MeV. For each energy the full-energy efficiency was calculated at various distances from the detector's cap. The results were analyzed according to Equation (3). As an example Table 2 gives the results of h_0 and the correlation coefficient for one energy (0.846 MeV) for the 49 detectors. It was found that for a constant thickness and various radii the value of h_0 is proportional to the radius of the detector, R_d, i.e. it is given by the equation:

$$h_0 = A \cdot R_d \tag{6}$$

Where A is a constant, which depends on the thickness of the detector. It was found, Figure 4, that for a constant radius, A increases asymptotically with the thickness of the detector-H_d.

$$A = \alpha \cdot (1 - b \cdot e^{-c \cdot H_d}) \tag{7}$$

The parameter α is the plateau value, A_{max}, in Figure 4. It was found that the parameter c is almost proportional to the ratio μ/ρ, where μ is the linear attenuation coefficient and ρ is the density of the detector material (μ/ρ is called the linear mass attenuation coefficient). It was found that for the range of 0.186–1.836 MeV the parameter c is given quite accurately by:

$$c = 4.29 \cdot \mu / \rho \tag{8}$$

Together with Equation (6) it leads to:

$$h_0 = a \cdot (1 - b \cdot e^{-4.29 \cdot (\mu/\rho) \cdot H_d}) \cdot R_d \qquad (9)$$

The concept of the virtual point detector can be stretched to radioactive bulky samples by integration over the whole sample. Debertin and Helmer [4] did it for a sample in the form of a thin disk of radius *R* at the distance *d* from the detector's cap and obtained:

$$\varepsilon(d) = \varepsilon_p(0,d) \cdot \frac{d^2}{R^2} \cdot \log(1 + \frac{R^2}{d^2}). \qquad (10)$$

Here, $\varepsilon(d)$ is the efficiency of the disk, and $\varepsilon_p(0,d)$ is the efficiency of the point source containing the same total activity as the whole disk and located on the axis of the symmetry of the detector at the distance *d* from the detector's cap.

For a cylinder, thicker than a thin disk, Equation (10) should be integrated, with the term for the photon absorption either included or neglected. The latter is justified for high energy photons because, for energies above 200 keV, the geometric effect is considerably larger than the absorption effect. In order to simplify this integration, Alfassi et al. [11] studied the possibility that the model of the virtual point detector can be used also for disk sources, i.e. fulfilling Equation (5) and to determine how h_0 depends on the radius of the radioactive disk. They used cylinders of small thickness of either powder or aqueous solutions of $Th(NO_3)_4$ and ^{232}Th daughters, thus having photons from 0.077 MeV to 2.615 MeV. Figure 5 shows examples of the experimental data indicating that photons of various energies in disk sources of various radii fulfill Equation (5) with high accuracy. All the lines have a correlation coefficients (Pearson squared) larger than 0.99 Table 3 summarizes the values of h_0 for the various radii of the disks for various photon energies.

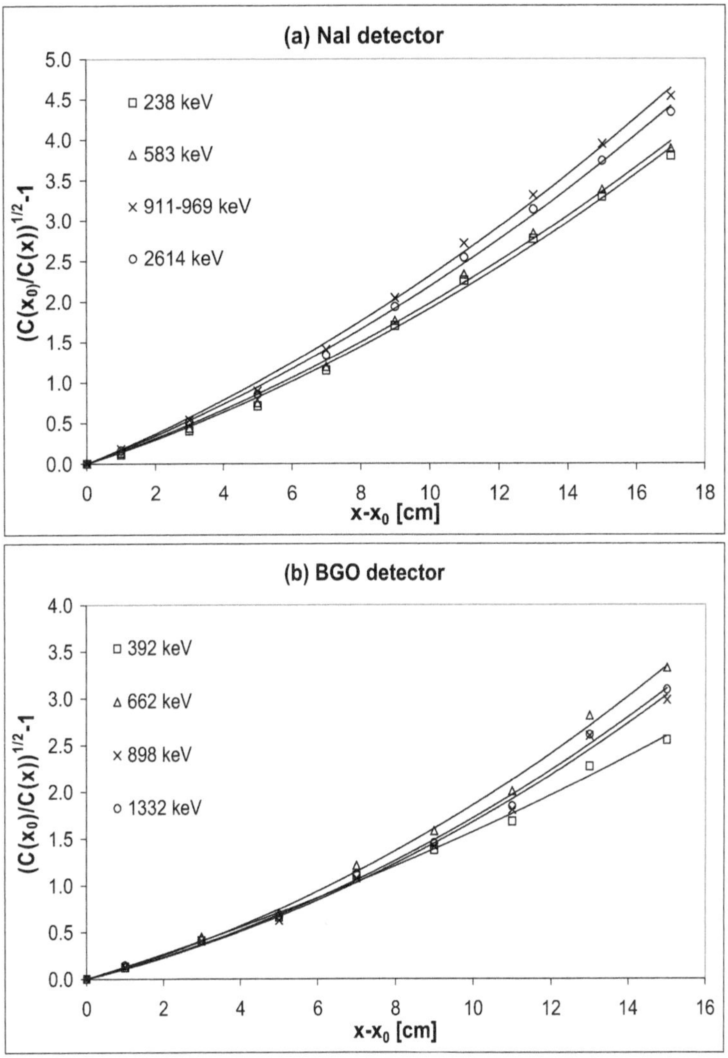

Figure 3. Fitted quadratic functions to experimental results for some photon energies, measured with: (**a**) a 75 × 75 mm NaI(Tl) detector (**b**) a 75 × 75 mm BGO detector

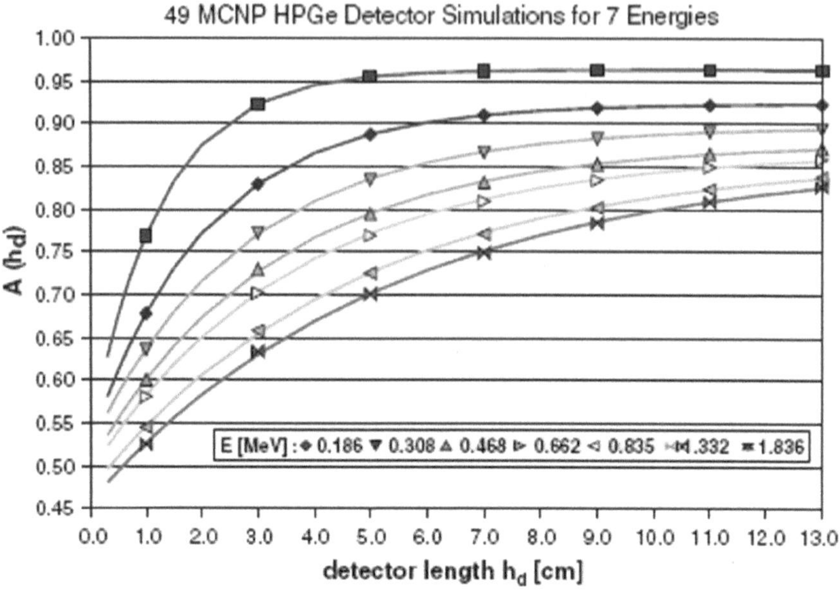

Figure 4. Plot of A versus H_d for seven different energies for results from MCNP simulations (*points*) and the corresponding non-linear fitted data (*lines*)

TABLE 2. Virtual detector point distances h_0 for 49 different cylinder-shaped detector geometries at 0.468 MeV (in brackets appear the correlation coefficients)

Rd (cm)	hd (cm)						
	13	11	9	7	5	3	1
13	12.3 (0.9964)	12.2 (0.9964)	12.1 (0.9963)	11.9 (0.9962)	11.4 (0.9961)	10.6 (0.9962)	8.6 (0.9970)
11	9.9 (0.9975)	9.8 (0.9975)	9.7 (0.9974)	9.6 (0.9973)	9.2 (0.9972)	8.5 (0.9971)	7.0 (0.9975)
9	7.7 (0.9985)	7.7 (0.9984)	7.6 (0.9984)	7.5 (0.9983)	7.2 (0.9982)	6.7 (0.9980)	5.6 (0.9981)
7	5.8 (0.9992)	5.8 (0.9992)	5.7 (0.9992)	5.6 (0.9991)	5.4 (0.9990)	5.0 (0.9988)	4.2 (0.9987)
5	4.1 (0.9998)	4.1 (0.9997)	4.0 (0.9997)	4.0 (0.9997)	3.8 (0.9996)	3.5 (0.9995)	2.9 (0.9993)
3	2.6 (1.0000)	2.6 (1.0000)	2.6 (1.0000)	2.5 (1.0000)	2.4 (0.9999)	2.2 (0.9999)	1.8 (0.9998)
1	1.2 (0.9999)	1.2 (0.9999)	1.2 (0.9999)	1.2 (0.9999)	1.2 (1.0000)	1.2 (1.0000)	0.8 (1.0000)

It was found that for a small radius, h_0 increases with the energy until it reaches a plateau. However, for larger radii, the plateau value is reached at very low photon energy. Thus, for the largest studied disk (radius 6 cm), the plateau value is almost reached even at 200 keV, Figure 6. It was found that for the small-radius disks, the increase in h_0 with increasing photon energy can be well described by the equation

$$h_0 = \alpha \cdot \exp(-\beta \cdot \mu) \tag{11}$$

Where α and β are constants depending on the radius of the disk and μ is the total absorption coefficient for photons with a specific energy in Ge.

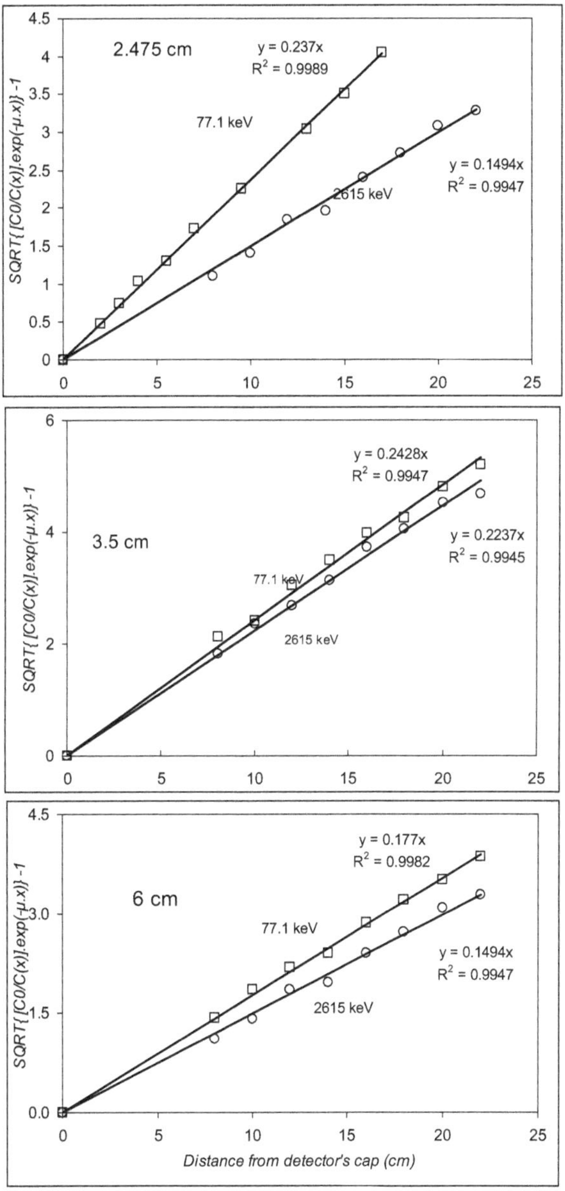

Figure 5. Experimental data plotted according to Equation (5) for disks with radii of 2.475, 3.5 and 6 cm

TABLE 3. Values of h_0 for the various full-energy peaks and disk diameters

Energy (keV)	Radius of disk (cm)								
	0.305	0.47	1.0	1.5	2.0	2.475	3.5	4.5	6.0
77.1	1.47	1.5	2.11	2.92	3.13	3.45	4.12	5.65	5.97
87.2	1.83	1.82	2.4	2.92	3.56	3.62	4.22	5.82	6.57
129.1	2.15	2.59	2.97	3.42	3.7	3.77	4.34	6.11	6.67
209.4	2.86	2.97	3.16	3.79	3.87	3.91	4.38	6.18	6.83
238.6	3.17	3.24	3.03	3.55	3.79	3.94	4.38	6.23	6.87
270.3	3.07	3.19	3.56	3.92	3.98	4	4.39	6.25	6.9
300.1	3.1	3.23	3.7	3.96	3.96	4.01	4.38	6.3	6.91
328	3.11	3.25	3.79	4	4.02	4.06	4.39	6.35	6.92
338.4	3.09	3.28	3.83	3.98	3.99	4.03	4.4	6.3	6.92

Figure 7 shows that h_0 increases with the radius of the disk. The increase is almost linear with the radius of the disk.

The findings that the concept of virtual point detector can be applied also to disk sources simplify the calculation of the efficiency of sources of cylindrical shapes [12]. Alfassi and Groppi [12] studied the count-rates of cylinders containing aqueous solutions of $^{232}Th(NO_3)_4$ and ^{232}Th daughters as a function of their heights. Similarly to previous study of Vesic and Anicin [13] who studied the count rates obtained from ^{232}Th-doped sand in cylinders of varying heights and radii. For a constant radius they found that the total number of counts as a function of height (or total volume) is a function describing saturation, i.e. a function which goes asymptotically to a limiting value. The limiting value increases with the radius of the cylinder. They found the same effect for several γ energies in the range of 239–2,615 keV from the mixture of sand with ^{232}Th and its daughters. They suggested by visual inspection that this dependence "suggests exponential saturation", i.e. $N(h) = N_\infty \cdot (1 - e^{-\mu \cdot h})$ where N_∞ and μ are constants and N(h) is the count rate for a cylinder with height h. They explained it as due to absorption of parallel beams of photons.

However, there are two factors that contribute to the observation that the count rate due to a γ emitting cylindrical source, with a constant radius, is less than proportional to the height of the cylinder (the volume of the sample); geometry and absorption. These are the two factors that cause the count rate due to a disk with thickness dh within the cylinder in distance h from the detector to decrease with the increase of h. The geometry factor is due to the isotropic nature of the γ emission and the decrease of the solid angle in which the detector sees the disk with the increase of h; the solid angle decreases with the increase of h. The second factor is due to the self attenuation of the photons due to scattering/absorption by the sample between the point where the photon was emitted and the detector; the scattering/absorption increases with the increase of h. The two factors lead in the same direction (decrease of the count rate with increase of h) but have different dependencies on h and the energy of the photons. The second factor depends strongly on the energy of the photons, while the first one is almost independent of the photons energy. For low energy photons, the second factor is the dominant one, however, for high energy photons the first factor is the dominant one.

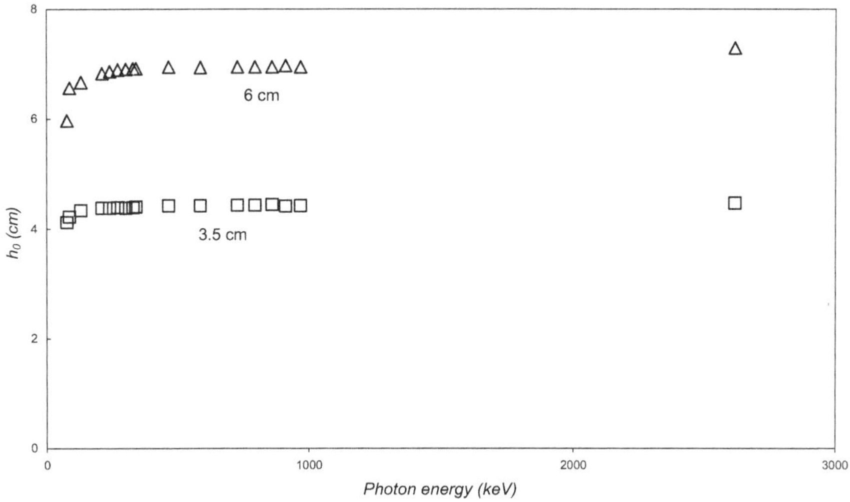

Figure 6. Dependence of h_0 on the photon energy for radioactive disks of radii 3.5 cm (■) and 6.0 cm (▲)

Vesic and Anicin based their equation assuming that the second factor is the dominant one but this is not true even for their lowest energy of 238.6 keV from ^{212}Pb (a member of the ^{232}Th chain).Their detector has an height (thickness) of 3.9 cm. If we assume that the virtual point detector distance is one half of the detector thickness (4), the count rate decrease due to the geometric factor for a point source at 10 cm from the detector compared to the point source positioned on the detector's cap is: $[(10 + 1.95)/1.95]^2 = 37.6$. For 239 keV photons the mass absorption coefficient is $\mu/\rho = 0.12$ cm^2 g^{-1}. Using for the sand $\rho = 1.5$ g/cm^3 the self absorption factor at distance of 10 cm is: $\exp(1.5 \cdot 0.12 \cdot 10) = 6.05$. Thus, the decrease of the count-rate due to the solid angle geometric factor is about six times larger than due to the self absorption. This ratio depends on h, but a factor of about 5 is correct for all distances between 4 and 16 cm. If the virtual point distance is lower, e.g. only one quarter of the thickness of the detector, then the geometric factor will be higher while the self absorption will remain the same, leading to larger dominance of the geometric factor. For 2,615 keV the geometric factor is about the same while the self-absorption will be only 1.82 (μ/ρ = 0.04), leading to the geometric factor being dominant by a factor of 20.

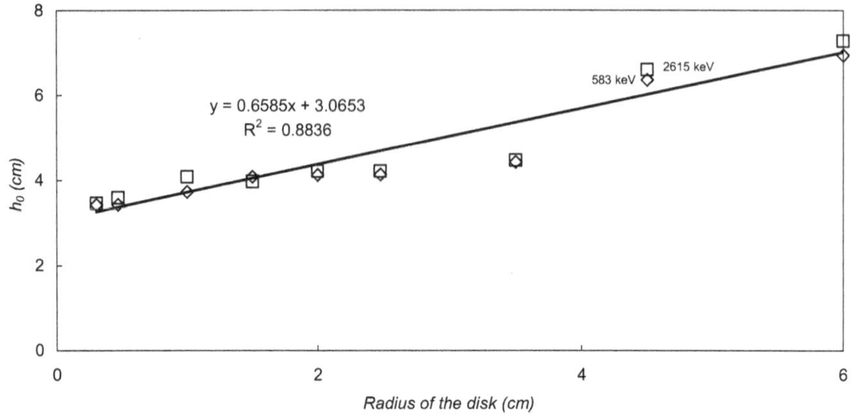

Figure 7. Dependence of h_0 on the radius of the disk for photon energies 583.1 keV (■) and 2,614.6 keV (▲)

Assigning the distance between the virtual point and the cap of the detector by h_0, the contribution to the total count rate from a disk of thickness dh in the cylinder at distance h from the bottom of the cylinder, which is positioned on the detector cap, is given by:

$$c(h)dh = A \cdot (\frac{h_0}{h + h_0})^2 \cdot e^{-\mu \cdot h} \cdot dh \quad (12)$$

Here A is a constant, which depends on the concentration of the radioactive material, the radius of the cylinder base and the characteristics of the detector. The count rate due to the whole cylinder with height H will be given by:

$$C(H) = \int_0^H c(h)dh \quad (13)$$

The integral, substituting c(h) from Equation (12), gives an analytic solution only in the form of an infinite series. However, in two cases it is possible to make some approximations which will lead to simpler results.

One case is when the bottom of the cylinder is far from the detector cap. This distance should be at least twice the virtual distance h_0, i.e. $h \gg h_0$. In this case we can use the approximation:

$$\frac{h_0}{h+h_0} = (1+\frac{h}{h_0})^{-1} = e^{\frac{h}{h_0}} \tag{14}$$

However, positioning the cylinder far from the detector will reduce the count rate, usually to values too low to yield sufficient accuracy of the radioactivity of environmental samples.

The second case is when the energy of the photons is sufficiently high such that the geometry factor is the dominant contributing factor and the absorption term can be neglected (Eγ > 250 keV), In this case the cylinder can be positioned on the detector's cap. Neglecting the scattering/ absorption term the integration yields:

$$C(H) = \frac{A \cdot h_0 \cdot H}{h_0 + H} \tag{15}$$

For a very short cylinder ho >> H and then $C(H) = A \cdot H$; meaning that C(H) is proportional to the cylinder height. The constant A can be assigned as C_0, the linear count rate density on the detector face, which has the units of counts/cm. Equation (15) is clearly leading to an asymptotically constant value with the increase of H, since for H >> h_0, Equation (15) yields: $C(H_\infty) = C_0 \cdot h_o$.

To validate the adequacy of Equation (15) to describe Vesic and Anicin's data and Alfassi and Groppi's data, Equation (15) can be transformed to:

$$\frac{1}{C(H)} = \frac{1}{C(H_\infty)} + \frac{1}{C_0} \cdot \frac{1}{H} \tag{16}$$

Plotting 1/C (H) versus 1/H should yield a straight line. In order to plot Equation (16) for different γ energies, which have different intensities and counting efficiencies, in the same plot we can normalize Equation (16) by multiplying the equation by C (H_1), where H_1 is the lowest height for which the measurement was done:

$$\frac{C(H_1)}{C(H)} = \frac{C(H_1)}{C(H_\infty)} + \frac{C(H_1)}{C_0} \cdot \frac{1}{H} \tag{17}$$

Figure 8 gives the plot of $C(H_1)/C(H)$ versus 1/H. which according to Equation (17) should be a straight line, for some of Vesic and Anicin's data both for the high and low photon energies (2,615 and 238 keV) for the largest and smallest radii (9 and 5 cm). Figures 9 and 10 give the same plot for some of Alfassi and Groppi's data, each figure is for two photon energies for the same cylinder. As can be seen in Figures 8–10 the fitness to straight line is very good, in both cases the correlation coefficient (R^2) is larger than 0.995. This agreement validates our explanation for the dependence of the total count rate on the cylinder height.

Our Equation (17) has also another advantage on Vesic and Anicin's equation besides being physically correct. For the largest radii (9 cm) they did not measure the limiting saturation count rate, either due to insufficient material or to too high dead time. Their equation $N(h) = N_{\infty} \cdot (1 - e^{-\mu \cdot h})$ does not allow finding both N_{∞} and μ by linear regression and for this case they must use non-linear regression. The linear plot of Equation (17) does not require the knowledge of the saturation value, and both C_0 and h_0 (which yields $C(H_{\infty})$, since $C(H_{\infty}) = C_0 \cdot h_o$) are obtained by linear regression.

For a constant volume of sample, it is obvious that both a very tall small radius cylinder and a very shallow large radius cylinder will lead to a small count rate, due to the large distance between most of the sample and the detector.

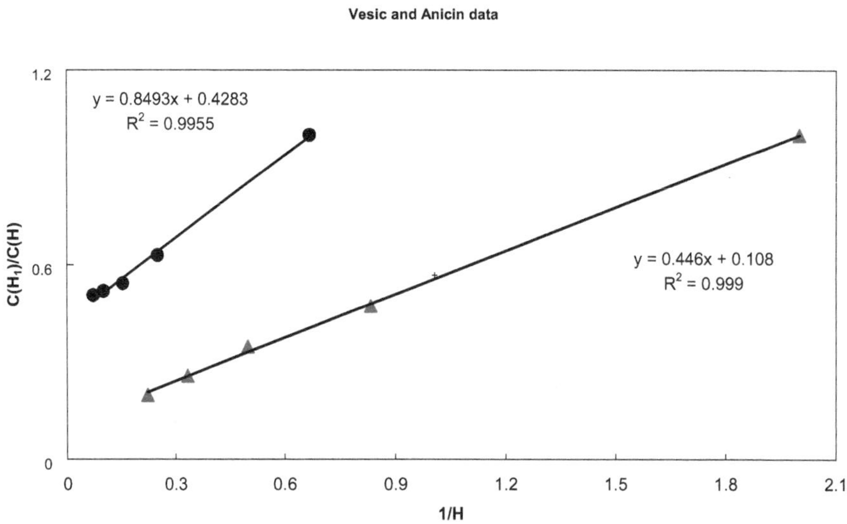

Figure 8. Analysis of Vesic and Anicin's data according to Equation (17) for R = 5 cm E_γ = 238 keV (●) and for R = 9 cm E_γ = 2,615 keV (▲)

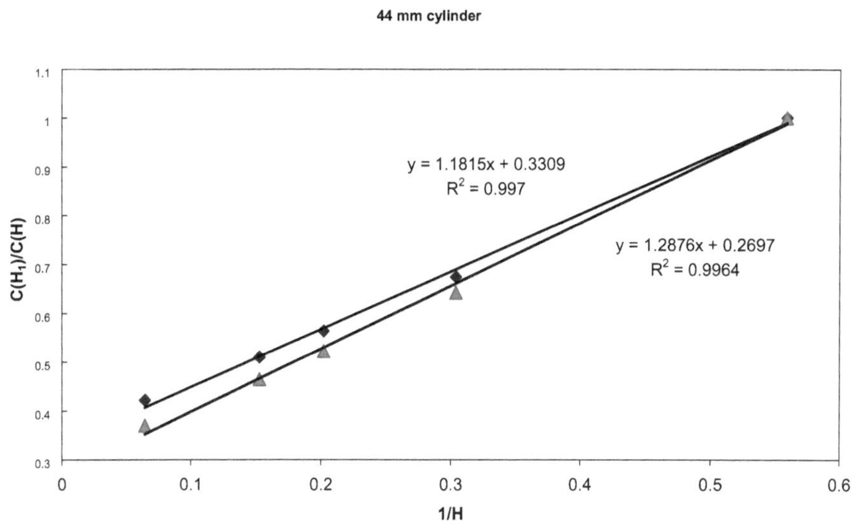

Figure 9. Analysis of Alfassi and Groppi data for according to Equation (17) for R = 4.4 cm and for the photon energies of 967 keV (•) and 2,615 keV (▲)

These two extreme cases indicate that between these cases there should be optimal dimensions of the cylinder, i.e. for each volume there will be optimal radius and height leading to maximum count rate. Equation (15) can be used in order to calculate the optimal dimensions of the cylinder. However, C_0 depends on the radius of the cylinder and the first problem is to find the relation between C_0 and the radius of the cylinder. Equation (1) can be written for a small radioactive source of area S on the detector cap in distance x from the center of the cap in the following form:

$$c(x) = \frac{\alpha \cdot S}{x^2 + h_0^2} \tag{18}$$

Where α is a constant proportional to the area activity concentration. The count rate due to a ring in distance between r and r + dr from the cap's center is given by:

$$c(r, r+dr) = \frac{\alpha \cdot 2 \cdot \pi \cdot r \cdot dr}{r^2 + h_0^2} \tag{19}$$

Integration of Equation (19) between 0 and R, where R is the radius of the cylinder base leads to a similar equation as Equation (10) [4]:

$$C_0 = 2 \cdot \pi \cdot \alpha \cdot \log(1 + \frac{R^2}{h_o^2}) \tag{20}$$

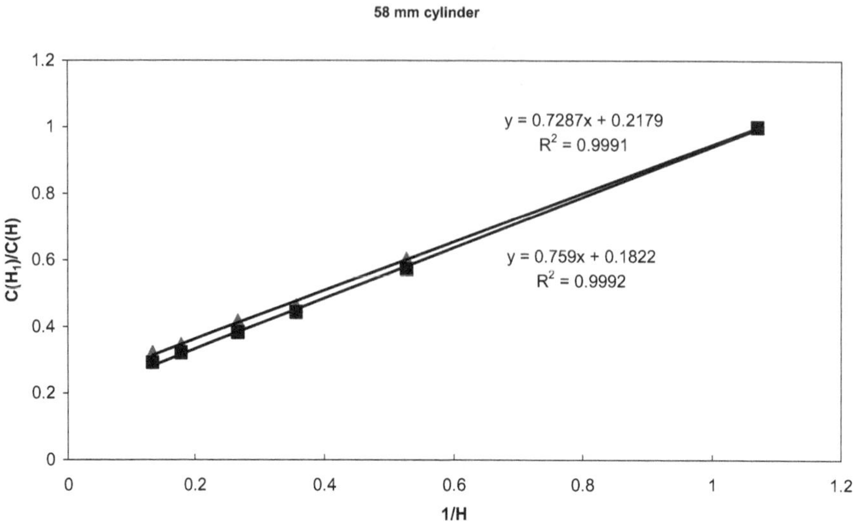

Figure 10. Analysis of our data for according to Equation (17) for R = 5.8 cm and for the photon energies of 238 keV (●) and 583 keV (▲)

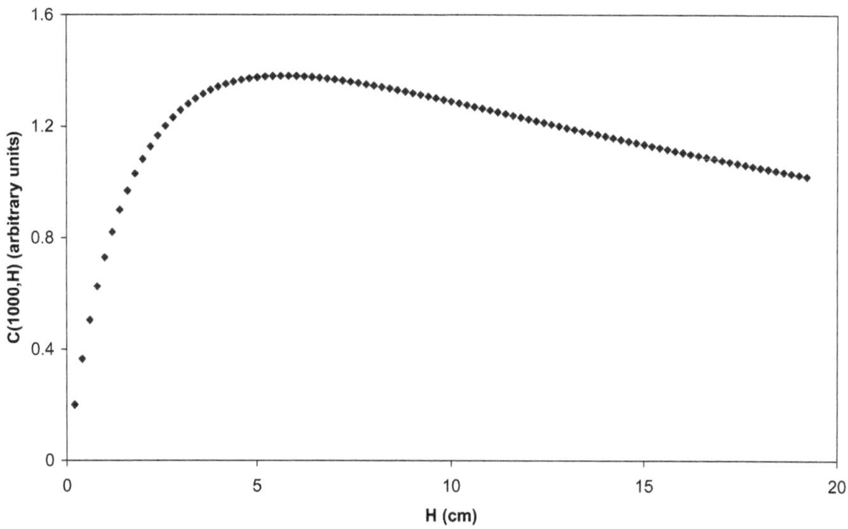

Figure 11. Dependence of C(H, V) on H for V = 1,000 cm^3

Hence:

$$C(H) = 2 \cdot \pi \cdot \alpha \cdot \frac{h_0 \cdot H}{h_0 + H} \cdot \log(1 + \frac{R^2}{h_o^{\ 2}}) \tag{21}$$

In order to find the optimal dimensions of a constant volume cylinder (V), it is sufficient to deal with the height of the cylinder, since for constant volume; R is a function only of the cylinder height. Substituting R^2 in Equation (21) by:

$$R^2 = \frac{V}{\pi \cdot H} \tag{22}$$

Leads to the following equation for the count-rate due to a cylindrical volume V and cylinder H:

$$C(H,V) = 2 \cdot \pi \cdot \alpha \cdot \frac{h_0 \cdot H}{h_o + H} \cdot \log(1 + \frac{V}{h_o^2 \cdot \pi \cdot H}) \tag{23}$$

To find the optimal H which will lead to maximum count rate for the volume V, C(H, V) should be differentiate with respect to H and the derivative should be equated to zero. It should be remembered that h_0 was found to be linearly dependent on R and hence on H [10]. However, even assuming h_0 to be independent of R, equating dC(H, V)/dH to zero does not yield an explicit equation for the optimal H. Hence, the simple way to find the optimal H is to plot Equation (23) for varying values of H, as was done in Figure 11 for V = 1,000 cm^3 and $h_0 = 3cm + 0.5 \cdot R$. Thus for $h_0 = 3cm + 0.5 \cdot R$ it can be found that the optimal H is 3.0 cm for V = 100 cm^3, 7.2 cm for V = 1,000 cm^3 and 15.6 cm for V = 10,000 cm^3. It should be pointed out that around the optimal H the change of C(H, V) with H is very small. Thus for V = 1,000 cm^3 allowing C(H, V) to be in the range of ±10% allow H to be in the range of 3.8–14.4 cm, Figure 11, indicating that for a first approximation we can take R = H, which yields:

$$H = \sqrt[3]{V / \pi} \tag{24}$$

Leading to 3.2 cm, 6.8 cm and 14.7 cm for the heights of cylinders with volumes of 100 cm^3, 1,000 cm^3 and 10,000 cm^3, respectively.

The optimal H depends on the variation of h_0 with the radius of the cylinder. If the dependence is $h_0 = 3cm + 0.25 \cdot R$ (i.e. R is multiplied by 0.25 instead of 0.5) the optimal values for the three volume mentioned above will be 2.5, 5.8 and 11.7 cm, respectively. The suggested first approximation will give not less than 95% of the maximal C(H, V) even in this case.

References

1. Alfassi ZB, Groppi F, Bonardi ML, Presler O, German U (2006) A note on the interpolation and extrapolation of gamma detection efficiency curve as a function of the distance. J Radioanal Nucl Chem 268:639–640
2. Debertin K, Ren KJ (1989) Measurement of the activity of radioactive samples in Marinelli beakers Nucl Instrum Meth Phys Res A 278:541–549
3. Notea A (1971) The Ge(Li) spectrometer as a point detector. Nucl Instrum Meth 91:513–515
4. Debertin K, Helmer RG (1988) γ- and x-ray spectrometry with semiconductor detectors. Elsevier, New York
5. Presler O, Pelled O, German U, Leichter Y, Alfassi ZB (2002) Off center efficiency of HPGe detectors. Nucl Instrum Meth Phys Res A 484:444–450
6. Presler O, German U, Pelled O, Alfassi ZB (2004) The validity of the virtual point detector concept for absorbing media Appl Radiat Isotopes 60:213–216
7. Presler O, German U, Pushkarsky V, Alfassi ZB (2006) Virtual point detector: on the interpolation and extrapolation of scintillation detectors counting efficiencies. Nucl Instrum Meth Phys Res A 565:704–710
8. Cline JE (1978) A technique of gamma ray detector absolute efficiency calibration for extended sources. Proceedings of the American Nuclear Society Topical Conference on Computers in Activation Analysis and Gamma Ray Spectroscopy Mayaguez, Puerto Rico, CONF. 780421
9. Alfassi ZB, Pelled O, German U (2006) The virtual point detector concept for HPGe planar and semi-planar detectors. Appl Radiat Isotopes 64:574–578
10. Mahling S, Orion I, Alfassi ZB (2006) The dependence of the virtual point detector on the HPGe detector dimensions Nucl Instrum Meth Phys Res A 557:544–553
11. Alfassi ZB, Lavi N, Presler O, Pushkarski V (2007) HPGe virtual point detector for disk radioactive sources Appl Radiat Isotopes 65:253–258
12. Alfassi ZB, Groppi F (2007) An empirical formula for the efficiency detection of Ge detectors for cylindrical radioactive γ sources Nucl Instrum Meth Phys Res 574A: 280–284
13. Vesic D, Anicin IV (1989) Some practical aspects of gamma ray spectroscopy of voluminous cylindrical sources with germanium detectors. Nucl Instrum Meth Phys Res A 276:216–222

THE LOCALIZATION OF A SMALL NEUTRON SOURCE IN A HOMOGENEOUS MEDIUM

SERGEI DUBINSKI, OREN PRESLER AND
ZEEV B.* ALFASSI
Department of Nuclear Engineering, Ben Gurion University, Beer Sheva, 84105, Israel

Abstract. The possibility of localization of an unknown neutron source in various bulky homogeneous media (box) was studied. For the planar case two ^{3}He detectors on the opposite faces of the box were used. A constant polypropylene shield around the box and the detectors was used to eliminate the varying contribution from the environment, to increase count rates of the detectors and to protect the experimentalist. It is shown that the location of a single small neutron emitting source in a large box can be found to a better than 7% by using two neutron detectors positioned on parallel faces of the box, coplanar with the source. The localization requires measurement of the count rate of both the unknown source and an extra source positioned on one of the faces of the box. The localization is based on the finding that the ratio of the count rates of the two detectors is an exponential function of the distance of the source from one of the detectors. In the case that the plane of the unknown source is not known, four detectors are required and an iteration method is used for localization of the source plane.

Keywords: Localization, neutron source, homogeneous medium, multi-detector measurement

1. Introduction

The purpose of this research is the localization of a small neutron source in an unknown homogeneous medium of known large size sample and subsequently finding its activity. Possible applications of our results are for example the measurement of radioactive wastes, finding small sources in

* To whom correspondence should be addressed. e-mail: alfassi@bgu.ac.il

glove boxes, the discovery of smuggled neutron emitting point sources as well as alpha sources due to (α, n) reactions, etc. The work was carried out both experimentally and by means of computational Monte-Carlo simulations.

It was established in the past [1–4] that there is a possibility to determine the location of a gamma radiation sources (by measuring their characteristic peaks) in homogeneous medium by the use of several detectors. But, the utilization of this method for the case of a neutron source is much more complicated. In the case of a γ source only the non-interacted photons can be measured due to the initial γ rays being mono-energetic, and to the measurement of the whole γ spectrum. In contrast, in the case of neutrons, most neutrons sources are not mono-energetic and the measurement of the neutron spectrum is very difficult. Most neutron detectors have considerably larger sensitivity for thermal neutrons and hence they yield mainly the flux of the thermal neutrons and not the more energetic ones. Hence, the detected neutrons actually interact through various scattering processes prior to the detection.

It was found [5] that the number of neutrons in a narrow beam in a homogeneous medium fall off exponentially with absorber thickness, but in case of real source (isotropic emission) the reflections from the environment that contribute to the count rate in the detector should be taken into consideration.

Some studies on the localization of a neutron source were made in the past. Antonopoulos-Domis and Tambouratzis [6] determined the presence of even plutonium isotopes (EPI) within sealed tanks by oscillating the suspect tank in a well counter. The well counter consisted of a paraffin cylinder and 12 ^{3}He detectors. The tank was rotated with a known frequency and the problem of localization was solved by least squares estimation.

Peurrung et al. [7] proposed the use of a moderator-free directional thermal neutron detector for identification and localization of neutrons sources even at distances up to 24 m. They placed neutron detector that is sensitive only to thermal neutrons inside a thermal neutron shield (cadmium box) and restricted the field of view using a collimator coated with a thermal neutron absorber. The experimental setup contained 23 ^{3}He proportional counter tubes placed in cadmium box with collimating array. This method will work only when some amount of moderator is present near the source or between the source and the detector.

Linden et al. [8] used a small scintillation detector, attached to an optical fiber to localize neutron source in a homogeneous water medium, by measuring of the flux and its gradient.

Later, Avdic et al. [9] measured scalar neutron flux and neutron current by optical fiber detector to localize a neutron source in a water tank.

2. Experimental Setup

The experimental setup consists of a rectangular box 460 × 200 × 200 mm^3 made of 5 mm thick Perspex, with two ^3He detectors on opposite sides of the box (i.e. at 180° one to another), at a distance of 485 mm (center to center). A small source (^{252}Cf or AmBe) is positioned at different places inside the box. Plates made of different materials, were inserted into the box. The source holder was made of polypropylene 2.5 cm thick, 30 cm height and 20 cm length (the source placed in the middle of the holder in height of 10 cm).

To eliminate the varying contribution from the environment, a constant polypropylene shield was placed around the whole area of the sample (box) and the detectors. Another purpose of the polypropylene shield is to increase the neutron count rate in the detectors, due to reflections. It was found that the increase in the neutron net count rate due to the reflector is up to a factor of 10. That make possible to detect weaker sources in reasonable time. This shield also allows the experimentalist to work close to the system.

Neutron calculations were performed with the Monte-Carlo code MCNP-4C [10], utilizing the cell flux tally (F4). The F4 tally is an estimator of the expected flux value in the cell. This tally, when weighted by the material atomic density and absorption cross section (F4 and FM4 combination), scores the number of neutrons absorbed in a real ^3He detector placed at the same flux [10].

The energy spectrum of the neutrons emitted by the AmBe source was taken from the literature [11]. ^{252}Cf spontaneous fission spectrum was taken directly from MCNP-4C libraries, according to Watt fission spectrum [10].

$$f(E) = C \cdot \exp(-\frac{E}{a}) \cdot \sinh(bE)^{\frac{1}{2}},$$

with the constants: a = 1.025 MeV and b = 2.926 MeV^{-1}.

3. Results

3.1. THE PLANAR CASE

In this case the source is located on a known line but the position on the line is not known. The neutron source was positioned at different locations on the plane connecting the two detectors at constant height which is about the center of the two detectors. A schematic diagram of the experimental system is shown in Figure 1. The two detectors were operated simultaneously, each connected to a separate multi-channel analyzer (MCA) via conventional electronic setup. The processing of the results was made by computing the sum over the spectrum, due to the neutrons. γ-rays have much smaller voltage in a ^{3}He detector then neutrons, and are rejected by the bias voltage of the MCA.

3.1.1. *Measurement with a single detector*

Experimental results and Monte-Carlo simulations show that for a single detector and AmBe or ^{252}Cf sources, in different moderating medium, the dependence of the count rate due to neutrons on the source-to-detector

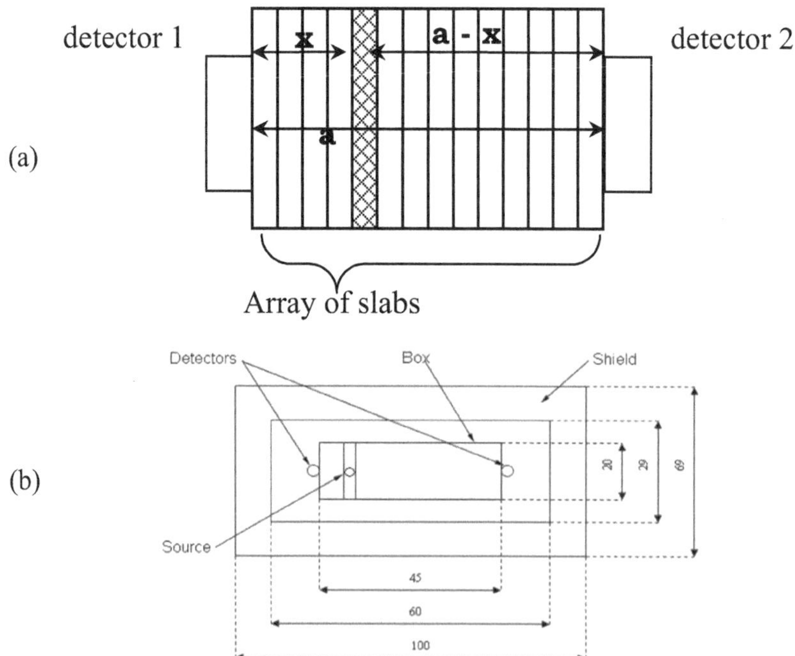

Figure 1. Diagram of the experimental system (**a**) side-view, (**b**) top-view

distance can be described reasonably, but not too well, by an exponential function, as can be seen in Figures 2a and 3a. In Figure 3a the value of the last point is lower than the value of the next-to-last point. This is due to absence of moderator between detector 2 and source in this point and hence the count rate drops. The detector is sensitive only to thermal neutrons, so without slowdown of neutrons the detector will count only slow neutrons originating from the source and neutrons reflected from the shield but not neutrons coming directly from the source.

3.1.2. *Simultaneous measurement with two detectors*

A better exponential dependence was found for the ratio of the count rates of the two detectors $R(x) = N_2(x)/N_1(x)$ (where in this case N_2 is the count rate of the detector positioned at the distance and N_l is the count rate at the detector positioned at distance 0). It can be seen in Figures 2 and 3 and Table 1, which gives the correlation coefficient for exponential dependence for a single detector N_2 and for the ratio of the two detectors R(x), that the exponential dependence is better for R(x) (correlation coefficient closer to 1). The difference of the agreement with exponential dependence between a single detector and the ratio of two detectors is more prominent for ^{252}Cf source than for AmBe source, probable due to the lower energy of neutrons or narrower energy spectrum, Figures 2 and 3. We can show clearly that the difference is larger for lower energy by MCNP calculation, Figure 4. In case of media with lower concentration of hydrogen the difference in R^2 of the single detector and the ratio of two detectors is negligible, but for high concentration hydrogenous media, the much better exponential agreement of the ratio than for a single detector was found also by the Monte-Carlo calculations. Hence we will use only the exponential dependence of ratio of count rates of two detectors rather than the counts of one detector.

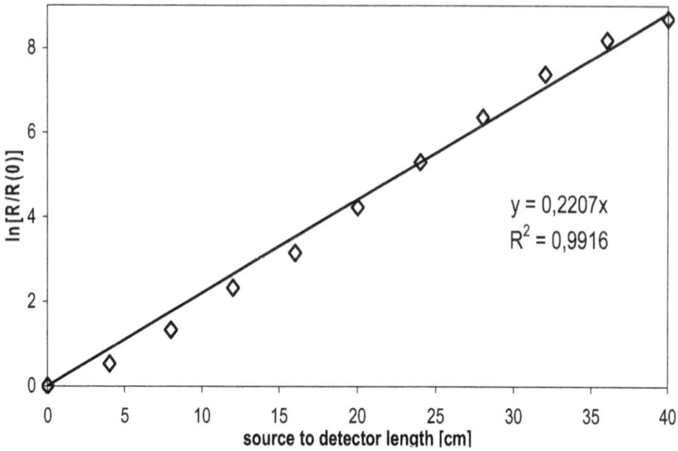

Figure 2. Ln normalized (**a**) Count rate of a single detector (**b**) Ratio of the count rate of two detectors as a function of source-to-detector distance for an AmBe source

TABLE 1. A comparison between the linear fit correlation coefficient of the natural logarithm of the detector 1 (N_1) and the counts rate ratio (R(x)), for different scattering medias within the measured bulky sample

	Exponential fit correlation coefficient R^2		
Source type	Scatering media	N_2	R(x)
AmBe	Air	0.9765	0.9968
	Paper	0.9948	0.9933
	Concrete tiles	0.9976	0.9991
	Kardboard	0.9985	0.9971
	Foamplast+perspex	0.995	0.9973
	Wood	0.9837	0.995
	Foamplast+Wood	0.9987	0.9989
	Paraffin	0.9689	0.9916
Cf	Polypropylene	0.8084	0.9921
	Perspex+foamplast	0.9898	0.992
	Grafite	0.9991	0.9991
	Paraffin	0.929	0.9878
	Perspex	0.9161	0.9889
	Wood	0.9939	0.9954

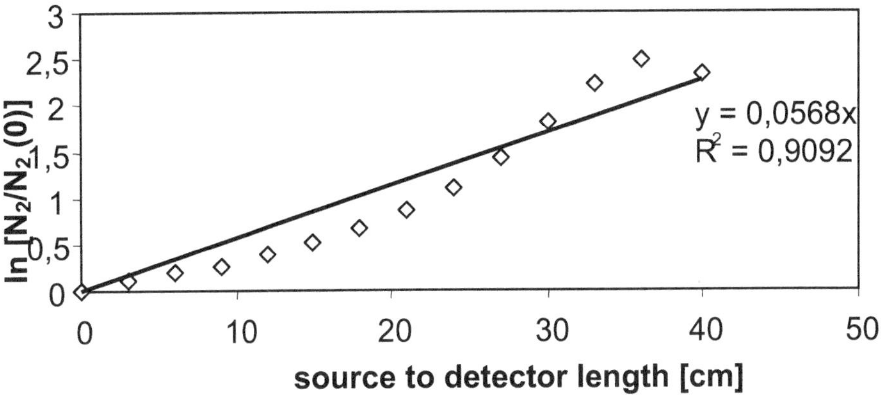

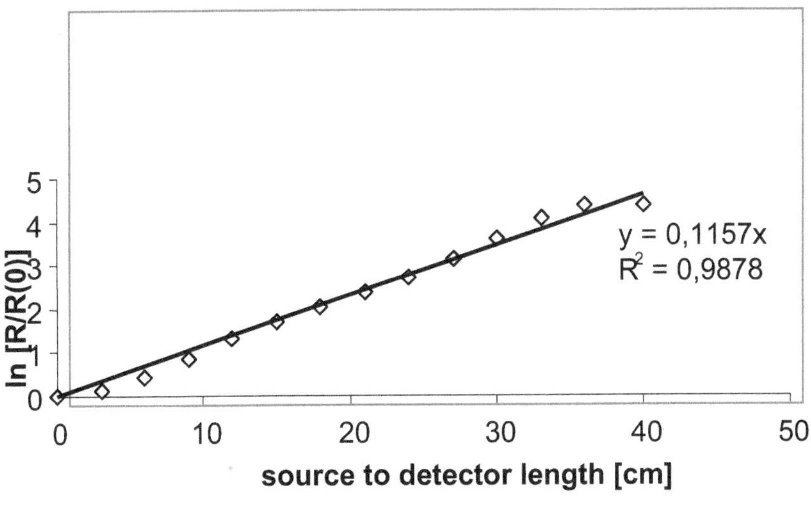

Figure 3. Ln normalized (a) Count rate of a single detector (b) Ratio of the count rate of two detectors as a function of source-to-detector distance for an Cf source

3.1.3. *Source localization*

The exponential dependence of the ratio can be written as:

$$R(x) = R(0) \cdot e^{\mu x} \tag{1}$$

The calculation of the location x can be done using Equation (1). If a is the length of the box, than from Equation (1)

$$R(a) = R(0) \cdot e^{\mu a}$$

However, the connection between R(a) and R(0) is the change of the naming of the detectors, detector 1 is now detector 2 and vice versa:

$$R(a) = R(0) \cdot e^{\mu a} = \frac{1}{R(0)} \tag{2}$$

Hence

$$R(0) = e^{-\frac{\mu a}{2}} \tag{3}$$

$$\mu = -\frac{2\ln[R(0)]}{a} \tag{4}$$

This equation can be developed not only for positions 0 and a but also for a general case. It can be written:

$$\frac{N_2}{N_1} = R(x)$$

$$\frac{N_1}{N_2} = R(a-x)$$

Since

$$R(x) = \frac{1}{R(a-x)}$$

Then

$$R(0) \cdot e^{\mu x} = \frac{1}{R(0) \cdot e^{\mu(a-x)}}$$

$$R(0) = e^{-\frac{\mu a}{2}}$$

$$R(x) = R_0 \cdot e^{\mu x} = e^{-\frac{\mu a}{2}+\mu x} = e^{\mu(x-\frac{a}{2})} \tag{5}$$

Equation (5) shows that R(x) is independent of source activity, and can be used in order to calculate x.

To find the experimental μ we can measure R(0), as Equation (4) shows the correlation between them, with an external source which will be located on the surface of the box.

From Equation (4):

$$x = \frac{1}{\mu} \cdot \ln\left[\frac{R(x)}{R(0)}\right] \tag{6}$$

Because the exponential parameter μ is a characteristic of medium and does not depend on the source activity substituting Equation (4) in Equation (6) yields:

$$x = -\frac{a}{2} \cdot \frac{\ln\left[\frac{R(x)}{R(0)}\right]}{\ln[R(0)]} = \frac{a}{2} \cdot \left(1 - \frac{\ln[R(x)]}{\ln[R(0)]}\right) \tag{7}$$

The value of R(0) with a known source cannot be measured unless we prior measure the contribution from the source in the box. Consequently, to find the location, a first measurement by the two detectors of the count rate of the unknown source positioned in an unknown place in the medium must be done. In the next step an additional external source is placed in position x = 0 (the source close to detector 1) and the count rate of the two sources together are measured by the two detectors. The count rate of the external source is calculated by subtraction of the count rate of the unknown source from the count rate of the two sources together.

The position x of the unknown source could be calculated by Equation (7).

Table 2 compared the measured x, from the actual position of the source with the calculated x from Equation (7) for the AmBe source. For ^{252}Cf source were received similar results.

It can be seen that the relative deviation in the source position between the calculated values to the measured one, relative to the size of the medium, is lower than 6.5% for every medium in the experiment. This is the linear error. The volume error will be (2*0.065)2 = 0.0169, since the dimension of the detector length is not studied. Thus it means that if we want to search for the source we have to search at most only 1.69% of the volume of the box. The same effect will be on the accuracy of the calculation of the activity of the source. The linear deviation in the source position in the box in absolute value is in all cases less than 2.11 cm. From Table 2 it can be seen that the deviation in the source position is larger when the source is positioned in the edges of the box. This is due to less scattering and slowdown of neutrons when the source is situated very close to the detector, and hence the accuracy in the measurement drops. Similarly for MCNP calculation the measured x (as given in the input data) was compared with the x calculated from Equation (7). The calculated source-detector distance is normalized to position 0 of the source. The results obtained from MCNP simulation and from experiments are in a good agreement with each other.

TABLE 2. The measured (x_{mea}) and calculated (x_{cal}) source-to-detector distance (cm) for an AmBe source

Wood			Paraffin			Air		
x_{mea}	x_{cal}	Δx/a	x_{mea}	x_{cal}	Δx/a	x_{mea}	x_{cal}	Δx/a
0.0	0.0	0.000	0.0	0.0	0.000	0.0	0.0	0.000
3.3	2.8	0.012	4.0	2.4	0.040	4.0	4.8	0.020
7.0	5.9	0.026	8.0	6.0	0.050	8.0	8.8	0.020
10.3	9.2	0.027	12.0	10.5	0.037	12.0	12.4	0.009
13.8	13.0	0.019	16.0	14.3	0.044	16.0	16.2	0.004
17.5	17.0	0.013	20.0	19.1	0.023	20.0	19.7	0.007
20.7	20.6	0.002	24.0	24.0	0.000	24.0	23.3	0.018
24.0	21.7	0.058	28.0	28.7	0.018	28.0	26.8	0.030
27.8	28.2	0.011	32.0	33.4	0.034	32.0	30.2	0.044
31.2	32.0	0.019	36.0	37.0	0.025	36.0	33.8	0.054
35.0	35.4	0.110	40.0	39.3	0.018	39.0	38.0	0.024
40,6	39.8	0.020						
Foamplast			Paper			Concrete tiles		
x_{mea}	x_{cal}	Δx/a	x_{mea}	x_{cal}	Δx/a	x_{mea}	x_{cal}	Δx/a
0.0	0.0	0.000	0.0	0.0	0.000	0.0	0.0	0.000
3.0	4.1	0.028	2.4	1.5	0.022	4.8	4.8	0.001
5.0	6.7	0.041	5.2	3.3	0.047	10.0	9.8	0.006
9.0	10.3	0.031	7.8	6.0	0.046	12.1	12.2	0.001
13.0	14.0	0.025	12.0	10.6	0.036	16.8	17.4	0.015
17.0	17.6	0.015	16.0	15.5	0.012	21.7	22.5	0.019
21.0	20.8	0.004	21.5	20.9	0.016	26.6	27.4	0.021
25.0	24.2	0.020	25.5	25.0	0.000	29.0	29.8	0.019
27.8	27.0	0.020	29.0	29.3	0.007	33.6	34.3	0.017
30.6	29.9	0.018	31.5	32.4	0.022	38.5	38.3	0.005
33.4	33.1	0.007	34.0	36.4	0.061			
36.2	34.6	0.041	36.5	38.0	0.039			
39.0	38.3	0.016	39.0	39.9	0.023			

This method is accurate as long we know the type of the neutron source. For completely unknown source in a box we have two more unknowns: the matrix and the type of the source which affects the neutron spectrum. In our method we ignore the activity of the source since we use the ratio of two detectors. The μ of the matrix is determined by the measurement of the external source located on the box face. The use of the μ of the external source assumes that the two neutron sources have a similar spectrum. This is the case for example when we look for a source of a known type in a glove box or for example in a measurement of nuclear waste, although a nuclear waste can have both sources of neutrons, both spontaneous fission and (α, n) reaction with ^{18}O and ^{19}F. However for a completely unknown source larger errors in the calculated location will be caused because of the error in ln R(0).

TABLE 3. MCNP's calculated distance of source from detector 1 cm within scattering shield and for various media

I. For AmBe source (R0 Cf source)

Media→	Perspex		Concrete tiles		Air		Fe	
x_{mea}	x_{cal}	$\Delta x/a$	x_{cal}	$\Delta x/a$	x_{cal}	$\Delta x/a$	x_{cal}	$\Delta x/a$
4	4.6	0.014	5.2	0.029	5.7	0.040	7.5	0.085
8	7.9	0.001	8.9	0.022	9.5	0.037	10.9	0.070
12	11.6	0.009	12.4	0.011	13.2	0.030	13.5	0.038
16	15.7	0.006	16.0	0.001	16.4	0.010	16.6	0.014
18.5	18.0	0.013	18.2	0.008	18.2	0.008	18.6	0.003
21	20.4	0.015	20.4	0.015	20.2	0.019	20.4	0.014
25	24.4	0.015	23.9	0.027	23.6	0.034	23.3	0.041
29	28.1	0.021	27.5	0.035	26.9	0.051	26.0	0.074
33	32.0	0.025	31.1	0.046	30.6	0.060	29.1	0.094
37	35.7	0.031	34.8	0.053	34.0	0.073	32.7	0.106
41	38.1	0.072	38.8	0.052	39.0	0.049	38.5	0.062

II. For Cf source (R0 AmBe source)

Media→	Perspex		Concrete tiles		Air		Fe	
x_{mea}	x_{cal}	$\Delta x/a$	x_{cal}	$\Delta x/a$	x_{cal}	$\Delta x/a$	x_{cal}	$\Delta x/a$
4	1.4	0.063	3.7	0.008	4.8	0.018	5.6	0.039
8	5.6	0.060	7.8	0.004	8.8	0.020	9.5	0.036
12	10.0	0.049	11.9	0.002	12.4	0.010	13.0	0.024
16	14.7	0.033	15.7	0.007	15.9	0.003	16.4	0.010
18.5	17.5	0.024	17.9	0.014	18.4	0.002	18.5	0.001
21	20.5	0.013	20.5	0.012	20.6	0.010	20.2	0.019
25	25.2	0.004	24.1	0.021	24.0	0.024	23.7	0.031
29	30.0	0.023	28.3	0.017	27.6	0.035	27.0	0.048
33	34.6	0.039	32.3	0.017	31.3	0.040	30.3	0.066
37	38.6	0.040	36.4	0.014	35.4	0.039	34.6	0.058
41	41.6	0.015	40.8	0.006	40.9	0.002	41.6	0.014

Table 3 gives the error generated by a wrong assumption of the type of the neutron source comparing the actual position of the source, as it was given in MCNP input with the calculated x from Equation (7) for the AmBe source by taking ln R(0) of the ^{252}Cf source, and for the ^{252}Cf source by taking ln R(0) of the AmBe source. The error caused by the unknown energy of the source depends on the position of the source. The larger errors were obtained for source close to one of the detectors up to 4 cm (10% in the case of ln R(0) of incompatible source and 8% for ln R(0) of the same source), and were about the same as for known source type in other points. In conclusion, even in the case of unknown source type Equation (7) may be applied to determine quite accurately the location of the source.

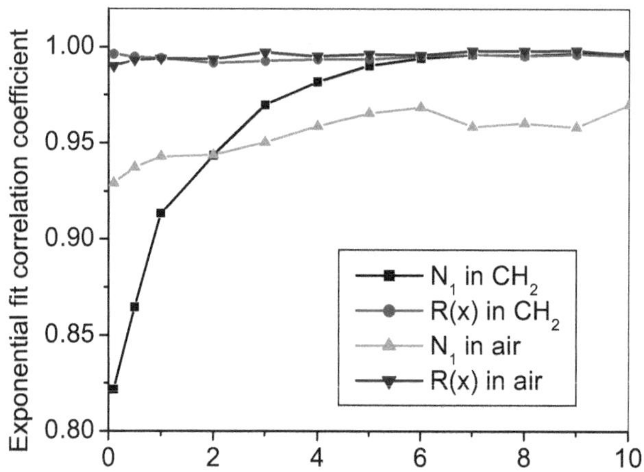

Figure 4. A comparison between the exponential fit correlation coefficient of the count rate of detector 1 (N_1) and the counts rate ratio (R(x)), calculated from MCNP simulations in media of paraffin or air for various energies of monoenergetic neutrons

3.1.4. *Compensation of the edge effect*

In order to cancel the effect of the edge of the box, where the error in the source localization is larger, we studied how positioning of polypropylene slabs of different thicknesses between the detectors and the box will affect the accuracy of the localization of the source. The purpose of these slabs is to thermalize the neutrons that go directly to the near-by detector. Due to the moderating character of the material between the detector and the source, the neutrons that do not go directly to the near-by detector are thermalized (moderated) by scattering reactions, while those that go directly to the detectors are mainly fast ones and therefore mainly undetected, since the ^3He detectors have much larger efficiency for thermal neutrons. Consequently, higher count-rate is expected with increasing source-detector distance for few centimeters from the detector. The count rate of a thermal neutron detector (such as ^{3}He) due to a point source, emitting mainly fast neutrons in a moderating medium, as a function of the detector-source distance, increases for small distances, due to moderation, and then decreases at larger distances, due to a smaller solid angle. Equation (1) indicates on a monotonic decrease of the count rate ratio R(x) with x, in contrast to the real increase with small x. This contradiction leads to wrong calculated distances for small x, smaller than the real ones, and even to negative calculated distance up to 2 cm from the detector.

Positioning of polypropylene moderating slabs between the detectors and the box will lead to maximum count rates in these slabs and monotonic decrease inside the box boundaries. The reference source used for measuring R(0) was positioned between the box and the slab. The accuracy is presented in Table 4 as the deviation of the calculated distance, x, from the measured one [Δx = x(calculated) – x(measured)], Table 4 gives also the ratio of the deviation to the sample size, $\Delta x/a$:the relative error in percents. The relative error is plotted also in Figure 5. It should be emphasized that even in the absence of the added slabs there is about 1 cm of moderating medium between the source and the detector, due to the thickness of the box walls and the source holder. Table 4 and Figure 5 show that the addition of the moderating slabs reduces the error in the calculation of the source positioning near the faces of the box. A 4 cm polypropylene slab reduces the relative error from 9% to 4.2%. The maximal absolute error is reduced from 3.5 cm to about 1.5 cm. The effect of the added slabs is not always a positive one. Table 4 and Figure 5 show also that while these slabs increase the accuracy of the localization at small distances, they have a deteriorating effect on the accuracy at the middle of the box. However, the reduction of the accuracy in the middle of the box is a minor one, only to very small extent, while improving considerably the accuracy of the localization in the edges. Figure 5 shows that the addition of moderator between the detectors and the box, in addition to reducing the relative error also moves the position with maximum relative error from about 3–4 cm to about 12 cm from the box wall . So if the results with the slabs show that the location is about 12 cm, it might be a good practice to count again without the slabs.

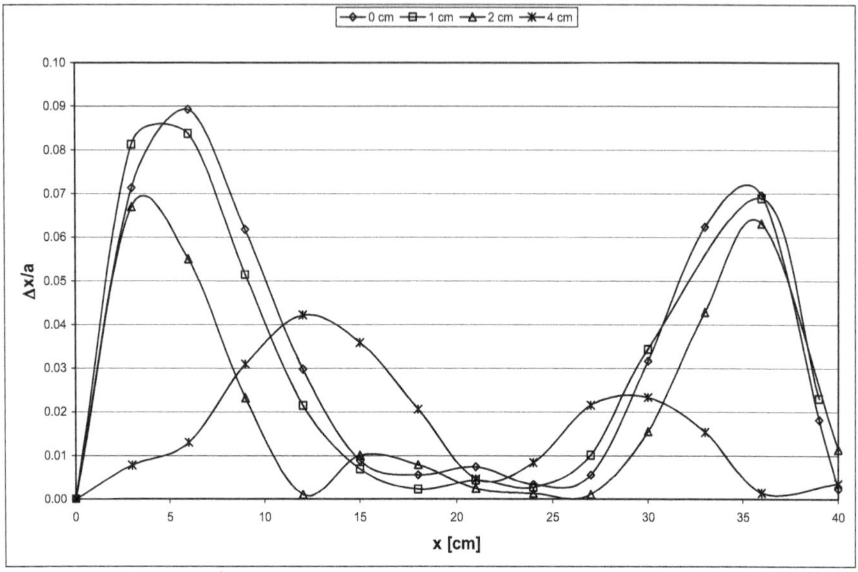

Figure 5. The relative error of localization as a function of the distance box face-source for various thicknesses of polypropylene between the detectors and the box

TABLE 4. The relative error of the localization of the source for various polypropylene slab thicknesses and source positions

Slab thickness	0 cm		1 cm		2 cm		4 cm	
$x_{measured}$ (cm)	x_{calc} (cm)	Δx/a	x_{calc} (cm)	Δx/a	x_{calc} (cm)	Δx/a	$x_{calculated}$ (cm)	Δx/a
0	0.000	0.000	0.000	0.000	0.000	0.000	0.000	0.000
3	0.150	0.071	0.250	0.081	0.323	0.067	2.689	0.008
6	2.432	0.089	2.653	0.084	3.801	0.055	6.518	0.013
9	6.533	0.062	6.941	0.051	8.074	0.023	10.233	0.031
12	10.811	0.030	11.141	0.021	12.041	0.001	13.685	0.042
15	14.648	0.009	14.723	0.007	15.402	0.010	16.435	0.036
18	17.778	0.006	17.910	0.002	18.316	0.008	18.823	0.021
21	20.703	0.007	20.827	0.004	21.096	0.002	21.186	0.005
24	23.867	0.003	23.892	0.003	23.948	0.001	23.666	0.008
27	27.220	0.006	27.406	0.010	27.043	0.001	26.141	0.021
30	31.262	0.032	31.372	0.034	30.620	0.015	29.068	0.023
33	35.490	0.062			34.714	0.043	32.387	0.015
36	38.782	0.070	38.754	0.069	38.523	0.063	36.057	0.001
39	39.726	0.018	39.920	0.023				
40	39.909	0.002			40.455	0.011	39.859	0.004

MCNP calculations, both for ^{241}Am/Be and ^{252}Cf sources (boxes of 40 × 40 and 40 × 20 cm) were performed. The results of the simulation for a ^{241}Am/Be source in 40 × 20 cm box are given in Table 5. It can be seen that for a ^{241}Am/Be source with a box similar to the experimental box dimension, the Monte Carlo calculations show similar results to those found experimentally. The most accurate source localization estimations are received with a polypropylene slab of 4–5 cm thickness. Thicker slabs show already a larger error, Table 2. Hence 4.5 cm slabs (equivalent to experimental 3.5 cm, due to box walls and source holder) seems to be the optimum moderator thickness. Additional MCNP simulations were performed to determine the optimal slab thickness for 40 × 40 cm box. It was found that optimal thickness is 6 cm, but almost the same accuracy was obtained with 5 and 7 cm slabs. Therefore, 5 cm slab was chosen for further experiments and calculations, which produces the same result as 6 cm in the MCNP simulation, due to box walls and source holder.

3.1.5. *The shield contribution*

The polypropylene shield, Figure 1b, serves several purposes:

1. Safety of the workers
2. Constant environment
3. Increase of the number of thermal neutrons reaching the detector

In order to study the increase in the count rates due to the shield, a thermal neutron absorber made of a Cd sheet was used in a series of experiments to prevent thermal neutrons reflected from the shield to reach the detector. A Cd foil 5 mm thick covered the box together with the ^{3}He detectors. However, fast neutrons still may returns, pass through the cadmium, thermalized in the box and counted in the detector. Similar MCNP simulations were also performed. For the simulation the box was kept in vacuum, so there were no returned neutrons. It was found that Equations (2)–(7) can still be applied to these results, but the number of counts drops dramatically up to a factor of 10.

3.2. THE 3D CASE

Actually there is no search for the location in three dimensions since the ^{3}He detectors are almost as long as the box and consequently there was no study of this dimension. The study was limited to the plane perpendicular to the ^{3}He detectors.

In the previous section the unknown source was in the plane of the two parallel ^{3}He detectors. However, in practice, the source position is unknown and it is most likely that the source position (plane) will deviate from the plane connecting the two ^{3}He. Therefore, the error in the source position estimation, caused due to the deviation of the source from the detector's centers plane, should be evaluated. Two factors may affect the accuracy of the localization method. The first factor is the error in the position of the source plane, which was not dealt with at all in the previous section. The second factor is the same as in the previous section, i.e. the error in the calculation of the source-detector distance. This leads to the question concerning the practice of our method, where do we have to position the calibrating source for measuring R(0) if the source plane is not the same as the detectors plane.

The box was subdivided to 130 rectangles by parallel lines. In one direction (perpendicular to the plane of the two detectors) there were 10 divisions (a–j) and in the perpendicular direction there were 13 divisions (1–13) as can be seen in Figure 6. In the center of each rectangle the neutron source was positioned and the count-rates of the two detectors was measured. Using Equation (7) the location of the unknown source was calculated. The external source, used for calculation of R(0), was positioned in various positions on the box face in the center of each rectangle.

Table 5a presents the absolute deviation between the experimentally measured source – box's face distance (the face that has the detector), x, and the calculated one for various source positions within the box. The deviation calculated for 40 × 40 cm sample, divided to ten planes (a–j) parallel to the plane between the two opposite detector caps. The reference source in this case was positioned at the detectors plane, and measured at distance x = 0.

Table 5b shows that the absolute error in the source localization increases with the increase in the distance between the parallel source plane and detectors plane. The deviation calculated for 40 × 40 cm sample, divided to ten (a–j) planes parallel to the plane between the two opposite detector caps. In other words the maximal deviation is obtained when the source was measured at the planes tangent to the sides of the box (planes a and j in Figure 6). The maximal absolute error obtained is 5.8 cm. It should be remembered that this is the error in the x coordinate besides an error of about 10 cm in the y coordinate.

The study of the effect of varying positions of the external calibrating source for the measurement of R(0) was done by positioning the calibrating source at different positions on the face of the box, measuring R(0) (by subtraction of the count rates of the unknown source which was measured prior to the addition of the external source) and calculating x from the various R(0) from Equation (7). Table 5b similarly to the Table 6.5a presents the absolute error in the localization of the neutron source when the reference source was positioned at the source plane, rather than at the detectors plane, at a distance of x = 0, Figure 6. In Table 5b, it is assumed that the source plane is known and the reference source is positioned at this plane, while the detectors remain at the same place, at the central plane. The maximal deviation in the source position estimation is 1.88 cm and in most cases the deviation is very small with respect to the sample (box) size. In other words, the conclusion is that if R(0) is measured in the source plane then the accuracy does not depend on the distance from the central plane.

Similar results were obtained by MCNP simulation for the same geometry. However, since the plane of the unknown source is not known, there is no prior knowledge where to position the external source. In order to solve this problem an iteration method was used.

The method can be done with only two detectors, but in this case in each step of the iteration both the detectors and the external source should be moved. If four detectors are used only the external source has to be moved in each step of the iteration. The other two detectors are positioned in the middle of the two faces of the box (not those at the top and the bottom faces).

Iterative method was applied to localize the unknown neutron source, by finding successive approximations to the position of the source plane, starting from an initial guess of unknown source plane In the first step the external source is positioned at x = 0. Then, the x coordinate of the unknown source was calculated, using Equation (7), from the results of the two detectors on the y-axis, detectors 1 and 2. Both R(x) and R(0) are measured with detectors 1 and 2, first the measurement of the unknown source alone, (R(x)), and then measurement of the unknown source together with a known external (calibrating) source; R(0) is the difference of the two measurements. In this first step R(0) was measured with the known (calibrating) source positioned at the center of the y-axis (on the central plane (initial guess), connecting the two detectors 1 and 2 (at x = 0.5 cm from detector 2). This R(0) is assigned $R(0)_x$. Then, the external calibrating source was moved to the x axis (right side of the box, y = 0) at the calculated x (to the nearest to calculated x cell) and R(0) is measured again using the two detectors on the y-axis, detectors 3 and 4. This R(0) is assigned $R(0)_y$ and was used to calculate y coordinate according to

Equation (7) using R(x) also of detectors 3 and 4. Now the calibrating source is moved to the y axis at the calculated y (to the nearest to calculated y cell). R(0)x is measured again and a new value of x is calculated. This process is repeated till either x or y is the same as in the previous iteration.

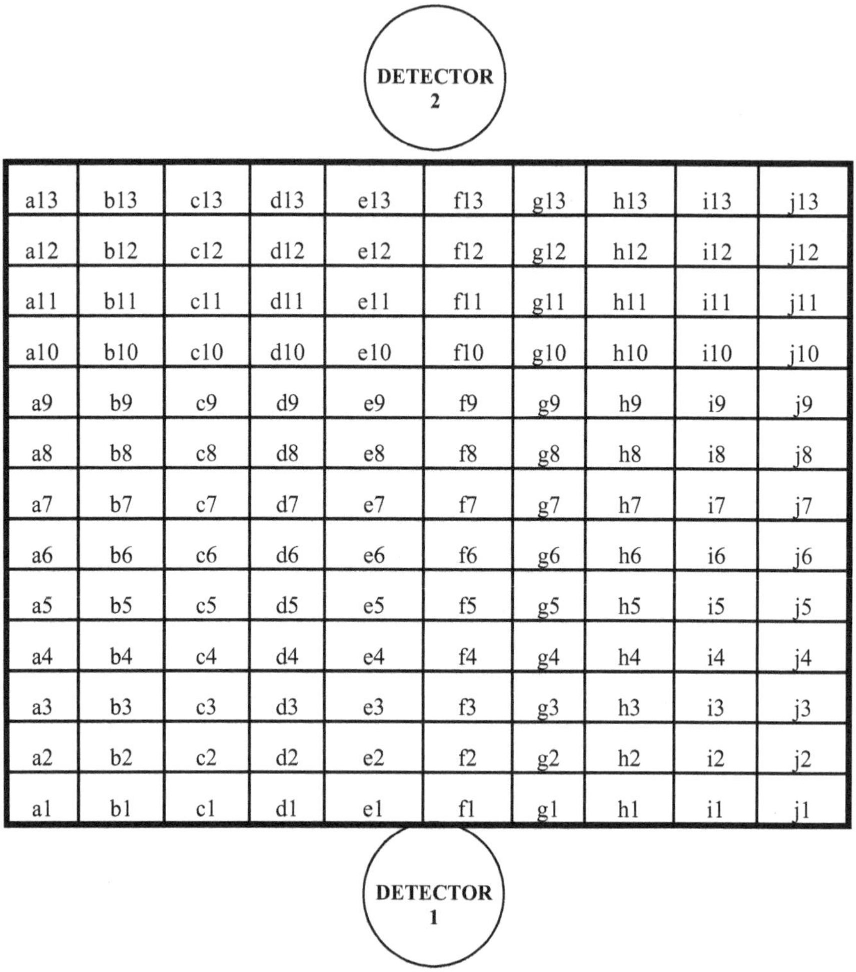

Figure 6. Schematic top view of the source positions measured in the 40 × 40 cm sample cells. The sample was divided to ten planes parallel to the detectors plane

TABLE 5a. The absolute deviation, in cm, between the measured and calculated source to detector distance x, for various source positions within the sample. The reference source in this case was positioned at the detectors plane, and measured at distance x = 0

Plane name	A	B	C	D	E	F	G	H	I	J
Distance from detectors plane (cm)	18	14	10	6	2	2	6	10	14	18
X(measured) (cm)	Δx (cm)	Δx (cm)	Δx (cm)	Δx (cm)	Δx (cm)	Δx (cm)	Δx (cm)	Δx (cm)	Δx (cm)	Δx (cm)
0	4.94	3.85	2.39	1.22	0	0.08	1.43	2.82	4.39	5.78
3	3.2	2.58	1.19	0.15	0.85	0.69	0.51	1.98	3.23	4.05
6	2.04	1.55	0.8	0.48	0.98	0.92	0.14	1.07	2.21	2.88
9	1.4	1.13	0.48	0.29	0.76	0.6	0.04	0.82	1.61	2.13
12	0.67	0.5	0.32	0.26	0.74	0.43	0.04	0.53	1.03	1.32
15	0.21	0.2	0.11	0.26	0.43	0.32	0.01	0.27	0.64	0.78
20	0.44	0.44	0.42	0.25	0.28	0.2	0.12	0.03	0.17	0.13
25	1.03	1.16	0.76	0.35	0.17	0.03	0.17	0.62	0.71	1.06
28	1.78	1.51	1.21	0.56	0.28	0.09	0.45	1.02	1.45	1.61
31	2.18	1.9	1.37	0.54	0.07	0.25	0.37	0.99	1.49	2.32
34	2.73	2.33	1.66	0.45	0.31	0.52	0.02	1.13	2.24	2.87
37	3.86	3.23	2.11	0.63	0.19	0.5	0.32	1.66	3.04	4.15
40	5.55	4.62	3.19	1.57	0.56	0.23	1.19	2.61	4.07	5.47

For box with dimension of 40 × 40, as it can be seen in Table 6 that the maximal possible accuracy is achieved after not more than four iterations of x and y. For larger boxes it might be necessary to make more iterative measurements. In the 40 × 40 cm^2 box case only one more measurement and calculation for y coordinate is required to receive larger accuracy than that of the first approximation.

Table 6 show an example of the iteration calculations for a box with dimension of 40 × 40 for the row b. For most points two iterations after initial guess calculation of x, were performed till x or y recurred. It could be seen that one iteration (two consecutive calculations of y and x) after initial guess, already yield major improvement to the accuracy. Similar results were obtained for the other rows.

Table 7 shows the distance between the real source position and the calculated one $\sqrt{(y_{cal} - y_{real})^2 + (x_{cal} - x_{real})^2}$ both for x and y calculated with central R(0) for both coordinates and the distances calculated after completing all iterations. One to four iterations required, depending on a source position.

TABLE 5b. The absolute deviation, in centimeter, between the measured and calculated source to detector distance x, for various source positions within the sample. The reference source in this case was positioned at the source plane (plane parallel to the detectors plane and which include the source), and measured at distance x = 0

Plane name	A	B	C	D	E	F	G	H	I	J
Distance from detectors plane (cm)	18	14	10	6	2	2	6	10	14	18
X (measured) (cm)	Δx (cm)	Δx (cm)	Δx (cm)	Δx (cm)	Δx (cm)	Δx (cm)	Δx (cm)	Δx (cm)	Δx (cm)	Δx (cm)
0	0	0	0	0	0	0	0	0	0	0
3	1.33	0.86	0.95	1.26	0.85	0.76	0.77	0.49	0.65	1.21
6	1.88	1.41	1	1.42	0.98	0.98	1.23	1.05	1.11	1.63
9	1.74	1.22	0.95	1.02	0.76	0.64	0.89	0.86	1.03	1.47
12	1.73	1.29	0.72	0.79	0.74	0.47	0.57	0.7	0.93	1.4
15	1.37	0.95	0.56	0.6	0.43	0.34	0.37	0.51	0.58	0.94
20	0.59	0.55	0.47	0.27	0.28	0.2	0.13	0.03	0.22	0.18
25	0.28	0.24	0.18	0.05	0.17	0.01	0.21	0.1	0.5	0.54
28	0.26	0.04	0.29	0.08	0.28	0.06	0.13	0.13	0.39	0.98
31	0.72	0.27	0.07	0.14	0.07	0.29	0.45	0.66	1.18	1.2
34	0.97	0.45	0.02	0.42	0.31	0.58	1.06	0.99	1.07	1.66
37	0.45	0.05	0.09	0.43	0.19	0.57	0.97	0.86	0.89	1.07
40	0.81	0.96	0.9	0.38	0.56	0.15	0.26	0.24	0.41	0.43

TABLE 6. Iteration calculations of the source-to-detector distance (cm) for row b

	y_{real}	x_{real}	Before iterations y	Initial estimate x	First y	Iteration x	Second y	Iteration x	Third y	Iteration x	After y	Iterations x
b01	4	0	10.95	3.79	6.04	1.54	**3.04**				3.04	1.54
b02	4	3	9.27	5.64	5.23	2.29	3.18	**2.29**			3.18	2.29
b03	4	6	7.97	7.58	5.31	4.68	3.33	**4.68**			3.33	4.68
b04	4	9	6.25	10.16	3.14	7.86	**3.14**				3.14	7.86
b05	4	12	4.85	12.41	3.22	10.63	**3.22**				3.22	10.63
b06	4	15	3.93	15.20	3.22	14.07	**3.22**				3.22	14.07
b07	4	20	3.29	19.61	3.29	19.52	**3.29**				3.29	19.52
b08	4	25	3.98	23.92	3.31	24.84	**3.31**				3.31	24.84
b09	4	28	4.77	26.40	4.13	27.90	3.24	**27.90**			3.24	27.90
b10	4	31	6.27	29.16	4.92	31.29	3.48	**31.29**			3.48	31.29
b11	4	34	7.75	31.70	5.31	34.44	3.50	**34.44**			3.50	34.44
b12	4	37	9.11	33.71	5.42	36.92	3.31	**36.92**			3.31	36.92
b13	4	40	10.42	35.46	7.28	37.61	5.48	39.07	3.52	**39.07**	3.52	39.07
error (cm)→			2.95	1.91	1.19	0.84					0.79	0.73

TABLE 7. Pre- and post-iteration distance calculations between the real source position and the calculated one

row, yreal→		a,y=0		b,y=4		c,y=8		d,y=12		e,y=16		f,y=20		g,y=24		h,y=28		I,y=32		j,y=36	
line	xreal	Distance before iterations	Distance after iterations	Distance before iterations	Distance after iterations	Distance before iterations	Distance after iterations	Distance before iterations	Distance after iterations	Distance before iterations	Distance after iterations	Distance before iterations	Distance after iterations	Distance before iterations	Distance after iterations	Distance before iterations	Distance after iterations	Distance before iterations	Distance after iterations	Distance before iterations	Distance after iterations
1	0	10.36	0.00	7.92	2.56	5.60	0.07	3.39	0.27	1.74	1.56	1.38	0.73	3.20	0.30	5.70	0.24	8.09	0.65	10.47	2.87
2	3	8.06	3.74	5.89	1.08	4.08	1.06	2.63	1.21	1.60	1.24	1.22	0.92	2.16	1.03	3.90	1.31	5.89	1.50	8.19	1.30
3	6	6.07	3.37	4.27	1.48	2.74	1.45	2.03	1.47	1.65	1.41	1.67	1.51	2.00	1.28	3.04	1.13	4.34	1.56	6.33	1.61
4	9	3.98	3.37	2.53	1.43	1.41	1.39	1.09	0.95	1.27	1.07	1.10	0.92	1.35	0.88	1.99	0.89	3.08	1.08	4.61	1.61
5	12	2.08	2.85	0.94	1.58	0.33	1.29	0.33	0.96	0.83	0.75	1.20	1.13	0.83	0.59	1.04	0.70	1.76	0.90	3.00	1.50
6	15	0.88	1.39	0.21	1.21	0.34	1.12	0.19	0.63	0.69	0.63	1.07	1.04	0.32	0.25	0.24	0.75	0.68	0.75	1.49	1.00
7	20	0.59	0.78	0.81	0.85	0.90	0.93	0.37	0.39	0.55	0.55	1.32	1.32	0.47	0.47	0.14	0.16	0.14	0.16	0.77	0.78
8	25	1.36	0.19	1.08	0.71	0.80	0.89	0.44	0.59	0.30	0.23	1.32	1.28	0.79	0.54	0.75	0.27	1.11	0.39	2.17	1.15
9	28	2.50	0.38	1.77	0.76	1.18	0.89	0.74	0.33	0.68	0.54	1.10	0.99	1.00	0.34	1.43	0.21	2.00	0.41	3.28	1.33
10	31	4.09	0.68	2.92	0.60	1.90	0.70	1.05	0.35	0.78	0.65	1.28	1.12	1.35	0.50	2.05	0.62	3.01	1.02	4.69	1.54
11	34	5.99	0.78	4.40	0.67	2.87	0.79	1.45	0.67	1.13	0.92	1.50	1.32	1.95	0.93	3.04	1.01	4.62	1.10	6.35	2.94
12	37	8.11	0.35	6.07	0.69	4.14	0.54	2.41	0.42	1.10	0.73	1.40	1.07	2.51	0.86	4.08	0.87	5.99	1.17	8.16	1.20
13	40	10.24	3.30	7.86	1.05	5.68	0.86	3.41	0.55	1.63	1.36	1.57	1.15	3.31	0.42	5.46	0.28	7.78	0.42	10.39	0.84
average		4.95	1.63	3.59	1.13	2.46	0.92	1.50	0.68	1.07	0.90	1.32	1.12	1.63	0.64	2.53	0.65	3.73	0.86	5.38	1.51

References

1. Presler O, Pelled O, German U, Leichter Y, Alfassi ZB (2002) Determination of a source in a box with two detectors. I. Non-absorbing media. Nucl Instrum Meth A 491:314–325
2. Presler O, German U, Alfassi ZB (2004) Location-independent determination of the activity of a point source in absorbing media. Appl Radiat Isotopes 60:221–225
3. Presler O, German U, Golan H, Alfassi ZB (2004) Determination of a source in a box with two detectors. The general case. Nucl Instrum Meth A 527:632–647
4. Pelled O, Tzroya S, German U, Haquin G, Alfassi ZB (2004) Locating a "hot spot" in the lungs when using an array of four HPGe detectors. Appl Radiat Isotopes 61:107–111
5. Glenn F. Knoll (2000) Radiation detection and measurement. 3rd edition, Wiley, New York
6. Antonopoulos-Domis A, Tambouratzis T (1996) Artificial neural networks for neutron source localization within sealed tanks. Ann Nucl Energy 23:1477–1488
7. Peurrung AJ, Reeder PL, Stromswold DC (1997) Location of neutron sources using moderator-free directional thermal neutron detectors. IEEE Trans Nucl Sci 44:543–550
8. Linden P, Karlsson JK-H, Dahl B, Pazsit I, Por G (1999) Localisation of a neutron source using measurements and calculation of the neutron flux and its gradient. Nucl Instrum Meth A 438:345–355
9. Avdic S, Linden P, Pazsit I (2001) Measurement of the neutron current and its use for the localization of a neutron source. Nucl Instrum Meth A 457:607–616
10. Briesmeister JF (Editor) (2000) MCNP – A General Monte Carlo N-Particle Transport Code, version 4c. Technical report LA-13709-M Los Alamos National Laboratory Los-Alamos NM
11. Compendium of neutron spectra and detector responses for radiation protection purposes (2001) Supplement to Technical Reports Series No. 318

PASSIVE SOLID STATE DOSIMETERS IN ENVIRONMENTAL MONITORING

MÁRIA RANOGAJEC-KOMOR*
Ruđer Bošković Institute, Bijenička 54, 10000 Zagreb, Croatia

Abstract. Environmental dosimetry systems have to fulfil the requirement to measure the man-made contribution to environmental radiation (1:10) under variable environmental conditions (UV sunlight, humidity, temperature). The recently developed SC-1 flat RPL glass dosimeters with FGD-202 reader for environmental dosimetry were compared to various high sensitivity TL dosimeters. All characteristics of RPL and TL dosimeters investigated fulfil the requirements of the new IEC 61066:2006 Standard for personal and environmental dosimetry. To reach international standards and to improve the environmental dosimetry methods there is a need for intercomparisons. The protocol and the aim of the intercomparison are discussed.

Keywords: Thermoluminescence dosimeter, radiophotoluminescence dosimetry, dosimeter, TLD reader

1. Introduction

The society today is faced with a problem of increasing man-made radiation. The starting basis for radiation protection is the exact knowledge of the doses of irradiation. The task is to carry out regular monitoring of individuals-professionals working with ionizing radiation, to control and measure doses in medical application on patients and staff and the monitoring in the environment. For solution of these tasks passive integrating solid state dosimeters (SSD) are widely used. The most often used solid state dosimetry methods are thermoluminescence (TL), radiophotoluminescence (RPL) and optically stimulated luminescence (OSL) dosimeters. In this paper the short principles, the elements and characteristics of RPL and TL dosimetry

* To whom correspondence should be addressed. e-mail: marika@irb.hr

systems and their application in environmental monitoring is discussed. Special attention is given to the importance of intercomparisons in environmental monitoring.

A review of the application of TL systems for environmental monitoring is published in [1]. The RPL dosimeter system described in this work is produced in Japan [2].

2. Principle of TL and RPL Dosimeters

The principle of TL [3] and RPL show some similarity however there are important differences which influence the application possibilities. Thermoluminescence or thermally stimulated luminescence is the emission of light during heating of a solid sample (insulator or semiconductor), previously excited by radiation. Radiophotoluminescence is the emission of light observed when some minerals or glasses, having been exposed to ionising radiation, are subsequently exposed to ultraviolet light. A short comparison of TL and RPL is shown in Table 1.

TABLE 1. Comparison of TL and RPL principle

	TL	RPL
Irradiation	Ionising radiation	
Detector	Various crystals	Mineral or glass
Effect	Excitation and trapping of electrons	Creation of stabile colour centres
Excitation (readout)	Heat	UV laser
Emission	Light	
Repetition of readout	Impossible	Possible as many times as necessary
Re-use	Possible after thermal annealing	

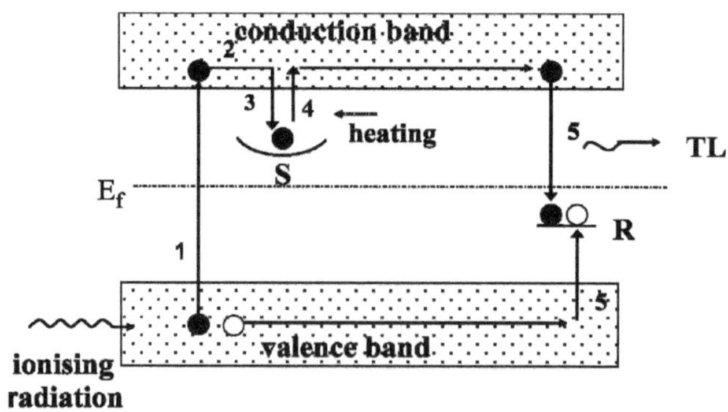

Figure 1. Simple model of the processes in TL phosphor S: potential electron trap; R: potential hole trap and recombination centre; E_f: equilibrium Fermi level

While for TL dosimetry many various crystals are suggested, for RPL the silver activated phosphate glass dosimeters are most commonly used. The simple schematic models of processes in TL and in RPL glass are s hown in Figures 1 and 2.

The TL process in Figure 1 can be described with the following steps:

1. Ionisation – produces free electrons in the conduction and free holes in the valence band.
2. Movement of electron in conduction band.
3. Trapping of electron in energy level S.
4. After heating electrons are released from the trap into the conduction band.
5. Recombination at recombination centre R and emission of light – TL response.

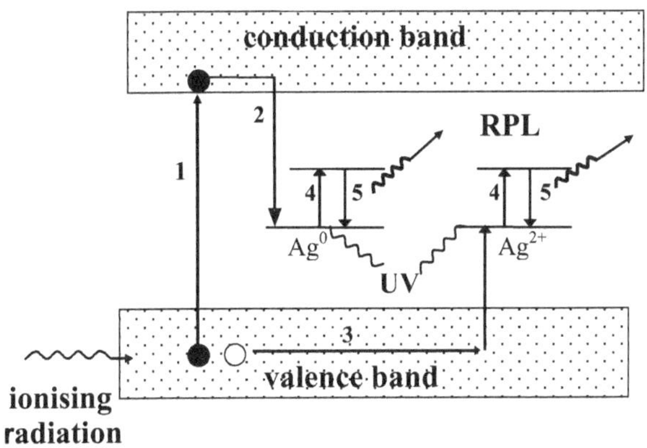

Figure 2. Simple model of the processes in RPL glass

The RPL process in Figure 2 can be described with the following steps:

1. Ionising radiation strips away electrons from the phosphate structure of the glass detector. In this way free electrons in the conduction band and holes (hPO_4) in the valence band are produced.
2. The free electrons react with Ag^+ activator in the glass and form Ag^0.
3. Ag^+ can react also with the holes ($Ag^+ + hPO_4$) resulting Ag^{2+}.
4. Ag^0 and Ag^{2+} are stabile luminescent centre, which after excitation with 320 nm UV light.
5. Return to lower energy level emitting an orange luminescence called RPL response.

In both processes the basic condition for dosimetry application is that the intensity of the emitted light (TL/RPL response) has to be proportional to the dose of ionising radiation.

In TL dosimetry the intensity of the emitted light as a function of the temperature is the thermoluminescence glow curve. The area under the peak or the peak height is the measure of the TL response. The shape, intensity and parameters of the glow curve differ for various TL phosphors and activators, Figure 3.

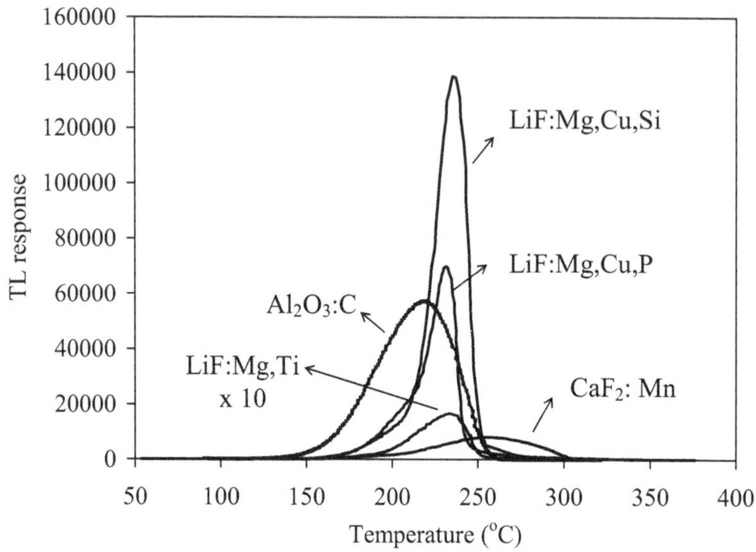

Figure 3. TL glow curves of various TL phosphors

3. Dosimetry Systems

The components of the passive solid state dosimetry systems are: the detector, the reader and the method of evaluation, i.e. the measurement cycle itself. All these three components determine the characteristics of the dosimetry systems.

3.1. DETECTORS

The commercially available TL detectors as well as the "home-made" detectors differ in physical shape (powder, disk, rod, chop etc.), in size and in chemical composition [1, 4]. Some of the most often used TL detectors are summarised in Table 2. TLD-100 and TLD-700 are the commercial names for natural and in ^{7}Li enriched LiF detectors activated with Mg, Ti (LiF:Mg,Ti) produced by "Harshaw" in US (today: Thermo Fischer Scientific). TLD-100 serves very often as reference material for comparison of characteristics. LiF phosphors doped with magnesium, copper and phosphorus or silica are high sensitivity detectors and therefore suitable for environmental dosimetry where low doses are expected [1, 5–8]. The characteristics of the detector are influenced by the chemical compound of the basic material, the quantity and quality of the activators, the method of crystallisation and production, etc. [9]. Various attempts can be found in the literature for radiophotoluminescent dosimetry from the early 1950s till today [10], however recently the Japanese silver activated phosphate glass detectors [2] are the most widely used for individual and environmental monitoring.

TABLE 2. Some of the most common TL detectors

LiF [5–8]	Borate [11, 12]	Calcium [13, 14]	Aluminium [15, 16]
LiF:Mg,Ti	MgB_4O_7:Dy	CaF_2:Mn	Al_2O_3:Mg, Y
TLD-100 (LiF:Mg,Ti)			
TLD-700 (^{7}LiF:Mg,Ti)			
LiF:Mg,Cu,P	$Li_2B_4O_7$:Cu, In	$CaSO_4$:Dy	Al_2O_3:C
LiF:Mg,Cu,Na,Si	$Li_2B_4O_7$:Cu, In, Ag		
LiF:Mg,Cu,Si	$Li_2B_4O_7$:Cu, Ag, P		

3.2. READERS

The TL reading process is based on the heating of the previously irradiated TL material from ambient temperature up to several hundred degree Celsius, while the emitted light is collected and measured quantitatively with photomultiplicator. The RPL reading process is the UV illumination (320 nm) of the previously irradiated glass detectors and the emitted light is also measured by photomultiplicator.

For TL dosimetry many commercial manual and automatic readers are available. TL reader combined with OSL reader is also widely used.

The most well-known RPL readers are produced in Japan and they differ according to the purpose of dose evaluation: for individual dosimetry, for radiotherapy measurements and for environmental monitoring. For environmental monitoring the SC-1 flat RPL glass dosimeters with FGD-202 reader were developed. In this work the study with this system is described. Nowadays a new RPL reader was developed by using an ultraviolet (UV) light emitting diode (LED) light source, appropriate optical filters and some electronic circuits based on a field programmable gate array which makes it possible to carry out fast signal processing and furthermore to change the system efficiently by rewriting the digital circuits [17].

3.3. MEASUREMENT CYCLE

The measurement cycle for TL and RPL consists of the following steps: annealing, package and storage, irradiation, readout and mathematical evaluation. All these steps influence the dosimetric characteristics of the system.

The *annealing* is a pre-irradiation heat treatment to release the traps and the colour centres for re-use of the detector. The annealing can cause variation in procedure therefore it is important to anneal all detectors involved in the measurement cycle identically. The annealing can be carried out in an oven or in the reader depending on phosphor characteristics. Today various ovens with computer controlled heating and cooling process are available.

The detectors mostly have to be protected against light, humidity, mechanical damage etc. during *storage* before or after monitoring as well as during the monitoring period. Therefore adequate *package* is very important for good accuracy of dosimetry. The detector become a dosimeter if it is placed in a holder. The holder in addition to the above protection has to ensure electronic equilibrium as well as can content various filter for energy dependence correction (Section 4).

Irradiation is carried out for two purposes: first for calibration in a well defined radiation field the other for the monitoring or dosimetry measurement in the unknown field. Passive solid state detectors are not absolute dosimeters. They have to be calibrated with a well-known dose (usual expressed as air kerma) of ionising radiation to measure the light output. For calibration usual ^{137}Cs or ^{60}Co is used, however for special radiation fields (heavy charge particles, low energy X-rays, etc.) special calibrations can be required.

The *readout* process is the measurement of the emitted light in various readers. For TL dosimeters it is a heat treatment in reader with an optimum heating rate (most often 10°C/s) to maximum temperature (T_{max}) defined by the detector to obtain the glow curve. The RPL readout process (UV excitation-light collection) is usual automatic.

To obtain dose values from the readout values i.e. TL or RPL response by *mathematical evaluation* the following factors has to be taken into consideration: calibration factor, correction factors (for reader, fading, energy dependence, sensitivity, background, etc.) and algorithm for calculation of dose value in terms of dose of interest. The calibration usual is carried out with dose expressed as air kerma and according to ICRU 39 [18] the dose in environmental monitoring has to be expressed as ambient dose equivalent, $H^*(d)$ for strongly penetrating radiation or as directional dose equivalent, $H'(d)$ for weekly penetrating radiation.

4. Characteristics of TL and RPL Systems

After completing the system the characterisation has to be carried out to determine the correction factors, to choose the best performances for the certain application. The following main characteristics have to be investigated: batch homogeneity, reproducibility detection threshold, sensitivity, linearity, fading, energy dependence.

The early RPL system was investigated in details according to European Laboratory Test Programme [19] Hsu et al. [20] compared the characteristics of RPL and TLD-100 dosimeters for individual monitoring, Ranogajec-Komor et al. characterised RPL (SC-1 flat glass dosimeters with FGD-202 reader) for environmental monitoring [21] according the requirements of international standard [22]. Both groups concluded that RPL has very convenient characteristics for these applications.

Numerous papers can be found in the literature about characterisation of various TL systems. In this work the earlier and recent experimental characterisation results of the author's laboratory [1, 23–29] are used.

In Table 3 some characteristics of RPL are compared to some TL dosimeters. The *batch homogeneity* can be expressed as the coefficient of variation (v) of the indicated dose value, E. *Reproducibility* is expressed as the statistical fluctuations of the indicated value in several measurement cycles. The *detection threshold* (or lowest detectable dose) is three times the standard deviation , s of the reading of ten unirradiated detectors.

TABLE 3. Characteristics of RPL and some TL dosimetry systems

Characteristics	RPL	TL
Batch homogeneity		
v (%)	1.2–3.7	2–4
Reproducibility		
v (%)	1.3–2.7	4–5
Detection threshold		LiF:Mg,Cu,P = 0.2
$3 \times s_E$ (μGy)	2.22	CaF_2:Mn = 7.0
		LiF:Mg,Ti = 61

v: coefficient of variation, s: standard deviation

The *sensitivity* is very important characteristics from the aspect of application. Dosimetry systems have to be chosen according the aim of application. High sensitivity dosimeters will be used for environmental dosimetry while for industrial application less sensitive dosimeters will be applied. The sensitivity of TL systems is expressed as the TL response relative to dose unit. Usual the relative sensitivity, $S(D)$ is used:

$$S(D) = F(D)_{TL\ mat}/F(D)_{TLD\text{-}100}$$

i.e. the dose response, *F*(*D*) of any TL material compared to the dose response of TLD-100.

Table 4 shows the relative sensitivity of some TL phosphors. The sensitivity difference can be seen also in Figure 1. It has to be taken into consideration that even the relative sensitivity can be changed if some parameters or part of the dosimetry systems (for example reader) is changed [23].

TABLE 4. Relative sensitivities of some TL detectors (with TOLEDO 654 reader)

Detector type (and origin)		S(D)
$Li_2B_4O_7$:Cu, In	(Serbia)	0.2
$Li_2B_4O_7$:Cu, In, Ag	(Serbia)	0.8
^{7}LiF:Mg,Cu,P	(Harshaw)	20
LiF:Mg,Cu,P	(China)	100
^{7}LiF:Mg,Ti	(Harshaw)	0.9
LiF:Mg,Ti (TLD-100)	(Harshaw)	1
Al_2O_3:C	(Russia)	65

The *linearity* is a basic characteristic of TL and RPL system. As it was explained (Section 2) the response of TL and RPL systems has to be proportional to dose, in ideal cases linearly proportional. In Figure 4 the linearity of several TL dosimeters is shown in a wide dose range.

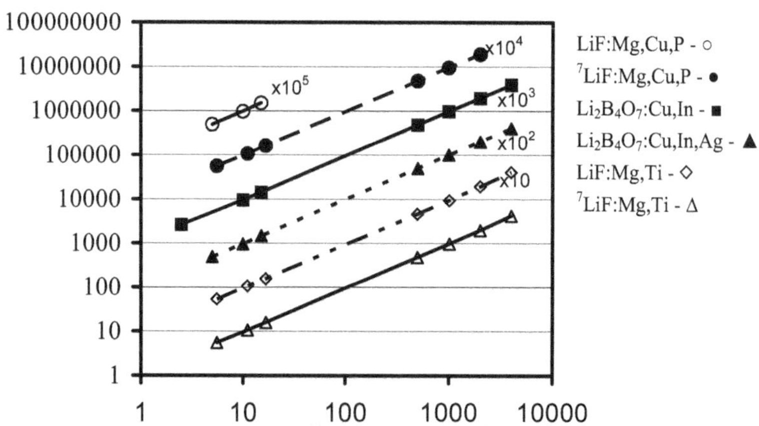

Figure 4. The dose response of some TL dosimeters

The linearity of RPL dosimeters in the dose range from 0.1 to 100 mGy of ^{137}Cs is shown in Figure 5. The standard deviations of the measured mean evaluated values were within 1%. The results (y and R^2) are in good agreement with the results in [20].

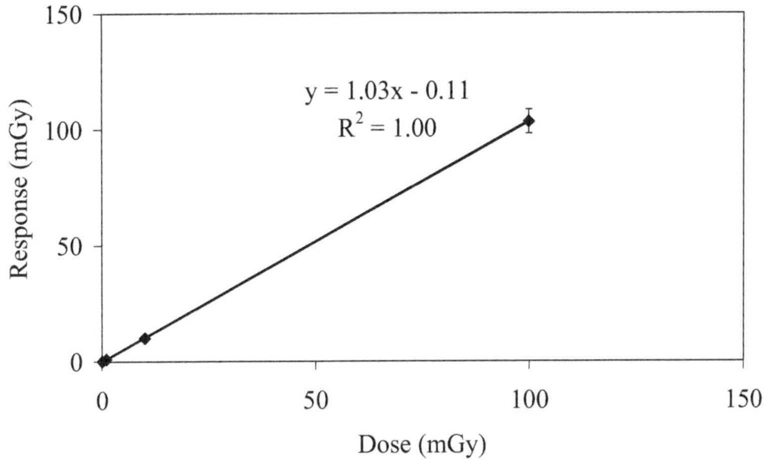

Figure 5. The dose response of RPL dosimeter

The stability of dosimeters under various climatic conditions as a function of time, i.e. the *fading* is very important in the environmental monitoring since in the environment mostly long-term exposures are the object of the investigations. Fading characteristics of newly developed materials are always investigated under different conditions (temperature, humidity, light effects). The fading of TL dosimeters depends first of all on the chemical composition of the detector and the dopants, the crystal structure, the thermal treatment during evaluation (pre-irradiation annealing, post-irradiation annealing/preheat, heating rate, etc.), the climatic and light conditions during exposure, usual fading in 1 month is about 3–30%. The fading correction can be carried out in different ways [1]. RPL dosimeters has an excellent fading characteristics, it has no fading [20].

The *energy dependence* has special importance in medical dosimetry because various and low energy X-rays are often used. In environmental monitoring the main component is the high energy cosmic radiation. In spite of this the energy dependence of various systems has to be determined and taken into correction.

Numerous studies were carried out on the energy dependence of various TL systems as well as the RPL system. An ideal dosimeter system would show the same response after irradiation with the same dose of radiation with various energies. Such dosimeter does not exist (even not theoretically), therefore the energy dependence has to be known and taken into correction especially because for calibration high energy ^{137}Cs or ^{60}Co radiation sources are used. There are different methods (filters, mathematical algorithms, the use of various calibration energies, etc.) for correction. In

Figure 6, the energy dependence of two high sensitivity LiF TL dosimeters and the RPL dosimeter is shown.

The energy dependence is influenced by the chemical compound of the dosimeters, by the composition of the activators, by the way of irradiation [1, 21, 23–27]. It can be seen in Figure 5 that in case of TL the dosimeters show underresponse in air and overresponse on phantom after irradiation with low energy X-rays and compared to ^{137}Cs.

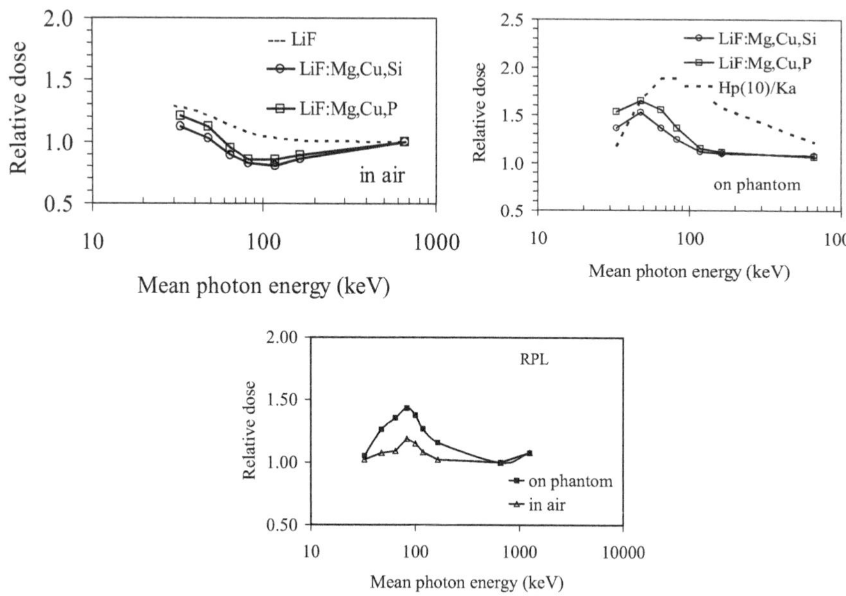

Figure 6. The energy dependence of TL and RPL dosimeters in air and on phantom

After characterisation of the SSD system the user can take the advantages and disadvantages of their own system into consideration, the possible errors can be estimate and the system can be applied in various field of dosimetry.

5. Environmental Monitoring with SSD

The responsibility of the society is to ensure adequate environmental quality. The potential increase of radiation dose in the environment as a consequence of human activities (weapon activity, possible facility accidents in nuclear power plant, medical, scientific and industrial application of various sources of radiation, illicit nuclear and radioactive agents) can be problem. To ensure the radiation protection in the environment the scientific community has the task to collect well documented data on radiation doses in the environment. The requirements for environmental dose measurements are not simple: it has to be measured the man-made contribution in the order of one tenth of natural environmental radiation and to follow the changes in natural environmental radiation. All these have to be done under variable and sometimes extreme environmental conditions (UV sunlight, humidity, temperature). The environmental radiation has two main components: the cosmic and the terrestrial radiation and varies according to location, altitude, season, etc. in the dose rate range 50–200 nSv/h.

TL and RPL systems are widely used for environmental monitoring. Their application to environmental dosimetry requires performance under laboratory and field conditions and performance testing according to national and international standards [1].

A standard is established for use as a rule or basis of comparison in measuring quantity, quality, value etc. Numerous existing standards and documents of relevance like international recommendations, technical reports etc. exist. It means that harmony is needed in standards. An overview of standards for individual monitoring can be found in the work of Alves [30] which contains useful informations for environmental monitoring also. As a small step to harmonisation of the international standards it was proposed by Ranogajec-Komor et al. [21] to change the IEC 60166 standard [22]: TLD systems for personal and environmental monitoring to a standard for solid state dosimeters (RPL and OSL).

An exemplar of national standard was the ANSI Draft Standard 13.29: "Environmental Dosimetry – Performance Criteria for Testing" which has foreseen test performance of TL environmental detectors under simulated extreme environmental conditions. The dosimeters investigated were placed in an environmental chamber that cycled twice through three 15 day periods of temperature and humidity conditions: −20°C; +50°C with 20% relative humidity; and +50° with 90% relative humidity [31]. Performance was measured by comparison with the criteria that would require the absolute value of the bias and standard deviation to each be less than or equal to 0.35, while their sum must be less than or equal to 0.50. All the TL dosimeter systems investigated (various LiF:Mg,Ti and LiF:Mg,Cu,P) satisfied these requirements, however they all showed higher fading than the quality control (QC) dosimeters. The QC dosimeters were treated on the same way (annealing, readout, irradiation) as the investigated dosimeters only they were not exposed to the extreme environmental conditions. The fading of TL systems observed in extreme temperature and humidity conditions called for further investigations and therefore inter-comparisons were organised.

To reach international standards there is a need for intercomparisons. Several large scale international intercomparison were organised by various organisations. The last published results for a large scale intercomparison were in 1999 [32]. The aim of these intercomparisons was the improvement of environmental monitoring methods and at the same time they also serve as periodic quality control for the participating laboratories.

Within these entire intercomparisons one lab was responsible for the organisation, for the field and laboratory irradiations and evaluation of the results (report). In the studies organised in frame of a Croatian–German–Hungarian [33, 34] and Croatian–Hungarian–USA [35–37] scientific cooperations the intercomparison was based on the joint-cross calibration in the laboratories involved and an exchange of the dosimeters within the respective national environment monitoring programs [1]. The objectives of the studies were to compare and to improve laboratory calibrations, to compare the results of different dosimetry systems at the same field site, to determine the field fading. The early RPL dosimeter system was also involved in the intercomparisons [1, 33, 34].

The dosimetry systems were compared through cross calibration and an exchange of dosimeters. Two measurement sites were selected in each country at which the environmental radiation field was different (sites in nuclear research institutes or nuclear facilities, private gardens, etc., sites at the see side and in mountains). The field doses at different locations, field fading under different climatic conditions, and the contribution of transit doses (because of air-mail) contributed to the knowledge of the properties of the TL and RPL systems for application to environmental dosimetry.

The following conclusions were made from these studies:

- The "new" (at this time) high sensitivity detectors were suitable for environmental monitoring.
- The large fading obtained in extreme temperature and humidity conditions was not observed in "normal" environmental conditions.
- The climatic conditions has no influence on fading and dose
  - Neither according the season.
  - Nor according the location.
- The fading correction does not improve the agreement between the dosimeter systems (which was within 10%).

The establishment of this international intercomparison allows a broad exchange of experience and dosimetric systems. As a consequence, the calibration and measuring procedures used in particular laboratories, as well as the quality of the dosimeter systems employed were improved.

Intercomparisons are organised from time to time according to the requirements of environmental dosimetry laboratories, the latest was reported in [38]. Recently a similar intercomparison is initiated as in refs. [33–37] using new developed high sensitivity TL and RPL detectors in comparison with the earlier investigated TLD systems.

6. Conclusions

Solid state dosimetry systems based on high sensitivity TL detectors and on RPL glass dosimeter are very suitable for long term environmental monitoring. In case of accidents they can be used also for shorter monitoring period. New systems are developed continuously. Before application the dosimeter systems has to be characterised according to national or international standards. Intercomparisons in environmental monitoring contribute to the knowledge of the properties of the TL and RPL systems for application to environmental dosimetry and help the laboratories to improve their environmental monitoring method.

Acknowledgement: The author gratefully acknowledges the cooperation of all coauthors and coworkers who participated in these studies. Special thanks for Chiyoda Technol Corporation, Japan for suport of the work with RPL dosimeters.

References

1. Ranogajec-Komor M (2003) Thermoluminescence dosimetry – application in environmental dosimetry. Radiat Safety Manag 2:2–16
2. http://wwwc-technolcojp/technol_eng/indexhtml
3. Bos AJ (2007) Theory of thermoluminescence. Radiat Meas 41:S45–S56
4. Kortov V (2007) Materials for thermoluminescent dosimetry: current status and future trends. Radiat Meas 42:576–581
5. Bos AJJ (2001) High sensitivity thermoluminescence dosimetry. Nucl Instrum Meth Phys Res B 184:3–28
6. Bilski P (2002) Lithium fluoride: from LiF:Mg,Ti to LiF:Mg,Cu,P. Radiat Prot Dosim 100:199–206
7. Tang K, Cui H, Zhu H, Fan Q (2007) Study of a new Lif:Mg,Cu,P formulation with enhanced thermal stability and a lower residual TL signal. Radiat Meas 42:24–28
8. Kim JL, Lee JI, Pradhan AS, Kim BH, Kim JS (2008) Further studies on the dosimetric characteristics of LiF:Mg,Cu,Si – A high sensitivity thermoluminescence dosimeter (TLD). Radiat Meas 43:446–449
9. Lee JI, Kim JL, Pradhan AS, Kim BH, Chung KS, Choe HS (2008) Role of dopants in LiF TLD materials. Radiat Meas 43:303–308

10. Hsu SM, Yang HW, Huang DYC, Hsu WL, Lu CC, Chen WL (2008) Development and physical characteristics of a novel compound radiophotoluminescent glass dosimeter. Radiat Meas 43:538–541
11. Prokic M (2001) Lithium borate solid TL detectors. Radiat Meas 33:393–396
12. Fernandes AC, Osvay M, Santos JP, Holovey V, Ignatovych M (2008) TL properties of newly developed lithium tetraborate single crystals. Radiat Meas 43:476–479
13. Dražić G, Trontelj M (1983) Sintered $CaSO_4$: Dy TL dosimeters. Int J Appl Radiat Isotopes 34:1633–1637
14. Ingle NB, Omanwar SK, Muthal PL, Dhopte SM, Kondawar VK, Gundurao TK, Moharil SV (2008) Synthesis of $CaSO_4$:Dy, $CaSO_4:Eu^{3+}$ and $CaSO_4:Eu^{2+}$ phosphors, Radiat Meas 43:1191–1197
15. Osvay M, Biró T (1980) Aluminium oxide in TL dosimetry. Nucl Instrum Meth 175:60–61
16. Akselrod MS, Kortov VS, Kravetsky DJ, Gotlib VI (1990) Highly sensitive thermoluminescent anion-defect α-Al_2O_3:C single crystals detectors. Radiat Prot Dosim 33:119–122
17. Ihara Y, Kishi A, Kada W, Sato F, Kato Y, Yamamoto T, Iida T (2008) A compact system for measurement of radiophotoluminescence of phosphate glass dosimeter. Radiat Meas 43:542–545
18. ICRU (1985) International Commission on radiation Units and Measurements, Determination of Dose Equivalents Resulting from External radiation Sources. ICRU Report 39, Bethesda, MD
19. Piesch E, Burgkhardt B (1984) Environmental monitoring, European interlaboratory test programme for integrating dosemeter systems. Commission of the European Communities, Luxemburg, EUR 8932
20. Hsu SM, Yeh SH, Lin MS, Chen WL (2006) Comparison on characteristics of radiophotoluminescent glass dosemeters and thermoluminescent dosemeters. Radiat Prot Dosim 119:327–331
21. Ranogajec-Komor M, Knežević Ž, Miljanić S, Vekić B (2008) Characterisation of radiophotoluminescent dosimeters for environmental monitoring. Radiat Meas 43:392–396
22. IEC (2006) International Electrotechnical Commission, Thermoluminescence dosimetry systems for personal and environmental monitoring CEI/IEC International Standard 61066:2006
23. Ranogajec-Komor M, Osvay M (1986) Dosimetric characteristics of different TL phosphors. Radiat Prot Dosim 17:379–384
24. Knežević Ž (2007) Influence of activators on energy dependence of thermoluminescence detectors. University of Zagreb, Faculty of Science, Zagreb, Croatia
25. Miljanić S, Ranogajec-Komor M, Knežević Ž, Štuhec M, Prokić M (2006) Comparative study of LiF:Mg,Cu,Na,Si and $Li_2B_4O_7$ TL detectors. Radiat Prot Dosim 119:191–196
26. Ranogajec-Komor M (2004) Thermoluminescence personal and medical dosimetry, Nato Advanced Research Workshop on Radiation Safety Problems in the Caspian Region (Proc Symp Baku Azerbaijan, 2003) (Eds. Zaidi MK, Mustafaev I) Kluwer, Dordrecht, The Netherlands, pp. 177–190
27. Miljanić S, Ranogajec-Komor M, Knežević Ž, Vekić B (2002) Main dosimetric characteristics of some tissue-equivalent TL detectors, Radiat Prot Dosim 100:437–442
28. Ranogajec-Komor M, Vekić B, Korenika Dž, Dvornik I, Piesch E, Burgkhardt B (1989) Standard test program and environmental monitoring with TL-dosimeters. II Yugoslav-Italian Symposium on Radiation Protection, Advances in Yugoslavia and Italy (Proc Symp, Udine, Italy, 1988) pp. 501–504

29. Ranogajec-Komor M, Muhiy-Ed-Din F, Milković Đ, Vekić B (1993) Thermoluminescence characteristics of various detectors for x ray diagnostic measurements. Radiat Prot Dosim 47:529–534
30. Alves JG (2008) Developments in standards and other guidance for individual monitoring. Radiat Meas 43:558–564
31. Ranogajec-Komor M, Klemic G, Sengupta S, Knežević Ž, Raccah F, Vekić B (1999) Investigation of the performance of ^{7}LiF:Mg,Cu,P under environmental conditions. Radiat Prot Dosim 85:217–222
32. Klemic G, Shobe J, Sengupta S, Shebell P, Miller K, Carolan PT, Holeman G, Kahnhauser H, Lamperti P, Soares C, Azziz N, Moscovitch M (1999) State of the art of environmental dosimetry: 11th international intercomparison and proposed performance tests. Radiat Prot Dosim 85:201–206
33. Ranogajec-Komor M, Vekić B, Piesch E, Burgkhardt BB, Szabó PP (1989) Intercomparison of solid state dosemeters within environmental monitoring programs, 30th Anniversary Symposium of Radiation Protection in the Boris Kidrič Institute of Nuclear Sciences, Radiation Protection-Selected Topics in Proc Symp Dubrovnik, Croatia, 1989, Eds. Ninković MM, Pavlović RS, Raičević JJ, pp. 385–390
34. Ranogajec-Komor M, Vekić B, Piesch E, Burgkhardt B, Szabó PP (1996) Intercomparison of solid state dosemeters within environmental monitoring. Radiat Prot Dosim 66:139–144
35. Ranogajec-Komor M, Klemic G (1997) Methods and advantages of intercomparisons of TLDs for environmental monitoring, 20th IRPA Regional Congress – The Second Regional Mediterranean Congress on Radiation Protection (Proc Congr Tel Aviv, Israel 1997) pp. 52–55
36. Ranogajec-Komor M, Uray I, Klemic G, Gabrić D (1999) Intercomparisons of new TLDs for environmental monitoring, IRPA Regional Symposium: Radiation Protection in Neighbouring Countries of Central Europe. (Proc Symp Budapest, Hungary, 1999) (Ed. Deme S), Roland Eötvös Physical Society, Budapest, Hungary pp. 504–511
37. Ranogajec-Komor M, Klemic G, Uray I (2002) Thermoluminescence dosimetry in environmental monitoring. IRPA Regional Congress on Radiation Protection in Central Europe (Proc Congress, Dubrovnik, Croatia, 2001) (Eds. Obelić B, Ranogajec-Komor M, Miljanić S, Krajcar Bronić I) CRPA, Zagreb, Croatia70-01
38. Duch MA, Sáez-Vergara JC, Ginjaume M, Gómez C, González-Leitón AM, Herrero J, de Lucas MJ, Rodríguez R, Marugán I, Salas R (2008) Long-term intercomparison of Spanish environmental dosimetry services Study of transit dose estimations. Radiat Meas 43:576–579

THE CHALLENGES FOR INVESTIGATION/DETECTION IN COMBATING TRAFFICKING OF RADIOACTIVE SOURCES IN ALBANIA

LUAN QAFMOLLA*
Institute of Nuclear Physics (INP)
SHYQYRI ARAPI
Institute of Public Health (IPH) Radiation Protection Office, Tirana, Albania

Abstract. The paper presents an overview of inalienable of radioactive sources and sporadic illicit trafficking of radioactive materials as new phenomena in Albania. The Institute of Nuclear Physics and Radiation Protection Office undertook the sealed radiation sources and radioactive thermoelectrically generators' situation in Albania. During the investigations of worker's group, some metallic scrap and spent radiation sources, which have penetrated from neighbor countries in illicit manner, were found. The most important output of this study is the evidence of the orphan/lost/found and conditioned radiation sources in Albanian territory during 1960–2007.

Keywords: Illicit trafficking, sealed radiation source, radioactive thermoelectrically generator, orphan, lost radioactive source

1. Introduction

1.1. THE EVALUATION OF SEALED RADIATION SOURCES SITUATION

The use of radiation sources in Albania has begun in the early 1960s mainly in geological researches as well in some military units for instrumental calibration purposes. These sources were mainly ^{226}Ra, ^{60}Co and ^{90}Sr.

In the mid of the mentioned years the first therapy unit with cobalt source of 150 TBq activity was begun to use at Tirana Hospital. After the establishment of the Institute of Nuclear Physics (INP), the use of

* To whom correspondence should be addressed. e-mail: l_qafmolla@hotmail.com

radioactive sources was expanded in different branches of medicine, industry, agriculture and researches. A large number of radioactive sources of alpha, beta, gamma and neutron emitters as ^{60}Co, ^{137}Cs, ^{90}Sr, ^{192}Ir, ^{241}Am, ^{238}Pu etc., were imported for different purposes from different countries mainly from France, United Kingdom and East Germany.

For 1970–1992, the INP has been in charge of the import–export, transport and inventory of radioactive materials in the whole country which have entered in Albania from other countries. Being based on this responsibility the INP has created the registry and inventory of all mentioned sources for this period [1].

In 1995 the Albanian Parliament approved the Radiation Protection Act and being based on this Act the Radiation Protection Commission (RPC) was established as national regulatory body in the field of licensing and inspection of all activities with radiation sources [2]. One year later the Radiation Protection Office (RPO) was established as executive body of the Radiation Protection Commission.

For the period 1992–2000 in Albanian, a decentralized policy for the import and export of radioactive materials was applied. After that period and especially after the Minister Council Ordinance (No. 56, dated 27.02.2000) for the control of import and export of radioactive materials by the licensed institutions, the RPO took the control of radioactive sources under its control and updated the database and the procedures of inventory of radioactive materials.

From the period 1960 up to now, in use of radioactive materials many irregularities were observed in cases of the lost of the control of radioactive sources. During the unrest events in 1997, some radioactive sources were stolen from military units and up to now the situation is ambiguous and unclear. During the period 1992–2000 some radioactive sources were imported and exported by local and foreign companies without obedience the legal procedures. These radioactive sources entered into the country together with their special equipments.

1.2. THE SITUATION OF RADIATION SOURCES DURING TRANSITION PERIOD

During the transition period (1990–2000), when the license and inspection control were not well established, probably some radioactive sources were out of the control of regulatory body, and actually they were lost or became orphan sources. In the same period a number of official companies, which used radioactive sources, were closed and therefore the radioactive materials were lost or are in unidentified situation. In some companies the sources may be stored in safety places, but this fact needs to be verified carefully under the investigation process by INP and RPO [3].

Regarding the radioactive sources, used by military units of Ministry of Defense, the agreement between the Ministry of Defense and INP was signed at 2003 and it was confirmed that all spent radioactive sources were sent to INP for conditioning and interim storage [4].

In the framework of this study, the INP/RPO have contacted in the officials of the Ministry of Defense and they confirmed that there are no lost or orphan radioactive sources. The Albanian Military Units had radioactive thermoelectrically generators (RTG) in use according to the non-official information [5].

During the control of radioactivity contamination in the metallic scrap at entry/exit points at customs, radioactive sources with medium or low activity were found. These sources were mainly used by army and/or for industrial applications and were exported by different Albanian companies. In 2004, three cobalt sources, which belonged to military units, were found in the north of Albania by police officers. In entry/exit points of the custom office at Tirana four items of ^{241}Am/Be neutron emitter sources with 0.4 GBq activities each and one item of ^{137}Cs with approximately 750 MBq activity were found in 2006. They all were used for industrial applications. It was evaluated that other radioactive sources as ^{60}Co and ^{90}Sr may be lost or stolen during the 1997. Another source of ^{137}Cs was found by INP staff which has not included at the inventory list of the radioactive sources used in Albania for period 1960–2008. It can be expressed that some of these sources have transported as metallic scrap to neighbor countries in illicit manner.

In Albania, the database of radioactive materials entered through legal procedures is stored in INP and a copy of the inventory is kept in RPO [1, 3].

2. Methodology

In the framework of the General Order Agreement 5 (RTR GOA-A/1) and the Statement of Work (SOW-02), Albanian institutions of INP and RPO were responsible for the sealed radiation sources and RTG in Albania during the period 1960–2006, with support from the Department of Energy of USA. An important output of this study was listing and putting in evidence of all orphan and lost radiation sources in Albania territory during the mentioned period.

To carry out the task described in Statement of Work SOW-02, a special working group was established with participation of the experts from INP, RPO and Geophysical Center. This group examined the most important documents related with import and export of radiation sources in Albania, carefully. The working group has checked all documents related with import and turned back at the suppliers of radioactive sources: invoices, reports, contracts, and registers as well all documents related with radioactive sources in use, storage, conditioned and suspected for period 1960–2007 in Albania (Table 1).

Being based on the inventory registers in INP and RPO, the spent radioactive sources were verified as well the total number of existing radioactive sources used in the public and private sectors. At the same time, the responsible persons who were in charge of radioactive sources of respective facilities were identified and their address were found in order to contact them for localization, identification and tracking the radioactive sources in next steps.

TABLE 1. The situation of radioactive sources in Albania (1960–2007)

Nuclide	Quantity (pcs)	Activity	Production year	Users	Imported	Licensed
Ra-226	3	0.5 mCi	1960	G.E.Tirana		
Co-60	1	500 mCi	1961	N.F.Fier		
Ra-226	2	0.8 mCi	1961	O.H.Tirana		
Sr-90	20	0.1 mCi	1964	Ch.D.Tirana		
Co-60	2	0.5 mCi	1968	O.H.Tirana		
Co-60	2	7 + 10 mCi	1968	O.H.Tirana		
Ra-226	1	5 mCi	1968	INP		Storage
Cs-137	1	2 mCi	1976	INP	Imported	
Cs-137	1	1 mCi	1976	INP	Imported	
Ra-226	Solution	Unknown	1976	A.D.Kucova		
Co-60	1	100 mCi	1978	FeNi, Prrenja	Imported	
Am/Be-241	1	2.1 Ci	1978	INP	Imported	In use
Pu-238	1	30 mCi	1978	INP	Imported	In use
Co-60	1	Unknown	1978	S.F. Maliq		
Am/Be-241	1	2.01 Ci	1978	Servcom	Imported	
Am/Be-241	1	100 mCi	1978	G.E.Tirana	Imported	Storage
Am/Be-241	1	100 mCi	1979	INP	Imported	In use
Cs-137	1	20 mCi	1979	INP	Imported	Storage
Am-241	4	Smoke det	1980	Port-Durres		Storage
Am/Be-241	4	Unknown	1980	M.C.Elbasa		
Co-60	1	600 mCi	1980	Ch.D.Tirana	Imported	
Co-60	1	580 mCi	1980	Ch.D.Tirana	Imported	
Co-60	1	8.3 mCi	1981	G.E.Tirana	Imported	
Sr-90	1	0.9 mCi	1982	INP	Imported	Storage
Cs-137	1	300 mCi	1982	Servcom	Imported	In use
Cs-137	2	50 mCi	1982	Servcom	Imported	In use
Co-60	1	560 mCi	1982	Ch.D.Tirana	Imported	
Co-60	1	580 mCi	1982	Ch.D.Tirana	Imported	
Co-60	1	560 mCi	1982	Ch.D.Tirana	Imported	
Co-60	1	580 mCi	1982	Ch.D.Tirana	Imported	
Co-60	1	560 mCi	1982	Ch.D.Tirana	Imported	
Co-60	1	580 mCi	1982	Ch.D.Tirana	Imported	
Cs-137	1	32.5 mCi	1983	INP	Imported	In use

Co-60	10	1–10 mCi	1983	Lezha Fact.	Imported	
Co-60	7	40 mCi	1983	M.CElbasan	Imported	
Co-60	8	123 mCi	1983	M.CElbasan	Imported	
Am-241	1	10 mCi	1983	INP	Imported	In use
Am/Be-241	1	40 mCi	1983	I.Soil.Study		
Cs-137	1	10,000 Ci	1985	INP	Imported	In use
Cs-137	6	15 mCi	1985	O.H.Tirana	Imported	
Cs-137	6	15 mCi	1985	O.H.Tirana	Imported	
Kr-85	3	100 mCi	1985	P.F.Lushnje	Imported	Storage/
Co-60	10	11 mCi	1985	Burel Fact.	Imported	
Co-60	1	5 mCi	1985	G.E.Tirana	Imported	
Co-60	1	5.2 mCi	1985	G.E.Tirana	Imported	
Am/Be-241	1	10 Ci	1986	Servcom	Imported	In use
Co-60	1	4.5 mCi	1986	G.E.Tirana	Imported	
Co-60	1	5 mCi	1987	G.E.Tirana	Imported	
Cs-137	1	150 mCi	1987	INP	Imported	Storage
Cs-137	1	20 mCi	1987	G.E.Tirana	Imported	
Pu-238	1	30 mCi	1988	G.E.Tirana	Imported	
Cs-137	1	2.2 mCi	1988	G.E.Tirana	Imported	
Co-60	1	8 mCi	1988	G.E.Tirana	Imported	
Co-60	1	5 mCi	1988	INP	Imported	
Co-60	1	8.7 mCi	1988	INP	Imported	
Co-57	1	50 mCi	1989	INP	Imported	
Co-60	1	1 mCi	1989	INP	Imported	
Pu-238	1	30 mCi	1990	INP	Imported	In use
Am/Be-241	1	10 Ci	1990	Servcom	Imported	In use
Co-60	1	3,000 Ci	1993	O.H.Tirana	Imported	
Ra-226	1	74 KBq	1994	W.AtlasFier	Imported	
Am/Be-241	1	666 GBq	1994	W.AtlasFier	Imported	
Am/Be-241	1	14 GBq	1994	W.AtlasFier	Imported	
Cs-137	1	74 GBq	1994	W.AtlasFier	Imported	
Cs-137	1	21 MBq	1994	W.AtlasFier	Imported	
Co-60	1	6,000 Ci	1996	O.H.Tirana	Imported	
Co-60	1	10,000 Ci	2003	O.H.Tirana	Imported	In use
Co-57	1	25 mCi	2001	INP		In use
Cd-109	1	20 mCi	2002	INP		In use
H-3	2	220 Ci	2002	INP		Storage
Cs-137	1	63 GBq	2002	Schlumber	Imported	In use
Am/Be-241	1	592 GBq	2002	Schlumber	Imported	In use
Am-241	1	18.5 GBq	2002	Schlumber	Imported	In use
Cs-137	1	200 mCi	2002	Schlumber	Imported	In use
Co-60	1	37 KBq	2002	Schrumber	Imported	In use
Cs-137	1	0.30 GBq	2003	Mapa Const	Imported	In use
Am/Be-241	1	1.48 GBq	2003	Mapa Const	Imported	In use
Am-241	1	Unknown	2003	Sarandopulu	Imported	
Co-60	1	20 MBq	2003	Kurum Int.	Imported	In use
Unknown	1	Unknown	2003	"OXY"USA	Imported	
Cs-137	1	20 Ci	2004	INP	Imported	In use
Ir-192	1	150 Ci	2004	I.Gas, Durres	Imported	Storage
Am-241	1	1.67 GBq	2004	C–C, Tirana	Imported	In use
Sr-90	1	20 MBq	2004	INP	Imported	In use
Co-60	1	10,000 Ci	2005	O.H.Tirana	Imported	In use
Cf-252	2	21 + 22 mCi	2006	Cement Fact	Imported	F.Kruja
Ir-192	1	154.8 Ci	2007	I.Gas, Durres	Imported	Storage
Co-60	1	0.5 m Ci	2008	KurumElbas	Imported	In use

It shows that 16 radioactive sources, used in public and private sectors are suspected as lost sources, Table 2.

TABLE 2. The radioactive sources suspected/lost during 1960–2007

Radionuclide	Quantity (pe)	Activity	Status
Ra-226	solution	Unknown	Suspected
Co-60	1	100 mCi	Suspected
Co-60	1	Unknown	Suspected
Am/Be-241	1	2.1 Ci	Lost
Am/Be-241	4	100 mCi	Conditioned
Co-60	1	8.3 mCi	Suspected
Co-60	7	40 mCi	Lost
Co-60	8	123 mCi	Lost
Am/Be-241	1	40 mCi	Found
Co-60	2	5 mCi	Suspected
Co-60	1	5.2 mCi	Suspected
Co-60	1	4.5 mCi	Suspected
Cs-137	1	20 mCi	Found
Pu-238	1	30 mCi	Suspected
Cs-137	1	2.2 mCi	Suspected

3. Results and Conclusions

Being based on the documents that we have studied, as well as at the data of SRS/RAIS programs which are shown at Table 1, 88 items of the radiation sources are under the control of official bodies. 26 items of the radiation sources are conditioned while 36 items are in use/storage. Ten items are turned back to suppliers while 16 items are suspected and need to be investigated on the near future, Table 2.

1. This study includes only suspected radioactive sources based on registers and documents of INP (1970–1992) and RPO (2000–2008). The situation before 1960 and 1990–2000 are actually unclear and doubtful.
2. The responsible persons who have in charge of radioactive sources of respective facilities were identified according to the documents. Their addresses were found in order to contact in next steps of localization, identification and recovery of radioactive sources.

3. The financial support of this investigation of the suspected/lost/found radiation sources needs to be ensured by Albanian governmental Institutions, INP and RPO and from other resources.

References

1. Report, Fundamental register of import/export of radioactive materials in Albania 1970–1992
2. Radiological Protection Act, Approved by Albanian Parliament, Tirana, Albania, Law No. 8025, Date 09.11.1995
3. SRS and RAIS database information and registration, CD data of INP/RPO, 1996/2003
4. Handling and Processing of Radioactive Waste from Nuclear Applications, INP, 2004
5. General Order Agreement RTR GOA-A/1 between INP/RPO and USA Department of Energy, October 2004

MACEDONIAN EXPERIENCE IN METAL SCRAP MONITORING AT BORDER CROSSINGS

TRAJČE TRAJČEV* (TRAYÇE TRAYÇEV)
Republic Institute for Health Protection – Radiation Dosimetry Department Skopje (Üsküp), Macedonia (Makedonya)

Abstract. Basic informations on ionizing radiation monitoring of the traffic of metal scrap materials trough the borders of the Republic of Macedonia. Techniques, procedures, instruments being used for monitoring. Collaboration with other institutions on this matter.

Keywords: Ionizing radiation monitoring, metal scrap materials, border crossing, dosimetry

1. Introduction

Since 2000, in Macedonia the radiation monitoring of the transport of metal scrap material is permanently developing along with the rising of the traffic trough the border crossings.

In 2007, total 7,975 of wagons and 824 of trucks were examined. The total mass examined at the borders is 369,617,911 kg [1].

Radiation Safety Directorate in Macedonia was established at 2002 and based on the Law of Radiation Protection and Safety (48/2002). Following the establishment of Radiation Safety Directorate, Radiation Dosimetry Department, within Public Health Protection Institute was instituted as the technical service for radiation dosimetric measurements.

* To whom correspondence should be addressed. e-mail: talepatale@gmail.com

2. Designated Locations for Ionizing Radiation Monitoring on Vehicles Transporting Metal Scrap Material

The transportation is carried out by railway and trucks. The designated locations are border crossings, custom's terminals, and melting companies gates. The trucks importing metal scrap materials are monitored for ionizing radiation at the border crossings. Some of the trucks are monitored at the Custom's terminals in Skopje and Tetovo. Almost in every city there are designated terminals for this type of monitoring.

Previously, the same procedure for railway wagons was carried out on a railway terminal near Skopje, where all wagons from all border crossings ware gathered. For better efficiency and better collaboration with the neighboring countries railway transport now is monitored at the entrances in the country.

Since this change has being made, in order to avoid traffic jam at the border crossings, monitoring teams are available 24 h in a day and 7 days in a week.

Railway wagons with metal scrap for export are controlled at the railway stations in Veles and Gradsko.

3. Procedures During Monitoring

Custom's Department, Macedonian Railways and Import/Export companies send request by fax for monitoring the vehicles to the Radiation Dosimetry Department.

Based to the data provided with this requests, Radiation Dosimetry Department's teams prepare certificates that contain:

- Certificate number
- Date of measurement
- Type of the material
- License number of the vehicle
- Type of transport
- Number of wagon composition (in case of railway transport)
- Mass of the material [kg]
- Company's name (exporter/importer)
- Origin of material, country
- Final destination
- Location of measurements
- Result of the measurement [μSv/h]
- Dosimetrist's name

3.1. OPERATIONAL VALUES

To estimate the ambient dose equivalent, value that is measured is dose rate given in Sv/h or R/h. For tracing ionizing radiation sources, survey meters containing scintillation detectors have advantage over the ones containing proportional, or Geiger-Miller's (GM) tubes, because of their fast response. In case of using this kind of survey meters, operational value is counts per second (cps).

3.2. PROCEDURES

Before starting the control, dosimetrists first establish background radiation by evaluating dose rate at distance from the vehicle approximately 10 m. Mean value of the background radiation in Macedonia is 150 nSv/h with highest level 250 nSv/h.

Alarm trigger of the instrument is set (manually or automatically) at position of approximately 20% of the background level in that particular surrounding. Ionizing radiation monitoring of vehicle's content is performed by positioning the survey meter's detector closest to the vehicle and sliding parallel to its sides over whole surface. Usually this is done by two dosimetrists on both sides of the vehicle in the same time, in order to avoid misdetection if one of the instruments is not functioning properly. Dosimetrists wear personal active dosimeters (pagers) attached on their bodies on the side closer to the vehicle. This is both, for the personal monitoring and for support of the survey measurements. If at some point of the wall of the vehicle higher number of counts per second is detected, then that spot is checked more then once to ensure the result. If this is the case, next procedure is to measure dose rate at this point by using instruments with proportional or GM tube. And, if it is possible (not obligation of this department) to identify the source and the energy of the photons with spectrum analyzer, the hot spot of the vehicle's surface is marked and Customs are provided with certificate and the entrance of this vehicle in Macedonia is forbidden. If the material in vehicle is of Macedonian origin, safe storage of radiation source is obligation of R. of Macedonia. Such cases are reported to Radiation Safety Directorate (RSD) which undertakes the next steps for safe storage.

3.3. IONIZING RADIATION SOURCES

Lightning Rod: If radiation sources are not detected, the Department gives import/export approval certificates for each vehicle. All data of the monitoring are kept in department's database.

3.4. PROBLEMS DURING MONITORING

During monitoring some problems may occur that can lead to misguides in further estimations. For example, there is case of electromagnetic induction in the survey meters from the cables of railway's power supply 25 kV, 50 Hz that produces electrical current trough the survey meter's detector, or its electronic components; which indicator articulates as higher dose rate of radiation and alarm is turning on.

Other cases are where background radiation is higher, or it only produces high number of photons but with very low energy. In this case there is possibility that the presence of some weak, shielded or screened sources may not be detected.

4. Equipment Department for Active Radiation Monitoring

- Thermo – basic unit FH 40G-L10 with integrated proportional detector (30 keV–4.4 MeV).
- Thermo teleprobe FH 40TG.
- Thermo – probe FHZ 512A with scintillation detector.
- Thermo – probe FHZ 612 with GM tube.
- Thermo – probe FHZ 732 GM, pancake for α, β, and γ radiation.
- Polimaster personal radiation detector (PRD) PM1703GN – pagers.
- Ludlum – monitoring instrument model 192.
- XRF spectrum analyzer and identifier ICS-4000.
- EXPLORANIUM spectrum analyzer and identifier GR-135 Plus.
- Personal dosimetry is performed with passive thermo-luminescent dosimeters, Thermo TLD100.

4.1. PANEL DETECTORS

Great contribution for radiation monitoring of the traffic of metal scrap material are the panel detectors. Border crossings equipped with this kind of instruments. The panel detectors were established at the points of Blace (recently reconstructed), Tabanovce (recently reconstructed), Deve Bair, Delčevo, Novo Selo, Bogorodica, Dojran, Medžitlija, Ḱafasan and Blato.

Another type of panel detector was established at the Airport of Skopje for cargo transport. Neutron panel detectors were placed in the passenger section at the airport of Skopje and at border crossing Bogorodica. But both of them are not in function. These devices are operated by Customs of Macedonia.

Employees at the border crossings, both from Ministry of interior and Customs are equipped with personal active dosimeters (pagers). None of the railway crossings is provided with panels. There is ongoing project for installing panels at railway border crossing Tabanovce, Figure 1.

References

1. Database of the Radiation Dosimetry Department – Republic Institute for Health Protection in Macedonia (2007) http://www.customs.gov.mk; http://www.drs.gov.mk/zakoni.html

RADIATION MONITORING AT THE BORDERS OF REPUBLIC OF UZBEKISTAN WITH THE USE OF PORTAL MONITORS

VITALIY PETRENKO*, BEKHZOD S. YULDASHEV, ULUGBEG ISMAILOV, NIKOLAY N. SHIPILOV AND ANVAR D. AVEZOV
Institute of Nuclear Physics, Uzbekistan Academy of Sciences, Tashkent, Uzbekistan

Abstract. According the program of radiation monitoring in Uzbekistan to prevent illicit nuclear trafficking main customs border checkpoints were equipped with radiation portal monitors. In total 30 checkpoints were equipped with 175 monitors, produced by Russian Company "ASPECT". Special radiation monitors were elaborated and manufactured in INP AS RU and installed in INP (main gates, research reactor and laboratories building) to prevent nuclear smuggling.

Keywords: Radioecology, radiation monitoring, radionuclide migration, NAA

1. Introduction

The increasing of threatening of the nuclear terrorism motivated the countries to find solutions to stop illicit trafficking of fissile and radioactive materials. Uzbekistan is located on the cross-roads between from the North to the South and from the West to the East. In old times the big part of Great Silk Road had passed through Uzbekistan. Presently there is a heavy traffic on this road; the vehicles from Europe, Russia, Kazakhstan, Iran, Afghanistan etc. All vehicles are possible carriers of nuclear smuggling because they are generally allowed to pass through border crossing points without any control. The cargos can only be inspected if there is sufficient suspicion that they were not declared clearly/truly by cargo sender. It is clear that the radiation control can decrease the probability of illicit trafficking of fissile and radioactive materials through the Uzbekistan, greatly. Before, custom and border officers on main entry points were equipped with radiation pagers which were attached to their belts. But

* To whom correspondence should be addressed. e-mail: petrenko@inp.uz

these devices were not effective enough because of high noise level originated from passing vehicles and because they have low detection sensitivity it was hard to monitor railway trains and pedestrians. It was the best way to install stationary portal radiation monitors at main customs border crossings or entry points to solve the problem. Their high sensitivity permits to detect signals exceeding background level on several percents in moving objects by recording these signals in computer and the object itself (vehicle, train or pedestrians etc.) by video-camera. Several thousands square centimeter volume of gamma detectors of "Yantar" produced by Russian company "Aspect" allows to get high sensitivity.

2. Installation and Operation of Portal Monitors at the Borders of Uzbekistan

Stationary systems for detection of fissile and radioactive materials "Yantar" are designed to detect radioactive and nuclear materials during continuous automatic monitoring of vehicles, trains, pedestrians and luggage at various checkpoints, nuclear power plants and nuclear cycle facilities. The basic set consists of pillars with detectors, electronic units and control panel. A video monitoring system, network devices, additional alarm devices, traffic lights, and drop bars are optional. Specifications of the system are as follows:

- False alarm rate – no more than 1/1,000
- Operation – continuous, automatic
- Uninterrupted operation after disconnection of 220 V power supply – no less than 10 h
- Service life – 12 years
- Bus with interface RS-485
- Protocol – MODBUS

Features of the system are as follows:

- Light and audible alarm indication
- Automatic adaptation to the varying natural background
- Archival storage of the alarm event data: date, time, detector count rate, type of channel (gamma or neutron)
- An optional video monitoring system provides a record of an alarm-causing object
- Gamma detector – based on organic plastic scintillators; neutron detector – based on proportional ^{3}He counters
- Operating temperature from −50°C to +50°C
- Compliance with EMC requirements for nuclear instruments
- Lightning protection of power and signal lines

- Possible integration of up to 16 systems of different types into a single information network without extra hardware and software
- Possible remote access

Radiation portal monitors are installed at one or two lanes of the road or railway for radiation level control of vehicles, pedestrians and trains. Special radiation monitors to control the transportation of metal scrap, the cargo in warehouses etc. are also installed. The passage of vehicles and pedestrians are under video-cameras surveillance and information of radiation levels and video are transmitted to the room of officer on duty. In case of detection the signal exceeding background level, light and sound alarm was triggered and the data on radiation levels along the vehicle (railway car) length both on gamma radiation and neutrons were recorded on server memory. In case of pedestrians' monitoring the pedestrian is recorded by his face. In railway station case, destination and number boards is also recorded. The radiation monitoring program to prevent illicit nuclear trafficking in Uzbekistan is supposed to equip main border checkpoints by using such monitors. For monitoring the pedestrians radiation monitors of Yantar-1P, Yantar-2P and Yantar-U are used. The radiation monitors of Yantar-1A and Yantar-2A are installed for radiation monitoring of vehicles and Yantar-1ZH and Yantar-2ZH monitors are installed for railway checkpoints. All these portal radiation monitors have gamma and neutron detection channels. Up to date, 30 checkpoints were equipped including 19 vehicles (total 118 monitors), 10 railways (total 40 monitors) and one for International Tashkent Airport (total 12 monitors). The military-Custom Institute is equipped with two monitors; Institute of Nuclear Physics is equipped with three monitors. The total amount of monitors is 175.

The institute of Nuclear physics of Uzbekistan Academy of Sciences provides the operation of these radiation monitors, technical assistance and consultancy in case of alarm signals and regular technical maintenance. Besides, in Institute of Nuclear Physics were elaborated and manufactured the radiation monitors KRIK [1] of which the operational principle differs from the other ones. The experience in operation of radiation monitors have shown that the majority of alarms were innocent and caused by NORM radionuclides mainly K-40 from various industrial products such as building materials, ceramics, mineral fertilizers. In some cases contaminated materials were detected, for example at one of the custom check-point Yallama alarm was caused by a truck with molybdenum oxide shipment. The analysis of shipment had revealed that part of it was contaminated by U-235, U-238, Pu-240 and some uranium fission products, Figure 1. The origin of shipment was Kazakhstan. Another detection of contaminated material happened at customs check-point Alat and caused by the truck with zinc powder shipment. Detailed analysis had shown that Cs-137 radioisotope was the main source of the contamination, Figure 2. On November 29, 2007 at 1.47:32 radiation monitors at railway station Nazarbek registered more than 20 times exceeding radiation background signal while passing the train #2306 from Kyrgyzstan to Iran, Figure 3. Another example on the same day; at 12.30 INP RU employees together with Tashkent province custom officers revealed the car # 64032139 which has the dose rate on surface exceeded 60,000 μR/h while the normal background dose rate is 12μR/h.

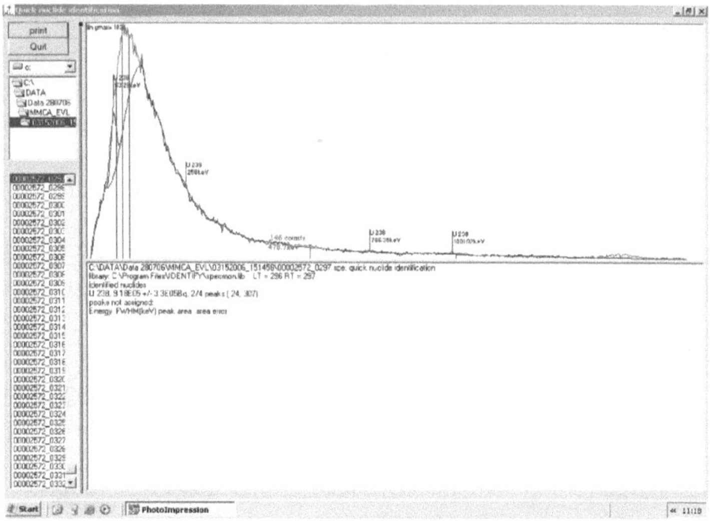

Figure 1. Gamma spectrum of Yallama shipment sample measured by MCA-16 and NaI(Tl)

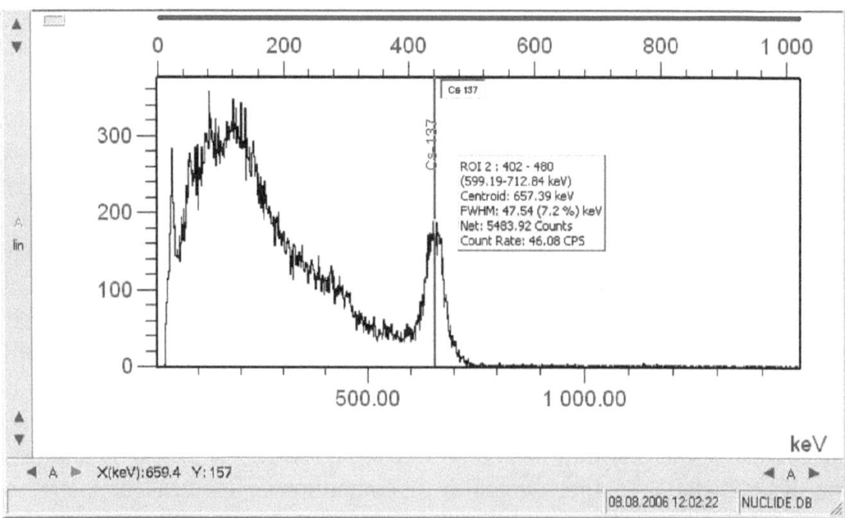

Figure 2. Gamma spectrum of Alat shipment sample measured by IdentiFINDER

The investigation of the case showed the following results:

1. The main source of radioactivity was observed in the middle of the car and in its lower part.
2. According to the measurements of radiation monitors and distribution of radioactivity along the car the radioactivity sources can be found.
3. By analyzing the obtained spectra ^{137}Cs, ^{231}Pa, ^{231}Th, ^{144}Ce, ^{134}I, ^{75}Se, ^{89}Kr, ^{235}U, ^{220}Ra, ^{67}Ga and other radioisotopes were identified, Figure 4. The measurement results obtained by using semiconductor detectors also confirmed the presence of ^{238}U decay products and ^{137}Cs. It's confirmed by elevated level on neutron radiation.

Figure 3. The contaminated wagon with scrap

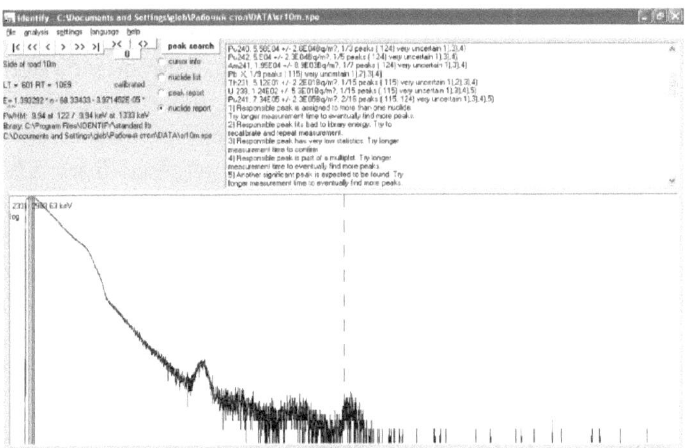

Figure 4. Gamma spectra taken on wagon surface

3. Radiation Detectors Elaborated in INP AS RU

The working principle of modern monitors is based on comparison between measurements of natural background without sample and with sample. In system, an alarm signal is triggered for values over the background. Monitors designed have two main weak spots leading to false alarms:

Object sensor: Performance of an object sensor depends on season, traffic intensity and a number of other factors.

Natural background: Background continuously changes for some technogenic reasons as cosmic radiation and others.

To reduce the effect of such disadvantages, a devised method is applied in monitoring procedures. In this case, the comparison between the signals from two detectors placed at some small distance from each other along the radioactivity traffic line which are separated by lead shielding, is considered, Figure 5. Working parameter of these two detectors is different from each other. The close located detectors allow excluding background deviations of cosmic component, and the difference between the signals from detectors must be equal to zero or close to a constant value, ideally. The difference between signal rates does not change or changes insignificantly in case there is no radioactive material. In case there is a radioactive material, the signal rate from the first detector increases and the difference between signal rates increases while the second detector is lead-shielded. Thus, two different peaks of the radioactive materials – of different polarity – are observed.

An alarm signal is set off of according to the logical circuit responds of these two peak values and the polarity difference over the normal value. During the testing of the proposed method Cesium-137 source with activity of 80 kBq was moved along the detectors position line at natural background of 577 counts/s. Each detector measured the radioactivity level at every 10 cm with time-step of 1 s and determined the difference between the signals. The peaks with different polarity were precisely determined. The background was artificially increased from 577 to 5,775 counts/s, that is more than ten times, but the difference between the signals was still same as before. The same Cesium-137 source was transported along the detectors position line for a period of 3 weeks continuously. During this period not more than two false alarms per week were observed. This result is significantly better than the ITRAP requirements of concerning monitors, which is very important for high traffic. Thus, if 25 false alarms are acceptable for 10,000 passing, because our testing is carried out for 5 s it means 86,400 passing in a week and therefore only one false alarm for 40,000 passing. Thus, these tests demonstrate that false

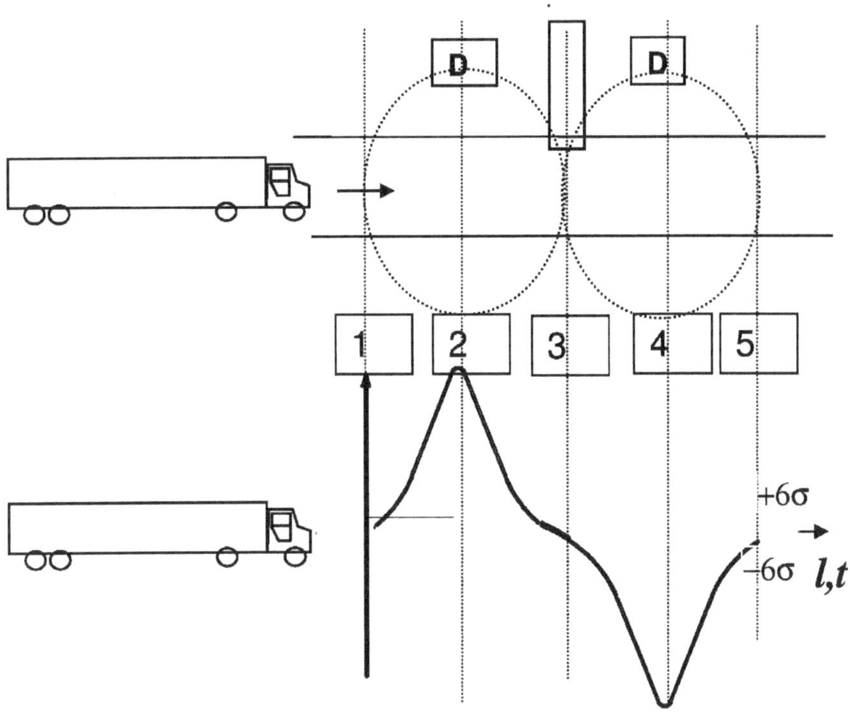

Figure 5. The different counting method of radioactivity detection in moving objects

alarms do not appear in ten times background changes and the system confidently detects approximately 80 kBq of radioactivity transportation. Such of these tests were conducted by means of modifications of KRIK monitors with plastic scintillators designed in INP AS laboratories in real conditions. An alarm signal from detectors was transferred to a counter placed in the computer. Analysis of signals from detectors and a decision making upon alarm was performed by an industrial computer by means of specially designed software.

In the Institute of Nuclear Physics, to detect nuclear fissile materials by use of instantaneous gamma quanta, which accompany to every spontaneous or induced decay of a nucleus, was suggested. Since the intensity of gamma radiation following the spontaneous neutron decay is much lower than the gamma radiation accompanying an alpha-decay, the former gamma radiation has not been used for passive analysis. We have examined the use of high multiplicity of this gamma radiation for coincidence detection. The Weapon Grade plutonium Pu-239 [2] contains approximately 6% of Pu-240, therefore, spontaneous decay activity of 10 g of Pu-239 is about 280 Bq. Here, neutron yield of spontaneous decay is about 600 neutrons/s, neutron yield of (α, n) reaction is about 380 neutrons/s. A coincidence of instantaneous gamma quanta was detected by using two organic-scintillator system. Every scintillator has an area of 1,200 cm^2. Neutrons were detected by using a set of ^{3}He counters with a total area of 1,200 cm^2. The background coincidences level can be determined by natural cascade processes. During studies it was found that the dependence of background coincidences is in ratio with the square times of the distance as well as of lead shielding thickness. The experimental results obtained in INP Laboratories allowed us to evaluate the background level for the system with a coincidence of 0.35 s^{-1}. That is a calculated signal of instantaneous gamma-quanta coincidences from 10 g of Pu-239 and is comparable with the background level of calculated neutrons signals. Based on the theoretical results of our study it can be concluded that it is possible to detect fissile nuclear materials by implementing the proposed method. The radiation monitoring can be done successfully by using the proposed method. As a summarize, the monitor consists of two/four detectors (for pedestrian/vehicular monitors, respectively) based on organic scintillators with the area of 1,200 cm^2 each, coincidence circuit, counter module and industrial computer with appropriate software. Advantages of this radiation monitoring system are as follows: to detect the fissile nuclear materials in an object; to use only one type of gamma-ray detectors; low costs. Disadvantages of this system can be expressed as low sensitivity in detection of alpha-neutron sources and no isotopes identification due to incapable of differentiation of gamma-ray energy. However, alpha-neutron sources can be detected by background excess based on sufficiently effective detection of gamma quanta and neutrons by using 10 cm thickness scintillator. As a result we conclude that it is possible to implement this method for the detection of fissile nuclear materials.

4. Conclusion

The radiation monitoring program is implemented to prevent illicit nuclear trafficking at main custom border checkpoints equipped with numbers of monitors in Uzbekistan. Today, 30 checkpoints were equipped including 19 vehicles (118 monitors), 10 railways (40 monitors) and one International Tashkent Airport (12 monitors). Together with 2 monitors at Military Custom Institute and 3 monitors at Institute of Nuclear Physics of Uzbekistan-Academy of Sciences totally 175 monitors equipped in Uzbekistan. The difference counting method with two scintillation detectors is proposed for detection of radioactive materials in moving objects. In this method, gamma–gamma coincidence detection of prompt gamma quanta from nuclear fission is proposed to detect the fissile materials.

The Institute of Nuclear Physics is responsible to operate of these radiation monitors; to give technical assistance, regular technical maintenance and consultancy in case of alarm and stable conditions.

Acknowledgement: The authors acknowledge the organizing committee of NATO advanced training courses in Mugla, Turkey, May 2008 for their support in our participation in the meeting and paper presentation.

References

1. Shipilov NN, Fazylov MI, Podkovyrin AI, Karimov YN, Petrenko VD, Yuldashev BS (2005) Identification of radioactive materials in moving objects. Applied Radiation and Isotopes 63:783–787
2. Reilly D, Ensslin N, Smith H, Kreiner S (1991) Passive Nondestructive Assay of Nuclear Materials. LA-UR-90-732 Los Alamos National Laboratory Los Alamos, New Mexico

THE UKRAINIAN EXPERIENCE OF APPLICATION OF GUARANTEES OF NON-PROLIFERATION AND REQUIREMENTS OF THE ADDITIONAL PROTOCOL

OLEKSANDR VISKOV AND ARKADY BATRAK*
West State Inspectorate for Nuclear & Radiation Safety, SNRCU State Nuclear Regulatory Committee of Ukraine, Kyiv, Ukraine

Abstract. The Ukraine State Regulatory System was developed for the use of nuclear materials (NM) and sources of ionizing radiation (SIR) and is functioning effectively. High-quality research workers are engaged in addressing nuclear energy problems, use of SIRs in industry and medicine, mining and production of nuclear material and have succeeded in the establishment of competent nuclear regulations. These regulations and State supervision demonstrates that Ukraine is a reliable and safe partner in the use and control of nuclear materials.

Keywords: Ukraine, nuclear regulation, nuclear material, SIR, SNRCU

1. Security Challenges

It is realized that first of all, the created system of the state control over use of nuclear materials and sources of an ionizing radiation should resist to criminal use Source of Ionizing Radiation (SIR) and Nuclear Materials (NM).

The main dangers associated with criminal use of SIR include:

- The conditioned criminal acts of some persons or groups which can use potentially dangerous properties of SIR to achieve their illegal aims.
- The danger caused by the criminal negligence of responsible persons which leads to hit especially highly active SIR, to the people who has no knowledge about what threat these subjects in not qualified hands can bear representations.

* To whom correspondence should be addressed. e-mail: wsinrs@ukr.net

System of levers has been created with the purpose of counteraction to the threats. State regulation of safer use of ionizing radiation sources provides;

- Normative documentation: an establishment of normative criteria and the requirements defining conditions of use of ionizing radiation sources in the country.
- Licensing: delivery of sanctions to the activity connected with use of nuclear installations and ionizing radiation sources.
- Supervising: conducting of looking after implementation of normative requirements and terms of the given out licenses including the coercive actions enterprises, organizations and persons, about the using nuclear installations and SIR.

2. Authorization-based Principle of Using of "SIR"

The purpose of licensing-system in the field of the use of nuclear energy is:

- To maintenance only the use of those ionizing radiation sources which level of safety is recognized by the international meetings, recognized requirements on the basis of an all-round estimation of all factors influencing safety, including maintenance of physical protection.
- To maintenance the realization of activity in use of nuclear energy only by responsible persons who can guarantee execution of requirements of the legislation, norms, rules and standards of nuclear and radiating safety.

Condition of licensing for carrying the nuclear and radioactive material on manufacture and use of SIR is confirmed documentarily to able to carry out the given activity.

Such requirements (license conditions) have been developed and introduced in practical use in current of last 10 years. Today license requirements arrange such kinds of activity in Ukraine, as manufacture of ionizing radiation sources, their uses in medicine, transportation of ionizing radioactive sources and behaviors in radioactive wastes.

3. State Regulatory Authority

The State Nuclear Regulation Committee of Ukraine State Nuclear Regulatory Committee of Ukraine (SNRCU) is the official state authority in Ukraine which carries out the supervision of all aspects of use of SIR. The government of Ukraine delegated SNRCU to give sufficient powers for realization of such activity.

- SNRCU makes realization of state policy in the field of the use of nuclear energy, providing of implementation of requirements of nuclear and radiation safety.
- regulations of the SNRCU approved by the Cabinet of Ministers of Ukraine dated 27.12.2006 with number of No. 1830.

SNRCU reports directly to the Government of Ukraine.

One of the SNRCU first-priority tasks is to implement the measurements to exclude possible SIR losses or theft, comply with physical protection requirements, exercise efficient state supervision and ensure high-quality SIR account and control.

In principle, the use of SIR is subjected to the authorization based on the Laws of Ukraine "On Use of Nuclear Energy and Radiation Safety", "On Authorizing Activity in the Area of Nuclear Energy", "On Human Protection against Ionizing Radiation" and "Procedure for Licensing Individual Activities in Nuclear Energy" approved by Cabinet Resolution No.1782 of 6 December 2000.

4. The Inspection Activities of SNRCU

With the purpose of reception of the full and adequate information on activity of the enterprise or the organizations on manufacturing, use or transportation of ionizing radiation sources SNRCU carries out regular inspection checks:

- Upon receipt of the application and related documents for licensing, SNRCU inspectors make the survey of enterprise or organization. The purpose of the inspection is to identify the untrue information filed with the licensing documents.
- Planned inspections licensees are planned for regularly carrying out to determine the conformity of the activities of the licensee standards legislation.
- Unplanned inspections are carried out in case of any inspection identifies specific areas of activity or any gaps in security. In these cases more detailed and more frequent inspection are required.

5. Territorial (Local) SNRCU Authorities

Considering the large number and prevalence of sources on Ukrainian territory SNRCU has created inspectorate on nuclear safety on each of nuclear stations and also eight Regional Inspectorates.

- In 2006, setting up Regional State Inspectorate on Nuclear and Radiation Safety, the State Committee for Nuclear Regulation of. Ukraine.
- During 2007, recruiting the staff of State Inspectorates, their training and internships.
- SNRCU headquarter provided continuous methodological support for State Inspectorates.
- Inspectorates greatly increased the controlling over compliance with the rules, requirements and standards of nuclear safety throughout Ukraine and with the work associated with licensing of the use of SIR.
- The chiefs of State Inspectorate are concurrently deputy chief of State Inspector of Ukraine and they are powerful enough for the most effective monitoring of compliance with the requirements of existing legislation in use of nuclear energy.
- State Inspectorate for Nuclear and Radiation Safety conduct the survey, inspection, considering licensing and control the affairs of licensee compliance with the conditions of activities SIR at their subordinate territories, at the moment.

The formation of State Inspectorate allowed optimizing the licensing process, especially in remote regions.

In recent years, most enterprises that use SIR for non-medical purposes have obtained licenses. Enterprises that have no license are issued prescriptions to terminate the use of SIR, and enforcement actions are taken under the law.

The primary licensing of medical devices at institutions that use radiation sources for diagnosis and treatment is underway of applying a differentiated approach depending on the potential hazard of SIR. Ukraine has about 3,000 medical apparatus (excluding dental X-ray devices) that use SIR. There are 52 cancer centre that use high-level radiation sources, over 2,700 X-ray units and 57 computer tomography devices.

6. State Registration of "SIR"

It was decided to establish a single computerized system in order to provide general accounting and operational controlling of the status and location of SIR, documented by the Cabinet of Ministers of Ukraine of

04.08.1997 and No. 847. On the basis of the Ukrainian State, industrial enterprise “Izotop” was established a separate unit – State Register of SIR in 1998, acting as the main registration center. On March 29, 2007 Register was adopted in commercial operations:

- The main objective is to prevent the emergence of orphan sources in Ukraine.
- Automated system “Register” is created by adaptation the International Program RAIS to requirements of the national legislation. Actions on maintenance of protection of the information of “Register” in conformity to requirements of the national legislation according to the Instruction of protection of the information of “Register” are all executed.
- A document of using procedure of the Register and interaction of Register with the State Customs Service was developed; the other document required for the operation of Register was also developed.
- Registration in electronic form contains information of all sources; from their appearance on Ukrainian territory until their removal from Ukraine and/or the transferring to specialized enterprises for dumping radioactive waste and/or for X-ray devices to convert them into the non-radiate position.

7. Accounting for and Control of Nuclear Materials (NM)

The SNRCU and other central executive bodies ensure fulfillment of Ukraine’s international obligations by regarding the nuclear weapon non-proliferation/Cabinet Resolution No. 1830 of 27 December 2006 entrusted the SNRCU with the coordination of implementing the Agreement between Ukraine and the International Atomic Agency for the Application of Safeguards in connection with the Treaty of Non-Proliferation of Nuclear Weapons and Additional Protocol to Agreement.

In 2007, to implement the Safeguards Agreement and Additional Protocol, the SNRCU:

- Prepared and provided the updated information for Ukraine’s declaration as required by the additional Protocol
- Submitted quarter declarations of export of equipment and materials covered by Article 2.a.ix of the Additional Protocol to the Safeguards Agreement
- Arranged and provided five additional accesses in compliance with the Additional Protocol
- Prepared and submitted 207 reports to the IAEA on physical inventory and changes in nuclear material inventory

- Sent 20 preliminary notifications on export/import of nuclear materials
- Arranged and conducted 75 IAEA inspections
- Arranged the agreement of central executive bodies for 35 IAEA candidate inspectors to act in Ukraine
- Prepared and submitted the schedules for main equipment repairs and physical inventory in each nuclear material balance area and information on radiation doses received by IAEA inspectors in Ukraine

The SNRCU regularly cooperates with the IAEA in implementing the Safeguards Agreement and Additional Protocol. The implementation of the Safeguards Agreement in Ukraine is based on data of the State System for Accounting and Control of nuclear materials (SSAC). To ensure efficient performance of the SSAC in compliance with international obligations on nuclear non-proliferation, Ukraine developed a regulatory and legislative framework, which is continuously improved. The SNRCU also contributes to the improvement.

An information system is one of the most important components of SSAC. Experts have developed and maintained the state databank for nuclear materials which provides information on the amount and composition of nuclear materials needed in compliance with international agreements of Ukraine. If necessary, such information is also provided to state authorities.

There are 121 enterprises where account of nuclear materials is kept. These enterprises and facilities are geographically distributed by material balance areas: RKQ0–11, RKQ1–23, RKQ2–17, RKQ3–37 enterprises.

The enterprises provided data on the application of safeguards in 2007. Based on these data, the following reports were processed and submitted to the IAEA: 141 reports on changes in nuclear material inventory, Inventory Change Report (ICR) and 66 reports on nuclear material inventories Physical Inventory Listing + Material Balance Report (PIL + MBR).

IAEA inspections in Ukraine are associated with the agreement between Ukraine and the IAEA for the Application of Safeguards in connection with the Treaty on Non-Proliferation of Nuclear Weapons/IAEA inspections in Ukraine, SNRCU inspectors are responsible for the interaction with IAEA inspectors.

The primary declaration of Ukraine was prepared and submitted to the IAEA in 2006 as required by the Additional Protocol to the Agreement between Ukraine and the IAEA for the Application of Safeguards in connection with the Treaty on Non-Proliferation of Nuclear Weapons, which was ratified by the Law of Ukraine No. 3092-IV of 16 November 2005. After an initial analysis of the declared information, the IAEA stated that the quality and completeness of the primary declaration were quite satisfactory. The comments were incorporated; answers to requests and updated information for the Ukraine's declaration were prepared and sent to the IAEA in a timely manner (by 15 May).

The IAEA currently verifies Ukraine's declaration by analyzing the documents submitted, by comparing its data with open-source information and results of satellite photography of Earth and by verifying data through additional access of IAEA inspectors to Ukrainian enterprises. Since the implementation of the Additional Protocol, ten additional accesses to sites were arranged and provided in a timely manner as stated in the declaration with 2- and 24-h preliminary notification as required by the Additional Protocol.

Therefore, the implementation of the Safeguards Agreement and Additional Protocol to the Agreement enables Ukraine to assure the world community that it fulfils all obligations on nuclear non-proliferation and uses nuclear energy for peaceful purposes.

8. Radioactive Material Transport

Radioactive materials are transported in connection with their use in energy, industry, medicine, radioactive waste management and nuclear fuel transit across Ukraine [1].

In 2007, the SNRCU issued 116 permits for international transport of nuclear materials. In particular:

- Six permits for fresh fuel transport from Ukrainian NPPs to Russia
- Two permits for uranium ore concentrate transport from Ukraine
- One permit for transit of uranium ore concentrate from Czech Republic to Russia
- Eleven permits for transit of fresh fuel from Russia to Slovakia, Hungary and Bulgaria
- Two permits for transit of spent fuel from Bulgaria to Russia
- One permit for transit of spent fuel from the research reactor in Czech Republic to Russia
- Seventy nine permits for transport of the radioactive materials

Transport activities are licensed as required by legislation. By the end of 2007, 37 enterprises had licenses for radioactive material transport. The NAEK Energoatom, Easten Ore Mining and Milling Works, Ukrainian State Production Association Izotope, State Interregional Specialized Plants of State Radon Association, State Enterprise Ukrgeofizika, State Specialized Enterprise Kompleks and State International Airport Borispol undertake the greatest scope of radioactive material transport. In 2007, the

SNRCU issued seven licenses to legal entities dealing with radioactive material transport and reissued eight licenses.

In 2007, nine certificates on approval of packaging design and special shipment of radioactive materials were granted and reissued.

Compliance of shipment participants with legislation and safety rules for radioactive material transport ensures the safety of the public, personnel and the environment. There are no incidents in radioactive material transport in Ukraine in 2007.

9. Management of Radioactive Waste Resulting from the Use of Ionizing Sources

The State Radon Association deals with the management of radioactive waste generated in use of ionizing radiation sources in national economy. The Radon Association includes six State Interregional Specialized Plants (SISPs): Kyiv, Donetsk, Odessa, Kharkiv, Dnipropetrovsk and Lviv.

The primary objectives of radwaste management at Radon SISPs are:

- To collect and store the radwaste to prevent its adverse effect on people and the environment.
- To improve the effectiveness of these plants by considering the radiation safety.

The Kharkiv, Lviv, Odessa, Dnipropetrovsk and Kyiv SISPs receive low- and intermediate-level waste. The Donetsk SISP operates only a radwaste decontamination and transportation station.

The SISPs receive solid radwaste; biological waste contaminated with radioactive substances (biological radwaste) and spent ionizing radiation sources (IRS).

Biological radwaste is placed separately from solid radwaste in special storage facilities with layered cementation.

Spent IRS are stored in biological shielding as ordinary radwaste or in a special pit designed to store unshielded IRS.

According to the law, the Radon SISPs manage radwaste under licenses issued by the SCRCU. The licenses specify both the scope of authorized activity and special terms imposed on it to improve the level of safety.

10. Conclusion

The Regional Inspectorate cooperates with the Bureau of Guarantees and checks performance of the requirements about the accountability of nuclear materials. If the licensee for any reason disagrees with performance of legal requirements by the inspectorate, it will convey its concerns administratively. This system has allowed the identification many objects where NMs are stored. The creation of regional inspectorates have allowed consistent performance by the Ukraine with the protocols of IAEA.

References

1. The State Nuclear Regulatory Committee of Ukraine, Nuclear and Radiation Safety in Ukraine, Annual Report, 2007

MODERN CONDITION OF URANIUM PROVINCES IN KYRGYZSTAN (IN AREAS OF KADJI-SAI AND MIN-KUSH)

AINAGUL JALILOVA*, BEKMAMAT M. DJENBAEV, ALAI B. SHAMSHIEV AND BAKTIAR T. ZHOLBOLDIEV
Institute of Biology & Pedology of National Academy of Sciences of the Kyrgyz Republic 720071, Ave Chui 265, Bishkek, Kyrgyz Republic

Abstract. On the Kyrgyzstan territory there are 46 tailing dumps where stored more than 600 Mm^3 of tails. They are man-caused solids of fine-dispersed waste of reprocessing and contain radioactive nuclides, heavy metals compounds and toxic substances used as reagents in extraction processes depending on the reprocessed ore. Tailing dumps of radioactive waste are in the cities of Mailuu-Suu, Kadji-Sai, Min-Kush and Kara-Balta occupied nearly 3,600 m^2 of total area. In this connection we are confirmed that for rehabilitation of uranium tailing dumps, along with the engineering works, reconstruction of ecosystems and soil densification, it is necessary to conduct by means of phytomeliorative actions.

Keywords: Uranium province, tailing dump, Issyk-Kul

1. Introduction

In territory of Kyrgyzstan there are a large number of radioactive sources (about 1,200). The used sources are stored in the long-term storage facility which was built in 1965 as a typical RADON design replicating similar facilities in other republics of the former Soviet Union. Because of natural cataclysm such as earthquake, landslip, mud flows and erosion processes, the threat of the further pollution of territory by radioactive substances increased. As a result of these natural processes a line of uranium tail deposits was damaged. The majority of tail deposits and the warehouse premises they are poorly supervised. Kyrgyz Republic meets by serious problems of radioecological character connected with production and

* To whom correspondence should be addressed. e-mail: ainagulnas@mail.ru

processing of uranium in the country. Following the disintegration of USSR, ownerless conditions have appeared in territory of Kyrgyzstan as; 36 tail deposits and 25 mountain dumps of uranium manufacture:

- 16 Mm^3 of volume
- 6,500 ha of polluted territorial area
- 90,000 Ci of total radioactivity

Radioactive withdrawals, heavy metals and toxic substances pollute the atmosphere, soil, ground and underground water sources, plants etc.

The following actions require immediate considerations:

- Radioecological estimation
- Rehabilitation works
- Infrastructure regulating on radiating protection for the long-term period

In 2005 Kyrgyz Republic became the member of the International Nuclear Information System of International Atomic Energy Agency (IAEA-INIS). An Analytical Laboratory Network of and also a Calibration Laboratory supported by IAEA is considered:

- Creation of the network of radioecological monitoring and estimations; the project of KIG/9/003 together with the regional project RER/9/086, directing on research of rehabilitation of tail deposits formed as a result of the previous mining and processing activities
- Development of infrastructure regulating (regional approach, if possible)
- Rehabilitation of uranium and tail deposits (regional approach, if possible)
- An estimation of irradiation of population live in these areas
- Improvement of the radiotherapeutically service at the National Centre of Oncology
- Establishment of a control system for detecting nuclear and radioactive materials at customs

2. Background-Procedures

Many tail deposits were formed within the limits of occupied items; technogenic uranium province areas as of Mailuu-Suu, Min-Kush, Shekaftar and geochemical province areas as of Sumsar, Kadji-Sai, Ak-Tuz, Kant etc.

2.1. TECHNOGENIC URANIUM PROVINCE OF KADJY-SAI

Tail deposit with uranium dumps is in 2.5 km east from an inhabited settlement, but because of the natural factors such rain, ground water, landslip and mud flow, it is an important ecological threat to lake Issyk-Kul and near settlements. The volume of the saved tails (industrial wastes) is about 150,000 m^3. By establishing the radiometric shooting, the radiation level in the hollow of Issyk-Kul, the settlement of Kadji-Sai it is rather low and changes from 150 to 470 nSv\h. At the beach zones of the southern coast (The-sand) the radiation dose values change from 30 to 60 μR/h, but at some points it reaches to 420 μR/h [1, 5].

2.2. GEOCHEMICAL PROVINCE OF MIN-KUSH

It is a mountainous-steppe belt in height of 2,200–2,500 m above sea level. In area of the town Min-Kush is located with;

- Four tail deposits, such as Tuiuk-Suu and Taldy-Bulak.
- Four mountainous dumps.
- Total amount of radioactive burial is 1 million, 150 thousands m^3.

The ore complex was maintained in the years of 1958–1969. Tail deposit emplaced at the flat land by the area of 6–7 ha and located on slopes by steepness of 25–400 between the mountains. Tail deposits of Tuiuk-Suu are located on the mouth of the River Tuiuk-Suu which run into the River Koko-Meren and further run into Naryn and Syrdaria. The radiometric shooting of various places around uranium tail deposit of Min-Kush resulted from 27 to 60 μR/h, but in some points for example at tail deposits of Taldy-Bulak it changes between the values of 550–660 μR/h.

3. Results and Discussion

The analyses were carried out in the Radiometric Laboratory of the Institute of Biology & Pedology using a radiometer (Radon PPA-01M-03) and dosimeter (DKS-96).

3.1. TECHNOGENIC URANIUM PROVINCE OF KADJY-SAI

The different parts of Lake Issyk-Kul have different amounts of uranium, Table 1. It is caused by non-uniformity of evaporation and distillation processes of coastal zones occurring in different parts of the lake. The mean uranium concentration of Lake Issyk-Kul is $3.0 \times 10–6\%$ [1, 3].

TABLE 1. The contents of uranium in water of the lake Issyk-Kul

Place of selection of tests of water	U in water (%)
Cholpon-Ata (Northern coastal zone)	3.331×10^{-6}
Gulf of Tup (East)	3.1×10^{-6}
Tamga (Southern)	1.7×10^{-6}
Bay Kolsovka (Southern)	3.68×10^{-6}
Gulf of Ton (Southern)	2.3×10^{-6}
Gulf of Ribachi (Western)	4.32×10^{-6}

Soil: The analysis of soil samples showed that on tail deposits, in the top 0–20 cm stage, the uranium concentrations changes from 1.1 to 2.6×10^{-6} g/g and it increases up to 3.0×10^{-6} g/g with depth. There is a zone of tail deposits where the uranium concentration in the top is equal to 4.2×10^{-6} g/g average, and in the bottom of 40–60 cm it increases up to 35.0 $\times 10^{-6}$ g/g, about 8.3 times higher than the value on the top [2, 5].

The level of a radiating background on the surface of industrial and tailings zone is not high. More detailed researches of isotope structure of the soil in the sub-region also showed that in the top layers it is higher in comparison with the bottom layers. When comparing Ak-Terek sand with soil from different regions it can be seen that the activity concentrations is 3–20 times higher, Table 2.

TABLE 2. Background concentrations of alpha-active isotopes in soils

Place of selection	Layer (cm)	Activity of the soils on isotopes (Bq/kg)									
		U-238	±	Ra-226	±	Pb-210	±	Th-228	±	Ra-228	±
Kara-Oi	0–5	71.8	12.7	35.1	3.9	147.4	13.0	39.5	2.2	35.2	8.8
	5–10	50.8	7.3	37.7	3.4	64.6	11.4	49.0	1.9	60.1	7.5
	10–15	44.0	1.7	35.1	3.2	50.1	7.2	45.6	1.8	52.3	3.5
	15–20	51.7	7.4	46.1	3.5	50.2	7.7	49.9	1.9	53.6	7.7
Kichi-Aksu	0–6	71.5	14.3	51.0	3.4	88.5	18.4	69.1	3.6	72.4	7.2
	6–11	52.1	6.5	43.2	3.1	71.7	10.2	43.2	3.3	59.2	19.7
	11–20	54.9	7.3	45.4	3.5	68.6	7.6	64.3	3.8	64.1	7.5
Ak-Terek (sand)	0–3	260.0	30.0	103.0	8.0	169.0	30.0	915.0	57.0	846.0	70.0

TABLE 3. Natural radionuclide activity concentrations in water of inflow rivers and Lake Issyk-Kul

Place of selection of tests	Uranium (total) (Bq/L)	U-234/U-238	Gross alpha (Bq/L)	Ra-226 (Bq/L)
Lake of Issyk-Kul settlement Kara-Oi	1.79 ± 0.18	1.13 ± 0.05	1.80	0.013
River of Bulan-Segetu	0.09 ± 0.01	–	0.10	0.002
River of Kichi-Ak-Suu	0.17 ± 0.02	–	0.20	0.009
River of Tup	0.23 ± 0.02	–	0.23	0.016
River Kara-Kol	0.21 ± 0.02	–	0.25	0.005
Lake of Issyk-Kul settlement of Ak-Terek	0.56 ± 0.06	–	0.60	0.02
Kadji-Sai, streams No. 1 up to the rain	4.21 ± 0.42	1.49 ± 0.05	4.5	0.007
Kadji-Sai, streams No. 2 up to the rain	10.2 ± 1.02	1.30 ± 0.05	10.0	0.005
Lake of Issyk-Kul settlement of Kadji-Sai, mouth of the river	1.69 ± 0.17	1.52 ± 0.05	1.67	0.015

Water: The activity concentrations of natural radionuclides in waters of inflow rivers and Lake Issyk-Kul and also the activity concentrations at the bottom zones of tails following the rain are shown in Table 3.

3.2. GEOCHEMICAL PROVINCE OF MIN-KUSH

The forwarding inspections in the area of uranium tail deposit of Min-Kush showed that the natural ecosystems were damaged in this region.

The soil of Min-Kush vicinity was specified as;

- The uranium concentrations by differentiating from 3.3 to $17.5 \cdot 10^{-6}$ g/g is rather high on the average.
- The large danger is caused on the territory located above the concentrating factory where the uranium concentrations in the soil reach to 30–35 × 10^{-6} g/g at a surface.

As a whole, soils of a geochemical province of Min-Kush are substantially enriched with uranium since the concentration of uranium is five to six times higher than in the soils of Kyrgyzstan [4].

Some possible variants of a safe storage of withdrawals should include:

- Disassembly and carry of tail deposit in more safe place
- Repair of hydraulic engineering structures and maintenance them in working properly during all long period of operation (1,000 years)

Spring and irrigation ditch water which takes place on territory of the town Min-Kush, have 3.3–3.5 × 10^{-5} g/L of uranium concentration. For example river Koko-Meren has 1.0 × 10^{-5} g/L of uranium concentration.

3.3. INHABITED PREMISES

The radiating background in some apartment houses of the town Min-Kush have also been investigate and the results of measurement showed that in inhabited premises, the radiating background raises up to two times of the limited admissible concentration. The basic reason of small increasing in level of a radiating background is connected with slag used for construction from local coal.

As it is said above, more than 1 million cubic meters of radioactive withdrawals are stored in eight tail deposits and mountain dumps of the settlement of Min-Kush in the area of Naryn. The burial place was formed in 1969 after the closing of Kyrgyz mountainous ore beds. The local experts assert that in case of a serious violent flood of water the tail deposit at Tuiuk-Suu can be destroyed. And consequently the radioactive substances can mix through the river Koko-Meren and then the river Naryn and the Aral Sea. So the problem of the burial places in Min-Kush should be considered as regional, not locally. Kyrgyzstan, Tajikistan, Uzbekistan and Kazakhstan are four states that are affected by this problem. In case of destruction of tail deposits in the Tuiuk-Suu territory; about 26,000 men of Kyrgyzstan, about 2.4 million men of Uzbekistan, about 0.7 million men of Tajikistan and about 0.9 million men of Kazakhstan will be affected.

4. Conclusion

1. The general level of the external radiating background at sub regional territory is within the limits of norm, except at technogenic territories.
2. Three types of radioactive anomaly is marked on technogenic sites:
   - The natural anomalies of a radioactivity sources by radioactive brown coals
   - Technogenic anomaly which exceeds hundreds times of the background sources by a concrete wall of dumps of grey thin-granular of substance

- Technogenic anomaly in standing water, which the activity is ten times higher than the background

3. It is necessary to afforest the naked area for protection against intensive erosion.
4. There is no any physical protection of tail deposits.
5. The easy approach is to limit the settling of people in this area, and to establish sanitary protective zone.

Acknowledgement: All above mentioned data are the evident results of the IAEA KIG/9/003 project.

References

1. Djenbaev BM, Shamshiev AB, Jalilova AA (2006) Radiation assessment in biosphere territories of the Issyk-Kul, 4th CCMS/NATO Workshop in: Management of Industrial Water and Substances Research, Ionia, Greece
2. ICRP Publication No. 60, Radiation Protection 1990: Recommendations of the International Commission on Radiological Protection (ICRP), Pergamon, New York
3. IAEA Safety Series No. 115 (1997) The International Basic Norms of Safety for Protection from Ionizing Radiations and Safe Handling of Sources of Radiation, IAEA
4. Norms of Radiating Safety (1999) Ministry of public health services of Russia
5. Djenbaev BM, Jolboldiev BT, Jalilova AA, Shamshiev AB (2008) Radioecological features of the technogenic uranium biogeochemical province of the biosphere of Kadji-Sai (Kyrgyzstan), J. ProbBiogeochem Geochem Ecol, Semipalatinsk 6:2–5

INSTRUMENTS FOR DETECTING THE UNSANCTIONED DISPLACEMENT OF RADIOACTIVE MATERIALS

YURY SAPOZHNIKOV*, IRINA BUTKALYUK AND PAVEL BUTKALYUK
Chemical Department, Lomonosov Moscow State University, Moscow, 119991, Russia

Abstract. In 2006, gamma-ray detector, consisting of the number of parallel scintillation units with BGO crystals was theoretically developed by the group of environmental radioactivity from Lomonosov Moscow State University. High effective atomic number of BGO and the selected detector geometry contributes to the effective registration of gamma-quanta, which enter mainly in parallel to the symmetry axis of the system. It is shown that the realization of detector as a whole will make it possible to check gamma-ray emitters in vehicles and to solve other problems connected with the prevention to terrorist actions. Some promising trends of the detectors developed for low level activities are examined.

Keywords: Bismuth germanate crystals (BGO), Cherenkov detector, radioactivity in water, high energy β-particles, scintillation gamma ray detector

1. Introduction

The equipment for measuring low level activities and the devices for the direct measurements of radioactivity in the environment have been developed by the group of the Environmental Radioactivity, Chemistry Department of Lomonosov Moscow State University. In the beginning of the 1980s the Cherenkov detector for determination of high energy beta-emitters in natural waters was developed. This detector could be used, for example, for detecting the unsanctioned ingress of Sr-90 in aquatic environment. For the suppression of bioluminescence during the passing of the water through the sensitive volume of detector, the membrane filters with the pore size of 0.1 μm were used. The instrument was tested under

* To whom correspondence should be addressed. e-mail: yas@radio.chem.msu.ru

laboratory conditions and sufficiently high registration effectiveness of a number of high energy beta emitters was demonstrated. In 2006 directional scintillation gamma-ray detector, consisting of the number of parallel scintillation units with Bismuth Germanate (BGO) crystals was theoretically developed in our Laboratory. At the present time the design of one unit was carried out and the laboratory tests confirmed its correspondence to the calculated parameters. It is shown that the realization of detector as a whole will make it possible to check gamma-ray emitters in vehicles and to solve other problems, connected with the prevention to terrorist actions. In this study, some promising trends of the detector developed for low level activities revealing in the natural media are examined.

The threat of terrorism produces the necessity of developing the countermeasures, aims to the early warning and protection of possible terrorists acts. Radioactive safety is the primary interest now. The possible terrorist's acts could include the pollution of the drinking water sources by radioactive materials or the use of so called "dirty" bombs causing radioactive contamination and panic in the places of high dense population. In the first case the geochemically mobile radionuclides, such as Sr-90, could be used. In the second case it could be Cs-137 or the spent nuclear fuel which contains, in particular, both above mentioned radionuclides.

The use of simple and non-destructive detection methods is essential for quick diagnostic of the possible radioactive inputs to the environment. From this point of view the on-line detection of beta- and gamma-radiation is of a big advantage.

In our laboratory we developed Cherenkov detector for the in-situ determination of Sr-90 in natural waters. The method is based on the detection of high energy β-particles ($E_{\beta\text{-max}}$ = 2.27 MeV) of daughter Y-90.

For the remote detection of gamma-emitting radionuclides we developed the gamma-ray scintillation spectrometer of the directional action. Initially this spectrometer was intended for measuring the seawater radioactivity. Seawater is the main contributor to the total radioactivity by the natural radionuclide K-40. This device could also be used for measuring radioactivity in the fresh surface and ground water.

2. Experimental

2.1. CHERENKOV DETECTOR FOR DETERMINATION OF HIGH ENERGY β-EMITTERS

The Cherenkov radiation is emitted when high energy β-particles (Eβ-max > 0.267 MeV) pass through the transparent dielectric medium as water with the refractive index n = 1.33. Its intensity rapidly grows with an increase of β-particles energy. Registration efficiency of Cherenkov radiation depends on the maximum energy of β-particles, and on the shape of the β-spectrum i.e. on the portion of β-particle energy spectrum fallen in the high energy part of the spectrum. For measuring of the natural radioactivity in water the Cherenkov detector was developed. The detector consisted of the stainless steel tube with an outside diameter of 90 mm and wall thickness of 1 mm. Two photomultipliers (PMT) FEU-110 were fixed inside the tube at its opposite ends with flanges. The sensitive volume of detector could vary between 300 and 1,700 mL depending on the length of the tube and correspondingly on the distance between the photoelectric cathodes of PMTs, Figure 1. Water could be continuously pumped through the detector using the branch pipes welded onto the flanges.

To prevent the bioluminescence in the sensitive volume of detector, the natural water was filtered through 0.1 μm pore size membrane filters which allow cutting off the most part of planktonic organisms. Pulses from the anodes of PMTs were gained and entered through the shaper to the coincidence circuit with the resolving time of approximately 25 ns which allows suppressing the inherent noise of PMT to a considerable degree. The block scheme of the Cherenkov detector for measuring seawater radioactivity is given in Figure 2.

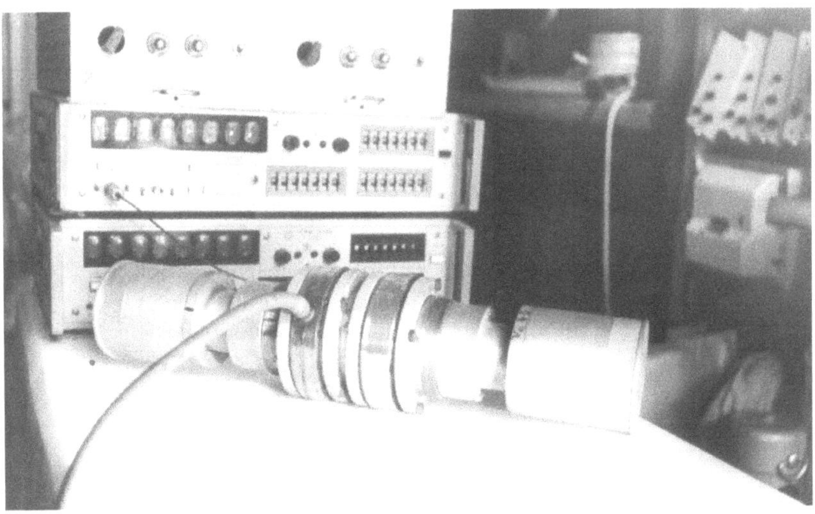

Figure 1. The exterior view of Cherenkov detector

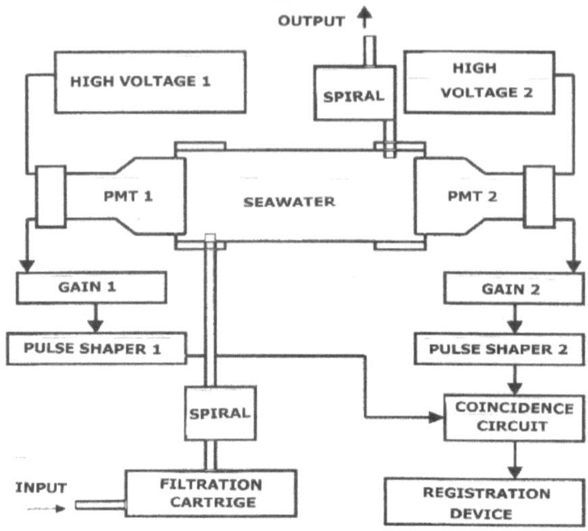

Figure 2. Block-scheme of the Cherenkov detector

2.2. THE RESULTS OBTAINED WITH CHERENKOV DETECTOR

The registration efficiency of K-40 and Y-90 in seawater reached to 13% and 40%, respectively. The possibility of determination of Cl-38, Na-24 and P-32 via Cherenkov radiation in seawater irradiated by the neutron source, was demonstrated under the laboratory conditions. The approaches to increase the effectiveness of this detector are outlined on the basis of the tests carried out. Instrument can also be used for measuring different beta-emitters in fresh surface water and groundwater.

2.3. DIRECTIONAL SCINTILLATION GAMMA-RAY DETECTOR

Usually to ensure the directivity of the detector functioning the systems of two or more scintillation crystals are used [1, 2]. In 2006 directional scintillation gamma-ray detector, consisting of the number of parallel scintillation units with BGO crystals with the ratio of length to diameter of 3:1 was developed, Figure 3.

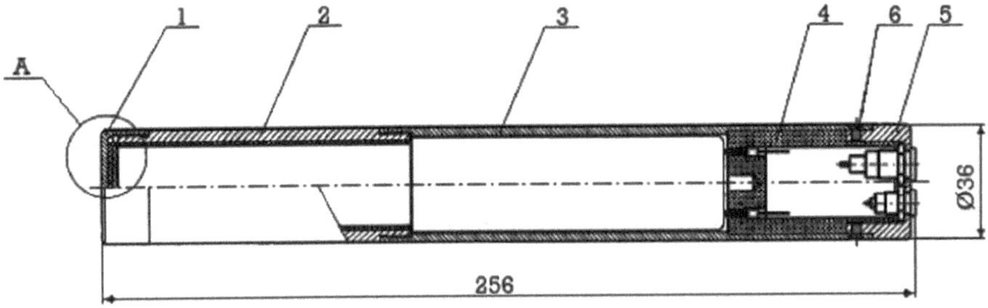

Figure 3. View of scintillation module with BGO crystal; *1* – lid of crystal case; *2* – case of crystal; *3* – case of PMT; *4* – case of the divider; *5* – lid PMT case

Entire detector consists of seven scintillation units located in the lead (passive shielding) with the open end, Figure 4. High effective atomic number of BGO and the detector geometry provide the effective registration of gamma-quanta which enter mainly in parallel to the symmetry axis of the system.

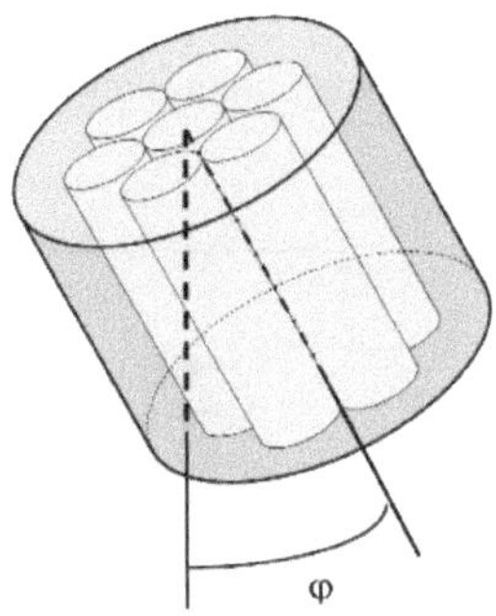

Figure 4. View of directional detector consisting of seven BGO units

For the suppression of Compton electrons scattered to the high angles, each of the unit must be included in the anti-coincidence circuit with respect to all remaining units.

2.4. RESULTS OBTAINED WITH DIRECTIONAL GAMMA-RAY DETECTOR

At the present time only one unit is designed. Laboratory tests confirmed its correspondence to the calculated parameters. Dependence of registration efficiency on angle from axis of one unit is presented in Figure 5. The thickness of lead (passive shielding) is approximately 5 cm. The radioactivity of the source of Cs-137 is about 0.6 MBq.

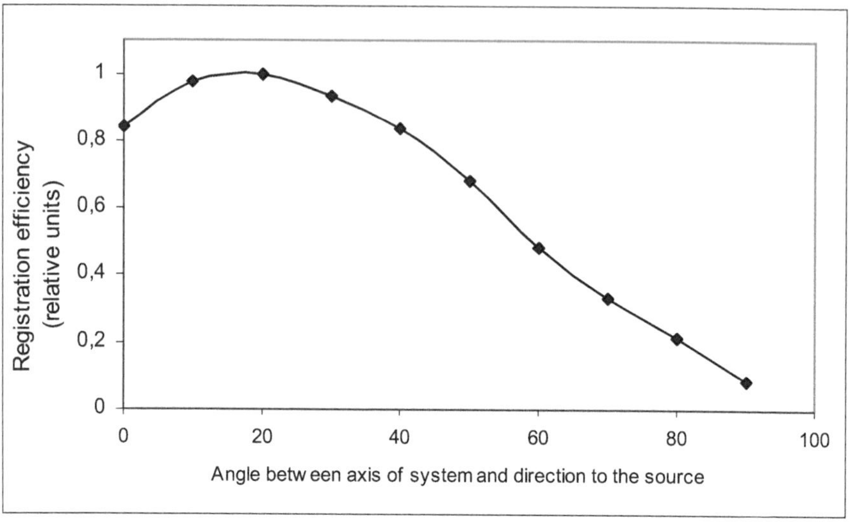

Figure 5. Dependence of efficiency of registration on the angle from axis of the system to source direction

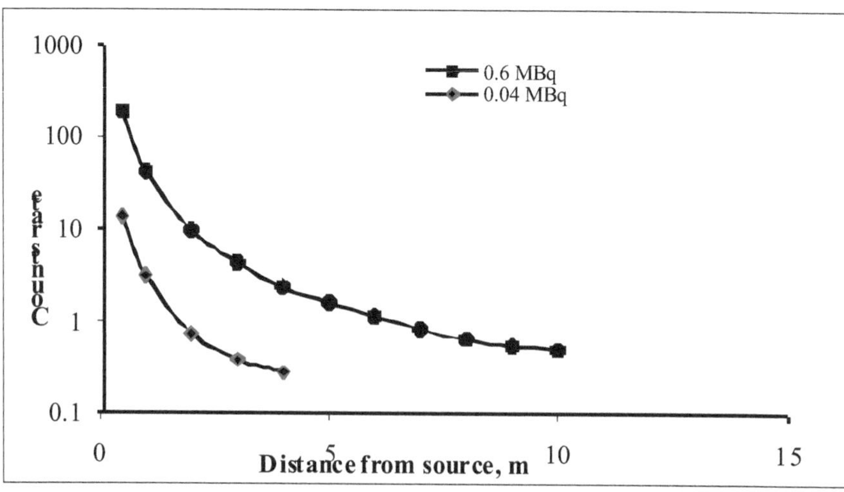

Figure 6. Dependence of counts rate on distance (m) from the sources

Dependence of registration efficiency on the distance from source to detector is presented in Figure 6. The thickness of lead (passive shielding) is about 5 cm with two Cs-137 sources: 0.6 and 0.04 MBq.

3. Conclusion

High efficiency of Cherenkov detector permits to measure the undesirable radioactive contaminations in natural waters. It allows us to detect the increasing beta-radioactivity levels in real time and could be used for continuous monitoring of drinking water.

At present we develop the instrument for the continuous preconcentration and measuring of the radioactivity concentration of Y-90 in water. The device is based on the use of transparent cartridge filled up by optically transparent sorbent for Y-90 extracted from water. Cartridge is located between two photocathode, registering quanta of light emitting due to Cherenkov process.

Directional gamma-ray detector was developed to reveal the radioactive sources on the distances up to 10 m. This instrument can be used, for example, for control of the vehicles moving through a blockhouse.

References

1. Sapozhnikov YA, Merkushev AM, Murzin VE (1984) The Cherenkov detector for measurement of the seawater radioactivity. In: Moscow University Scientists – for science & industry. Discoveries, inventions and results of scientific researches, proposed for the practical use. MSU Edition, Russia
2. Bowyer TW, Geelhood BD, Hossbach TW et al. (2000) In situ, high sensitivity measurement of strontium-90 in ground water using Cherenkov light. Nucl Instrum Meth A 441: 577–582
3. Orphan VJ, Muenchau E, Gormley J, Richardson R (2005) Advanced gamma-ray technology for scanning cargo containers. Appl Radiat Isotopes 63: 723–732
4. Shirakawa Y (2004) Development of directional detectors with NaI(Tl)/BGO scintillator. Nucl Instrum Meth 213: 255–259
5. Butkalyuk P, Sapozhnikov YA (2008) Express technique for Sr-90 determination in the seawater. All-Russian Conference of Chemical Analysis. 21–25 April 2008. Abstracts: Moscow-Klyaz'ma 109–111

PRACTICAL INSTRUMENTATION CONSIDERATIONS WHEN PLANNING A RADIATION MONITORING PROGRAM FOR THE FIELD AND THE LABORATORY

N. ANTHONY GREENHOUSE*
Berkeley Laboratories, 788 Mickinley Avenue, Oakland, CA 94610-3833, USA

Abstract. Very often the selection of appropriate radiation monitoring instruments is beset by budgetary constraints. In these situations it is helpful to select instruments which can cover both needs without unduly burdening either setting. A review of basic radiation survey instruments along with their application to monitoring in a laboratory setting or an outdoor setting was conducted. Their use in the Marshall Islands demonstrates that in general no special care need be taken for use of survey instruments in the field. Thus, with the possible exclusion of alpha monitors, the survey instrument collection can be universally applied as long as the detection limitations of some instruments are considered. Several instrument detector types will be discussed, including G-M counters, scintillation detectors, and pressurized ionization chambers.

Keywords: Radiation monitoring, detectors, Marshall Islands, survey instruments

1. Introduction

The operability of survey instruments from in routine operations, to accidents as massive as that at Cernobyl, to current threats from terrorist groups pose concerns for the radiation protection community today. This presentation will demonstrate that no special care need be taken for the use of radiation survey instruments in outdoor environments, as long as their limitations are considered.

The cost of appropriate radiation survey equipment is often a limiting concern prior to their purchase. These instruments are usually constructed

* To whom correspondence should be addressed. e-mail: nagreenhouse@envirhealth.com

to tolerate the extremes of both environments, but there are some concerns that still must be dealt with, e.g., their operation in very cold and sometimes in very humid environments. The following discussion bears upon getting the maximum utility per unit money spent [1–3].

Instruments and Their Applications: Portable survey instruments are usually designed with ruggedness in mind. As long as they are routinely calibrated and serviced by qualified staff, they can take a lot of abuse, and still be relied upon to perform well.

Background Radiation and Its Significance: Higher-sensitivity survey instruments can detect background radiation with varying degrees of success. A useful method is to make 30 or so measurements of background radiation levels, say after each formal calibration. Then calculate the sample mean and sample standard deviation, ±б, ±2б, and ±3б. Record these results on an electronic spread sheet, or equivalent data storage location.

The background parameters will be re-established if any of the following occurs:

- Instrument consistently does not meet pre-established parameters;
- Instrument receives extensive repairs; or
- A new instrument and detector combination is used.

Minimum Detectable Level or Concentration: Calculate and record the minimum detectable concentration (MDC) using the following formula for each detector individually:

$$\text{MDC} = \frac{3 + (4.65\ (\text{B}1/2)}{\text{T}\ \epsilon\text{Tot G}}$$

where:
MDC = minimum detectable concentration level in disintegrations/minute/ 100 cm^2;
B = background (total counts) in time interval, T;
T = count time (min) to be used for field measurements;
€Tot = total efficiency = counts per disintegration = €i × €s, where
€i = instrument efficiency and €s = source efficiency (unless otherwise determined)

For example:

€s = 0.5 for βmax > 400 keV (e.g., ^{90}SrY)
€s = 0.25 for βmax < 400 keV (e.g., ^{99}Tc, ^{204}Tl); and

$$\text{G} = \text{geometry} = \frac{\textit{Physical Detector Area}\ \text{cm}^2}{100}$$

The above formula calculates the activity level in dpm/100 cm^2 which can be detected at the 95% confidence level [4, 5]. Compare this value to the site guidelines to determine adequate sensitivity of the instrumentation.

2. Beta–Gamma and Alpha Surveys

Let us review basic detector types along with considerations of the desirable characteristics and limitations of each type. The Geiger-Meüller (G-M) detector has seen universal applications because of its simplicity and sensitivity to beta particles and secondary electrons from gamma rays. The G-M instrument is also likely to be the least expensive radiation survey device to purchase. Cylindrical G-M detectors are commonly used in survey instruments calibrated for exposure rate measurements at low levels. “Pancake” type G-M detectors are designed for contamination monitoring in lab or hospital environments, and are economical because of their sensitivity and ease of production [6]. Thin-walled G-M detectors such as the “pancake probe” can detect alpha particles. However, they are quite sensitive to natural background radiation and to ambient fields from beta–gamma emitters, and therefore should not be used to monitor for highly radiotoxic alpha emitters.

Ion chamber detectors tend to be not very sensitive in terms of measuring natural background levels, but are useful for exposure rate measurements in the mR/h to kR/h range (and higher). The sensitivity of chambers of modest size, around 100 cm^3, is sufficient for them to measure dose equivalent rates as low as 1 $\mu Sv\ h^{-1}$. A specialized form, the pressurized ion chamber (PIC), overcomes the sensitivity limitation by providing many more opportunities for interaction with incoming gamma rays, and is the basic detector for very sensitive exposure rate measurements in the μR/h range. Typically, PICs contain stable argon at 10 or more atmospheres pressure. Commercial PICs are expensive instruments, and will be useful primarily for cross calibration of more conventional survey instruments, especially in outdoor environments.

The second case, dose rate monitoring, is considerably more important from a health physics standpoint. But, also here the instrument designers have gained the advantage in making instruments that function very well with respect to scientific and legal limits to exposure of radiation workers and the general public. Ion chamber survey instruments are well suited to do this job [6]. The sensitivity of pressurized ionization chambers makes them very useful for exposure rate evalution in environmental settings. For example, Reuter-Stokes manufactures the RSS-112 chambers with readout apparatus which performs this job well. In fact, this instrument can reliably measure natural background dose rates in areas without the contribution from terrestrial nuclides in the uranium and thorium decay chains. These instruments are bulky and heavy which limits their utility when compared with true hand-held survey instruments. But, the survey team attached the detector module of a predecessor PIC, the RSS-111, to a boom on a jeep for extensive measurements of large contaminated areas at Bikini Atoll in the Marshall Islands. In this environment there is essentially no natural terrestrial nuclide contribution beyond that from ^{40}K. The contribution from cosmic rays and aquatic radionuclides added slightly to the natural background exposure rate. Coral atolls such as the Marshall Islands, when not contaminated with fission products are probably among the lowest terrestrial natural background radiation areas in the world, averaging about 3.5 μR/h.

3. Scintillation Detectors

Scintillation detectors are, in essence, solid state devices that are capable of great sensitivity. They, along with proportional counters, are also capable of providing signals in proportion to the energy deposited in them. Thus they can be incorporated into portable spectrometric instruments for use in building and field environments. The sensitivity of crystalline scintillators such as NaI:Tl, CsI:Tl, etc. detectors is roughly proportional to the mass and density of the detector. The sensitivity of survey instruments with crystalline scintillation detectors is remarkable, but these devices have several undesirable characteristics. One is that they are extremely energy dependent in terms of their measurement of radiation exposure rate or dose equivalent rate. They are usually less expensive than the previously mentioned pressurized ion chambers, but please recall that the PICs can measure exposure rates accurately at 60 keV and below. On the other hand, because of their energy dependence, scintillation devices must be specifically calibrated to a radiation field in question before an accurate assessment of the exposure rate can be made. If you will recall an earier slide on the mass absorption coefficient for NaI, this in effect duplicates the energy dependence of NaI:Tl scintillation detectors.

As an example, the northern Marshall Islands are contaminated today with ^{137}Cs among other radionuclides from atmospheric testing of nuclear explosives. The ^{137}Cs, at levels too low to cause significant external exposures, rapidly finds its way into coconuts and other food products grown by the resident people causing their internal doses to approach and even exceed maximum permissible levels. It was desirable, therefore to locate islets with lower levels of Cs to minimize the uptake by coconut trees and other plant in the human food chain. Scintillation detectors can be calibrated with good geometry to Cs in a laboratory environment, but these devices performed marginally in this field location because if the intimate association of the nuclide with the coral soil. The result is that the 0.667 MeV gamma peak became thoroughly compromised by the Compton Effect, thus defeating the advantage of the proportional nature of the scintillation detector. Most of the commercial survey meters had scales reading in "counts per minute" and "mR/h", but the exposure rate scales over responded by factors of 5–8 depending on the amount of overburden of clean soil. So, even though the instruments were quite sensitive (a valuable asset in this situation) they had to be cross calibrated against a pressurized ion chamber before relatively accurate readings could be made.

Plastic scintillators are made by dissolving toluene and/or other scintillating liquids with a plastic matrix such as polymethyl methacrylate (Lucite or Perspex). They are not as sensitive to gamma or x-rays per unit mass as their crystalline counterparts, but they can be made into very large and/or unusually shaped detectors for specialized purposes such as around accelerators. They are also good for detecting beta particles, muons, etc. Plastic scintillators have been utilized in muon radiography equipment to detect the presence of contraband plutonium and uranium in shipping containers on board ships and in tractor-trailers on land. This technology will hopefully inhibit the illicit transport of these materials by most commercial means.

3.1. CONTAMINATION VERSUS DOSE RATE MONITORING

Let us examine two generic types of survey instruments with respect to their function: contamination monitoring and exposure-rate or dose-rate monitoring. The former function is usually classified as the "good house-keeping" function. Here if any radiation is detected above "background", the monitoring staff or cleanup staff is called in to remove the contaminant. This is a classic use for a G-M survey instrument. This must, however, be contrasted with large releases of radioactive materials to the outdoor environment in accidents or by terrorist attacks. In these cases, the first order of business is to protect the public from over-exposure. Field monitoring and meteorological estimates must be effectively coupled to achieve this goal.

If it is desirable to measure exposure rates (or to estimate dose equivalent rates) in a building of in the field, then an ion chamber instrument is the most desirable device to accomplish this task. This is especially at occupational exposure levels. Pressurized ion chambers can be fruitfully used at low radiation levels, and scintillation instruments can also be used here as long as the user recognizes the potential shortcomings of this type of survey meter.

3.2. LIMITATIONS OF ALPHA SURVEY EQUIPMENT

The function where contamination monitoring instruments meet a utilitarian problem in the lab environment as contrasted with the outdoors is with alpha monitoring. In the indoor environment, surveys for alpha surface contamination can be tedious, but they can be accomplished with some dedication on the part of the technician. In the outdoor environment however, the problem is twofold. Alpha particles, of course, have a notoriously short range, and a spill outdoors involving alpha emitters will virtually always involve seepage into the ground, or interference by plants, or coverage by, say oil or soil, and other issues in the outdoor environment. Plus, the detectors, whether they are proportional counters or alpha scintillators are delicate in their own right. The scintillation detectors will be rendered null and void by a tiny puncture, and the alpha proportional counter will "die" from not much more. Lastly, the difficulty of obtaining a good signal has its own "human engineering" problems because of the necessity of holding the probe in very close proximity to the contaminant.

The only practical solution for detection of alpha-emitters outdoors is when the radioactive material has an intrinsic gamma ray associated with its decay process, or it has a short half-lived daughter or associated (contaminant) radionuclide with this benefit. In these situations the FIDLER, an acronym for "Field Instrument for Detection of Low Energy Radiation" [7], and its successor scintillation detectors can be godsends. In fact, they can be used to estimate the average overburden of soil or other materials that contain the contaminant. Also, with proper calibration, the FIDLER can approximate the contaminated area as well. Unfortunately, these specialized survey instruments are expensive.

A more recent version of the FIDLER utilizes a ~13 cm diameter × 1 mm thick NaI:Tl primary detector laminated to a 50 mm thick CaF2:Tl scintillators to compensate for background radiation. The two types of scintillators produce pulses with different decay times, thus a pulse shape discriminator can distinguish the source of the pulses, and when this discriminator is coupled to an "and gate" it can greatly minimize counts from higher energy background radiation. This device, in spite of its high cost, is a "must" in situations where outdoor contamination by relatively high specific activity alpha emitters is likely. FIDLER detectors and associated analyzer/count rate instruments are available Bicron Corporation in the United States as well as other manufacturers.

3.3. NEUTRON SURVEYS

Monitoring for neutrons is not likely to be a problem in the environment. But here again, neutron survey instruments tend to be bulky (and expensive) devices. The range of possible neutron energies covers much more than nine orders of magnitude. And, the fluence rates as a function of neutron energy must be multiplied by a Radiation Weighting Factor, wR, which vary by two orders of magnitude as a function of neutron energy [8, 9]. Therefore, it is very difficult to measure neutron dose equivalent rates as a function of neutron energy with a single instrument. However, it is possible in a practical sense to adequately assess neutron hazards in most workplaces, e.g., nuclear reactors, because neutron survey instruments are available to perform adequately over a limited range of neutron energies. They almost always involve a sizeable and thus heavy moderator surrounding a thermal neutron detector such as a ten BF3 proportional counter or a nine LiF scintillator. The bulk and weight of these instruments makes neutron surveys somewhat difficult, but in a single facility frequent surveys can be replaced with, e.g., activation detectors, as a part of the environmental monitoring program. The goal here is to be certain that, for example, the reactor and its radiation protection components continue to operate within acceptable limits.

3.4. GAMMA SPECTROMETRY

The use of survey instruments is often supplemented by portable gamma-ray spectrometers. This is a refined solution to the identification of gamma emitters in the lab or in the field, and modern spectrometers use germanium solid state detectors. More will be said about this in the next presentation.

3.4.1. *Environmental concerns*

Most, if not all, radiation detectors are protected from issues such as high humidity, dust, airborne chemicals, etc. which might interfere with their performance. One exception is that many ion chamber instruments contain air as the interaction medium, and the chamber is not sealed. The electronics are usually contained within a sealed metal box thus negating very humid conditions as a factor which will compromise satisfactory performance. It is important here to mention that most crystalline scintillators are very hygroscopic, and also the coupling between the scintillators and the optics between it and the photomultiplier tube (PMT or MPT) must remain intact. These devices are normally produced with rugged housings that prevent humid air from entering the detector compartment, and also provide support for the PMT and optics and the scintillation detector as well.

Cross-contamination is also possible whenever contamination monitoring instruments are used in contaminated areas. These instruments should be checked for contamination with (e.g.) a filter paper wipe which is then counted at a bench top scaler, or more coarsely with another survey instrument at the same location. Here again, ambient radiation levels should be taken into account since they will affect the adequacy of the second survey meter measurement. The results of these checks should, of course, be recorded in a log book.

The issue of proper operation at low temperatures can be more serious, as may be the case with most battery-operated electronic equipment. At more "normal" ambient temperatures the detector, associated electronics, and power source will operate as expected. But, as the temperature is lowered, the battery, which is an electro-chemical energy source, becomes frozen. And therefore, at low temperatures, it will not provide adequate energy to operate the survey meter. The best solution here is to make a remotely operated battery pack that can be worn under a parka and close to the operator's body. Care must be taken here because the hermetic seal on the instrument case will be broken. The use of water-tight feed-throughs is suggested to bring the power connections to an outside battery pack. It is unlikely that any user will have to work in an environment that is so cold that the survey instruments will not work because the detector and associated electronics will not function. But care must be given with the use of crystalline scintillation detectors in transit from warm (typically indoor) environments to cold environments and back again. These devices are single crystals and are subject to fracture when moved quickly between areas having high temperature differential, a process called "thermal shock". A scintillation detector having one or two thermally induced cracks will continue to operate, but with reduced resolution. There is no easy way, apart from disassembly of the PMT and scintillator assembly for examination, of identifying thermally iduced cracking of the crystal.

High humidity and rainy wet conditions can also be of concern. First, very high ohmage resistors are frequently used to change ranges in many survey meters. High humidity can result in water vapor deposition on these devices, thus altering their efficacy and thus the radiation intensity readings associated with them. Most instruments use hermetically sealed cases designed to prevent moisture deposition inside of them. Users should be aware of this possible error source, and take necessary actions to minimize its occurrence. One possibility is to install fresh desiccant packets in the instrument cases, and to change them at each formal calibration, or more often, if necessary. New instruments usually arrive at their owner's location with desicant packets in them. Often these are left in the instrument without regard to the fact that they require periodic replacement for their continued effectiveness.

Secondly, moisture deposition on the insulators of ion-chamber instruments will cause response errors if not a complete break down. This occurrence related to very high humidity is hopefully unlikely.

3.4.2. *Proper calibration and service*

Many organizations such as the International Commission on Radiation Units and Measurement (ICRU), the International Atomic Energy Agency (IAEA), the United States National Council on Radiation Protection and Measurements (NCRP), etc. have publications recommending the minimum frequency for calibration and servicing of radiation survey instruments. Most of these recommend that the instruments be calibrated at least semi-annually, and that calibration checks with a radioactive source be done prior to each use of the survey meter. Formal calibrations must be accompanied by battery changes, if necessary, and by operational checks of switches and electronics connected with proper operation of the instrument [10].

3.4.3. *Global positioning systems*

The possibility of terrorist activity, and the probability of the accidental release of large amounts of radioactive materials make a global positioning system a valuable asset. Global Positioning Systems (GPS) are often used in conjunction with radiation survey equipment to locate contaminated areas with moderate precision. In some cases GPS systems are included as accessories with commercial survey instruments. There is a major advantage to having GPS associated with a survey when a large open area requires coverage. The correlation of contamination levels and/or dose or exposure rate readings with locations is obvious when maps must be produced relating radiological issues to the involved areas. The addition of GPS equipment should be a serious consideration for an enhanced radiation protection program.

4. Conclusion

Commercially manufactured instruments are usually well constructed with a mind to proper operation indoors or out being taken into account.

However, there are some situations where the outdoor use of these instruments can cause some concern.

- Alpha Monitoring – In this case the short range of alpha emissions and the ease of stopping them in most materials makes monitoring for alpha emissions virtually impossible outdoors. A very acceptable alternative is the use of FIDLER-type scintillation detectors calibrated to measure the photons associated with alpha decay from the parent or from daughters and radioactive contaminants associated with the parent.
- Instrument use in very cold environments – Proper operation of most battery powered sources will be significantly degraded in low temperatures. The use of a remote power source kept near body temperature will correct this deficiency.
- Thermal shock must be avoided with crystalline scintillation detectors.
- It is necessary that portable radiation survey instruments be properly calibrated and serviced at acceptable intervals.

Acknowledgement: I am thankful to the organizining committee of the NATO ATC on providing financial support for my travel and stay.

References

1. International Commission on Radiation Units and Measurements, ICRU (1992) Measurement of Dose Equivalents from External Photon and Electron Radiations, Publication No. 47
2. International Atomic Energy Agency, IAEA, Construction and Use of Calibration Facilities for Radiometric Field Equipment, Technical Reports Series (1990) No. 309
3. National Council for Radiation Protection and Measurements, NCRP, Instrumentation and Monitoring Methods for Radiation Protection, Publication 57 (1978)
4. ANSI N323A-1997, American National Standard Radiation Protection Instrumentation Test and Calibration, Portable Survey Instruments. 1997
5. NUREG-1507, Minimum Detectable Concentrations with Typical Radioactive Survey Instruments for Various Contaminants and Field Conditions (1998) US Nuclear Regulatory Commission, Washington, DC
6. International Standard, ISO 7503-1 (1988) Evaluation of Surface Contamination – Part 1: Beta Emitters (maximum beta energy greater than 0.15 MeV) and alpha-emitters
7. Eriksson M (2002) On Weapons Plutonium in the Arctic Environment (Thule, Greenland) PhD thesis, Risø National Laboratory Roskilde, Denmark
8. International Commission on Radiological Protection, ICRP (1990) Recommendations of the International Commission on Radiological Protection, ICRP Publication No. 60
9. International Commission on Radiation Units and Measurements, ICRU(1980) Radiation Quantities and Units, Report No. 33
10. Radiation International Commission on Radiation Units and Measurements, ICRU (1970) Report. Radiation Protection Instrumentation and Its Application No. 20

GAMMA SPECTROMETRY IN THE FIELD

N. ANTHONY GREENHOUSE*
Berkeley Laboratories, 788 Mickinley Avenue, Oakland, CA 94610-3833, USA

Abstract. The portability and speed of personal computers and the availability of germanium detectors and Dewar's have made gamma ray spectrometry in the field a functional reality. Software developers continue to write appropriate interfaces making it a relatively easy task to do spectrometry outdoors, and often in environments that 20 years ago would have been prohibitive. This paper is intended to generate discussions on the applications of gamma ray spectrometry in environmental situations.

Keywords: Gamma spectroscopy, radiation, environment, GE and HPGe

1. Introduction

The practical use of gamma ray spectrometry in the field has been the desire of radiation protection personnel for many decades. Initially spectrometry was approached with single channel analyzers (SCA) and NaI:Tl scintillation detectors. However, the desirability of producing a complete spectrum over a wide range of energies resulted in the development of portable, battery powered multi-channel analyzers, initially having 128 channels (or less). Later, the development of portable computers became a reality. This permitted spectrometrists to utilize the PC's memory to expand the available channels, and the programmers to focus on appropriate interface software to simplify the use of this apparatus.

Later, the development of solid-state detectors such as lithium-drifted Ge greatly enhanced the spectrometers resolution. Once high-purity Ge (HpGe) detectors became available, the evolution of portable gamma spectrometry arrived at today's level.

* To whom correspondence should be addressed. e-mail: nagreenhouse@envirhealth.com

G.A. Aycik (ed.), *New Techniques for the Detection of Nuclear and Radioactive Agents*,

Currently portable gamma spectroscopy equipment utilizes solid detectors that can be categorized in two major classes based on the temperature at which the detector is operated. Scintillation detection systems operate at ambient temperatures, but Ge detectors must be cooled either by a liquefied gas (usually nitrogen) in a Dewar flask or with an electro-mechanical cooler. Packaging of the detectors such as NaI:Tl that operate at ambient temperatures are usually thought of as hand held instruments weighing only a few pounds making them truly portable. Cooled detectors used in portable spectroscopy systems are mostly high purity germanium (HpGe) detectors. The HpGe systems will include both the detector with cooling medium and associated electronics that comprise an operational system. This combination of hardware will easily weigh over 20 lb. Thus they are actually semi-portable! When contrasting these two classes of spectroscopy systems the user must understand that nuclide identification is the primary purpose of the equipment. The key attribute of a detector is that it must be able to identify complex spectra in the presence of natural radioactivity. That question can be answered with a single word, resolution. It's a well known fact that cooled (HpGe) detectors have superior resolution to that of scintillation detectors operated at ambient temperatures [1].

Along with the hardware associated with these spectroscopy systems there is typically a host of software and firmware provided for the user interface. This software is used for system configuration, calibration, spectral acquisition, system diagnostics, and spectral analyses. One of the must useful additions to the software capabilities is the ability to quantify radionuclides via mathematically calculated efficiencies for complex geometries. Considerable software engineering has been focused in this area over the last few years. This has resulted in a more intuitive interface for the users making the instrument more portable across many computer operating systems. Looking to the future, anticipating what the next generation of portable spectroscopy equipment will encompass is certainly not clear. Many factors are currently influencing the development of spectroscopy systems which focus on the needs of my country, the United States. One issue is the minimization of the possibility of a terrorist attack using radioactive materials. Portable spectroscopy equipment plays a fundamental role in the security of all of our homelands against this threat. For the most part portable equipment is used by personnel that respond to initial alerts based on intelligence or actual detection alarms. In this role, future ambient temperature detectors need to be developed so they have spectral resolution comparable to or approaching the current HpGe detector resolution. Another possibility is the future development of a light weight compact electro-mechanical cooling system for HpGe detection systems so these systems become more like the current hand held spectroscopy systems. In general, high resolution hand held spectroscopy systems are likely to be developed to enhance our readiness.

All of the future spectroscopy systems will be software driven requiring further software development. Enhanced or smart software systems will have to be developed with capabilities that can identify radionuclides with greater reliability, possibly locate the radioactivity emission point, image the area about the emission point, and reconstruct an un-attenuated spectrum from the acquired spectrum. It is also likely that new software will automatically calibrate system efficiency as related to gamma ray energy in order to quantify the identified radionuclides.

2. Some Uses of Field Spectrometry

2.1. NATURAL BACKGROUND ASSESSMENT

Gamma radiation is, of course, a part of natural background. At any facility it is necessary to know the background contributions per channel in order to determine low level additions to them from potential environmental contaminants. Natural background radiation intensity varies as a function of geographic location (even over relatively short distances), and as a function of time day and time of year. Weather patterns also have an influence on natural background radiation levels. The sunspot cycle controls the intensity of solar cosmic rays on the upper atmosphere with an 11 year period. So, periodic determinations of natural background spectra are necessary. This should be a routine use for portable and/or fixed-location spectrometers in the environmental monitoring programs at large nuclear facilities.

2.2. ROUTINE ENVIRONMENTAL MONITORING

Most nuclear facilities offer sufficient potential risks with their operation to warrant a well defined environmental monitoring program. Usually this will entail collection of environmental samples for analysis in the lab, and an environmental dosimetry program. Periodic use of a portable gamma spectrometer should also be a part of such a program. Spectrometry results should be incorporated into public exposure models used to verify that the facility is operating within acceptable limits. The following figures demonstrate possible areas where in-situ gamma spectrometry can be used to advantage. In Figure 1, the environmental elements such as “crops”, “soil”, or “animals” can be analyzed in the field. In Figure 2, some of the elements such as aquatic plants and animals, and sand and sediment would need to be sampled first. These samples could be transported to the lab for analysis, and/or they could be analyzed in the field. The same is true for air samples collected on filters or cartridges. The results of the sample analyses would then, of course, be incorporated into an environmental dosimetry program. The results would normally be radiation dose predictions for individuals located at the facility "fence line", or anywhere else where environmental transport mechanisms such as air and water would carry materials away from the facility.

2.3. SPECIALIZED STUDIES

Clouvas et al. used a portable spectrometer to analyze indoor radon in Greek buildings [2]. Enghauser and Ebara used their spectrometry apparatus to obtain accurate estimates of complex radiation shielding requirements for multiple nuclides having unique geometries [3]. Other such applications are available in the literature as well as on-line.

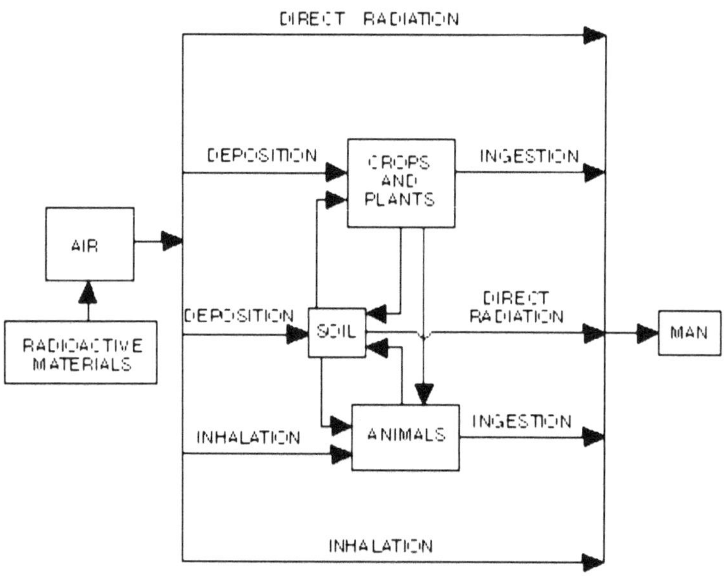

Figure 1. Simplified pathways for airborne releases to man [4]

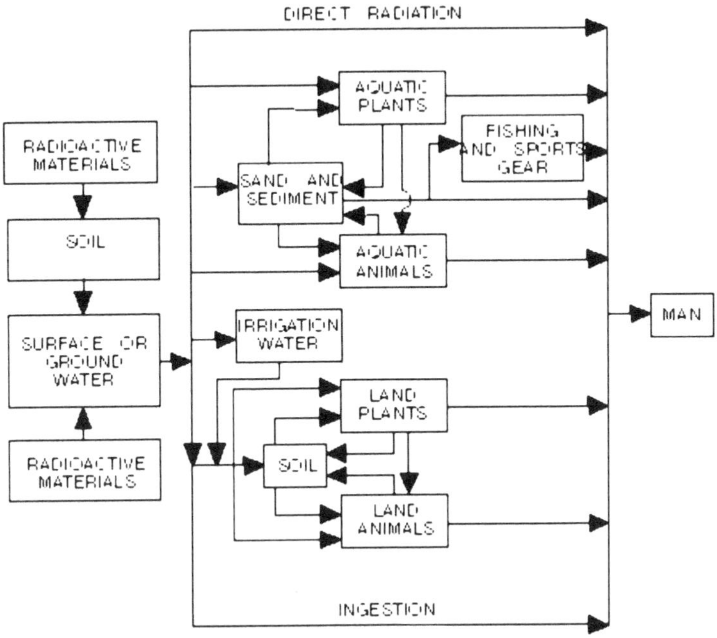

Figure 2. Simplified pathways for waterborne releases to man [4]

2.4. CORRELATION WITH METEOROLOGICAL PREDICTIONS DURING AND AFTER A SPILL

Let us assume a worst case at a nuclear reactor where there has been a loss of radioactivity from a fuel element with a release from containment. Of course, conventional radiation detection equipment can be used to identify the direction of the plume, but fission and activation products will be lost by radioactive decay and by scavenging on stable aerosols and surfaces. So, a portable gamma spectrometer can identify the gamma-emitters in the plume as a function of time after the accident.

Likewise, after a "dirty bomb" explosion, or other terrorist activity, the need to monitor and track gamma-emitters in the plume becomes all important. Of course, this can often be done with more conventional air sampling equipment, but as the plume disperses the sensitivity of an HpGe detector will be of value.

Meteorologists can usually predict plume dispersion accurately with time, but they need feedback to improve the accuracy of their estimates. Also, members of the public are potentially at risk in situations such as this, and protective measures, including evacuation need to be made with the best available environmental information. The field gamma spectrometer can be a very valuable asset in this situation.

2.5. EXERCISES DESIGNED TO SIMULATE RELEASES OF RADIOACTIVE MATERIAL(S)

Realistic exercised should be a part of the management program for any major nuclear facility. Such exercises should, of course, include the use of portable spectrometry equipment. Meteorological models should be tested, and plume directions and changes thereof should be incorporated into the exercise protocol. For example, some fluorescent chemicals can be released and measured at significant distances from the facility and later measured on filter papers in-house. The spectrometry apparatus will "simulate" the presence or absence of these chemicals in field air samples. The results of each such exercise along with recommendations for improvement of facility staff functions should be logged in an exercise report and included with the evolution of the emergency plan.

2.6. DECONTAMINATION AND DECOMMISSIONING

Portable gamma spectroscopy systems provide a practical way to characterize dispersed radionuclides in or on the soil at nuclear facility decommissioning and restoration sites, and in surrounding areas. The objective is to determine radioactivity of plant-related nuclides per unit area or unit volume of soil. Traditional methods generally involve gross (non-spectroscopic) field surveys, followed by extensive field sample collections for subsequent laboratory gamma spectroscopic analysis. However, field surveys with gross counting instrumentation do not identify specific nuclides, and therefore cannot discriminate between plant-caused activity and anomalous distributions of natural activity or global fallout activity. Gross counting techniques also cannot detect small amounts of problem nuclides in the presence of larger amounts of other natural nuclides. Any discrete sample taken for laboratory analysis will only identify what was at that one specific sample site. This means that for cases where the contamination is not uniform, some hot spot areas could be missed. In situ gamma spectroscopy, on the other hand, can effectively detect all of the gamma activity over more than 100 m^2 of area. For high energy gammas, it even detects radioactivity buried below the surface of the soil. With in situ gamma spectroscopy, there is a much higher probability that nothing will be missed.

With laboratory analyses, there is much labor involved, and a long turn around time for the analysis results. Samples must be collected and documented, labeled, and transported to a remote lab (with subsequent potential loss of chain-of-custody). Then the samples must be prepared, analyzed, and the reports sent back to the user. With in situ gamma spectroscopy, the results are available immediately, with equivalent or better accuracy, and with less labor.

2.7. SPECIALIZED SOFTWARE

Most commercial portable gamma spectrometry apparatus comes with its own operational and data analysis software. However, some commercial software packages have been available for gamma spectroscopists, apart from those which accompany new set-ups for use in outdoor environments. An example is "SNAP" from Eberline Instrument Company which is tailored for waste disposal analyses. Depending upon one's needs for this equipment, some research should be placed upon specialized software selection.

3. Conclusion

The availability of truly portable gamma spectrometry apparatus has been a welcomed development for persons interested in environmental analyses, tracking radioactive materials after an accidental release or terrorist activity, decontamination and decommissioning operations, and many specialized operations. The only difficulty at this point is the need for a Dewar and liquid nitrogen or an electromechanical cooling apparatus for proper operation of an HpGe detector. Scintillation detectors can be used in place of the cooled HpGe detector, but with a substantial loss of resolution. This loss may be adequate for some of the functions discussed here, but the tremendous advantages of HpGe over conventional scintillation detectors are obvious [5, 6].

References

1. Smith R J (2006) Portable Gamma Spectroscopy – A Brief Look at the "State of The Art" and a Vision of the Next Generation. Westinghouse Savannah River Company, Aiken, South Carolina, USA
2. Clouvas A, Xanthos S, Antonopoulos-Domis M (2003) A combination study of indoor radon and in situ gamma spectrometry measurements in Greek dwellings. Radiat Prot Dosim 103:363–366

3. Clouvas A, Xanthos S, Antonopoulos-Domis M (2006) Simultaneous measurements of indoor radon, radon-thoron progeny and high-resolution gamma spectrometry in Greek dwellings. Radiat Prot Dosim 118:482–490
4. Radiological Characterization of Surface Soil, Professional Enrichment Program Course Materials, Health Physics Society, 35th Annual Meeting, June 24, 1989; http://www.canberra.com/literature/972.asp
5. Englehauser M, Ebara S (1997) Quantitative Portable Gamma Spectroscopy Sample Analyses for Non-standard Sample Geometries Sandia National Laboratory, Albuquerque, MN, 87185
6. International Atomic Energy Agency Technical Meeting on Nuclear Spectrometry Methods for in-situ Characterization of Materials. IAEA Headquarters, Vienna, Austria, 19–23 May 2008

GAS-FILLED AND PLASTIC SCINTILLATION DETECTORS: ADVANTAGES AND DISADVANTAGES

MOHAMMED K. ZAIDI* AND SYED F. NAEEM
Idaho State University, Pocatello, ID. 83209-8060, USA

Abstract. The radioactivity detectors used in radiation monitoring operations are discussed with special emphasis to their advantages and disadvantages when used in a field and or laboratory setting. They are highly sensitive and need care while being used in the field. Their sensitivity is improved when used in a laboratory setting because of the correct predictability of the local environment. The physics of these detectors is discussed relative to their use in lab and field settings.

Keywords: Radiation, detectors, gas-filled detectors and plastic scintillation detectors

1. Introduction

1.1. BASIC RADIATION PHYSICS REVIEW

Nuclei of atoms are composed of protons and neutrons that are held together by a strong nuclear binding force. An element *X* having atomic mass *A* with a number of protons *Z* is symbolically written as ($_{Z}X^{A}$) The number of neutrons can be determined by subtracting the number of protons from the atomic mass. Unstable nuclei formed as a result of nuclear reactions can undergo transformations in the form of radioactive decay to reach their ground state or into stable nuclei. The transformations can be isomeric or isobaric. Radioactive isotopes of a given atomic number differ only in number of neutrons in nuclei, whereas number of protons differs in nuclear isobars but with the same atomic mass.

The activity of a radioactive source or radionuclide is defined as the number of nuclei of a source that disintegrate per unit time with the emission of massive particles or mass less electromagnetic particles called

* To whom correspondence should be addressed. e-mail: zaidmoha@isu.edu

photons. The radioactive decay process is random, but with a certain probability for a radioactive atom to decay within a given time interval. Half-life ($t_{1/2}$) is the amount of time in which a radioactive source loses half of its activity from the radioactive decay process. Each radionuclide has a physical half-life that can range from fraction of a second to millions of years or longer. The System International (SI) unit of activity is called the Becquerel (Bq), which is defined as one disintegration per second (older units are the Curie (Ci), which is 3.7×10^{10} Bq). Various forms of radioactive decay occur in the form of alpha decay (α-decay), beta decay (β-decay), gamma emission (γ-emission), electron capture (EC), and internal conversion (IC).

1.1.1. *Alpha decay*

Radionuclides containing excess nucleons A > 209 are generally unstable and they or their progeny emit radiation in the form of α-decay. The α-particle is a ^{4}He nuclei comprised of two protons and two neutrons as shown below in the following α-decay transition:

$$(Z, A) \rightarrow (Z - 2, A - 4) + \alpha \tag{1}$$

Common α-emitter sources such as ^{241}Am, ^{241}Po, and ^{242}Cm are frequently used in the nuclear laboratory. The α-particles are generally emitted with an energy of 4–6 MeV, are doubly positive charge (+2e) and travel only a few centimeters (cm) in air because of their high rate of energy loss in matter.

1.1.2. *Beta decay*

Nucleons having excess number of neutrons or protons or because of weak interactions can decay by the emission of β-particles (fast electrons or positrons) and the decay process is called β-decay. The β-decay process shows a continuous energy spectrum within few tens of keV to a few MeV for the β-particles. The Majority of the β sources are not pure β emitters. That is, they first decay by β-emission into excited state of a nucleus (^{137m}Ba) that may further decay via γ transitions to the ground state of a daughter nucleus. These delayed photons that resulted after β-decay is called beta-delayed (β-delayed) photons as shown in Figure 1. Some of the common pure β-emitters are ^{3}H, ^{14}C, ^{32}P, ^{99}Tc, and ^{90}Sr/^{90}Y.

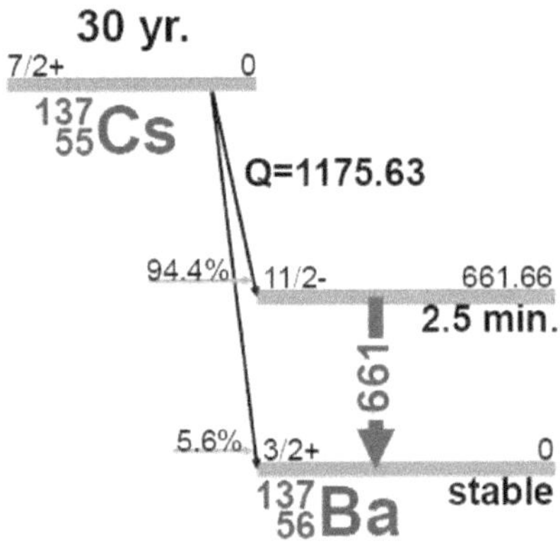

Figure 1. ^{137}Cs decay scheme [1]

A neutron rich nucleus emits an electron (e^-), and a antineutrino ($\bar{\nu}$) particle and increases the nuclear charge by one as shown in the following transitions,

$$n \rightarrow p + e^- + \bar{\nu} \tag{2}$$

$$^3H \rightarrow {}^3He + e^- + \bar{\upsilon}\text{, etc.} \tag{3}$$

Similarly, a proton rich nucleus such as ^{22}Na emits a positron particle (β^+), and a neutrino (ν) particle,

$$^{22}Na \rightarrow {}^{22}Ne + e^+ + \nu \tag{4}$$

1.1.3. *Gamma emission*

Electromagnetic radiation in the form of γ-photons is emitted when an excited state of a nucleus directly decays to its stable daughter nucleus or to another excited energy level. The energy of these emitted photons typically ranges from a few hundred keV to a few MeV. A γ-photon of 661 keV is released during the decay process of ^{137}Cs as shown above in Figure 1. An isomeric transition during the radioactive decay process occurs when one or more photons are emitted from the excited state of the daughter nucleus that further decays down to its ground level (661 gamma is coming from ^{137m}Ba).

Isomers are represented by the symbol "m" next to the atomic number of an element such as ^{60m}Co, which is an isomer. ^{60m}Co whose excited meta-stable daughter state decays directly into its stable daughter ground state with the emission of 58 keV photon and into multiple excited daughter states that further decays via β-delayed photons emitting 826, 1,332, and 2,158 keV γ-emission. ^{60m}Co nuclear decay diagram is shown in Figure 2.

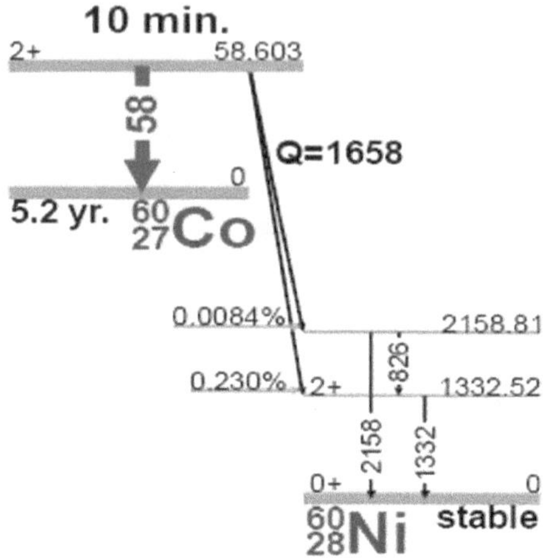

Figure 2. ^{60m}Co decay scheme [1]

Positrons annihilation are another source of γ-emission. Positrons quickly annihilate with the surrounding electrons produce two 511 keV photons and in order to conserve momentum these photons travel in almost in opposite directions.

1.2. RADIATION DETECTION AND SOURCES CALIBRATION

Detection efficiency of a radiation detector is determined by its absolute efficiency ε_{abs} and intrinsic efficiency $\varepsilon_{int.}$ [2] The absolute efficiency can mathematically be determined using the following expression:

$$\%\varepsilon_{abs} = \left[\frac{cpm - BKG}{dpm}\right] \times 100\% \tag{5}$$

where: *cpm* = counts per minute or number of pulses recorded per minute, *dpm* = disintegration per minute or number of radiation quanta emitted by source, *BKG* = background counts per minute.

Detection efficiency of a radiation detector is generally dependent on the geometry of the detector that is, the distance from the source to the detector and the solid angle of the source with respect to detector. Intrinsic efficiency ε_{int} can be expressed mathematically as;

$$\varepsilon_{\text{int}} = \varepsilon_{abs} \cdot \left(\frac{4\pi}{\Omega} \right) \tag{6}$$

$$\Omega = 2\pi \left(1 - \frac{d}{\sqrt{d^2 + a^2}} \right) \tag{7}$$

where: d = source-detector distance, a = detector's surface radius. Ω = solid angle of the detector seen from the source in steradians (Steradians are a measure of the angular 'area' subtended by a two dimensional surface about the origin in three dimensional space, just as a radian is a measure of the angle subtended by a one dimensional line about the origin in two dimensional (plane) space. Steradians are equivalently referred to as 'square radians.' A sphere subtends 4π square radians (steradians) about the origin. By analogy, a circle subtends 2π radians about the origin. Numerically, the number of steradians in a sphere is equal to the surface area of a sphere of unit radius. i.e., area of sphere = $4\pi r^2$, but with r = 1, area = 4π Likewise, numerically, the number of radians in a circle is equal to the circumference of a circle of unit radius, i.e., circumference = 2πr, but with r = 1, circumference = 2π. As one would expect, steradians (square radians) can be converted to square degrees by multiplying by the square of the number of degrees in a radian = 57.2957795°. For example, the number of square degrees in a sphere is equal to $4\pi \times (57.2957795)^2$ = 41,253 square degrees (rounded to the nearest square degree). For those who prefer to work in square degrees, it is helpful to remember that the number of square degrees in a sphere contains the digits 1 through 5, with no repeats. Steradians occur virtually anywhere in physics where a flux through a three dimensional surface is involved. For example, the ubiquitous factors of 4π that keep popping up in formulas derived in electromagnetics really just represent the scaling, or normalizing, of whatever is being described to the angular area subtended by a sphere. Not surprisingly, steradians find heavy use in antenna engineering to characterize such properties as the 'directivity' of an antenna relative to an 'isotropic' radiator, one that radiates uniformly in all directions through the surface of an imaginary sphere [3].

The radiation is detected due to the interaction of radiation with material in the detector. The material in most commonly used detectors is gas. They have three parts: detector, amplification system and a measuring device. The output pulse of the detector is shaped and amplified and then converted. The most commonly employed to detect ionizing radiation are as described in the following sections.

2. Ion Collection

The way ions can be collected at the electrodes is categorized by the six-region curve which is depended on the applied voltage. The different regions, shown in Figure 3, are as follows [2, 4]:

I. **Recombination Region**: The applied voltage is very low in this region however, the size of the pulse generated increases with increasing voltage. This region is not useful for radiation detection purposes because the number of ions collected in this region is less than the number of ions produced as a result of interaction between ionizing radiation and gas molecules in detector's chamber.

II. **Ionization Chamber (IC) Region**: This portion of the curve is essentially flat as the number of pulses generated in this region does not change with voltage increase. Every ion created is essentially collected and the voltage is not high enough to cause further ionization to the migrating ions.

III. **Proportional Region**: The number of ion pairs collected in this region is greater than the number created and the pulse size increases proportionally for a given type of radiation as the voltage increases. The gas amplification, the creation of new ion pairs by the migrating ions on their way to the detector's electrodes, factor reflects the multiplication of the number of ions collected versus created. The gas amplification factor changes for different types of radiation but it is consistent within specific radiation types.

IV. **Limited Proportional Region**: Gas amplification also occurs in this region however, it is not linear with change in voltage. This region is not useful for radiation detection purposes as it will not give meaningful results.

V. **Geiger-Muller (GM) Region**: Pulse size in this region increases slightly as voltage increases but the change in pulse size is not noticeable over relatively large changes in the applied potential. The applied potential is so high that every event occurring within detector's chamber generates more ion pairs as they migrate towards respective electrodes, this effect is referred to as Townsend avalanche. The pulse size, as a result of avalanche, does not depend on the number of ion pairs produced. It is important to specify that the pulse size is the same, no matter what type of energy of radiation caused by ionization. GM region therefore, is useful to monitor radiation but it is not possible to determine the type of incoming radiation because of avalanche process.

VI. **Continuous Discharge Region**: This region follows GM region however, it is not useful for radiation detection purposes because there is a continuous discharge of electricity following ionization takes place in the detector's chamber.

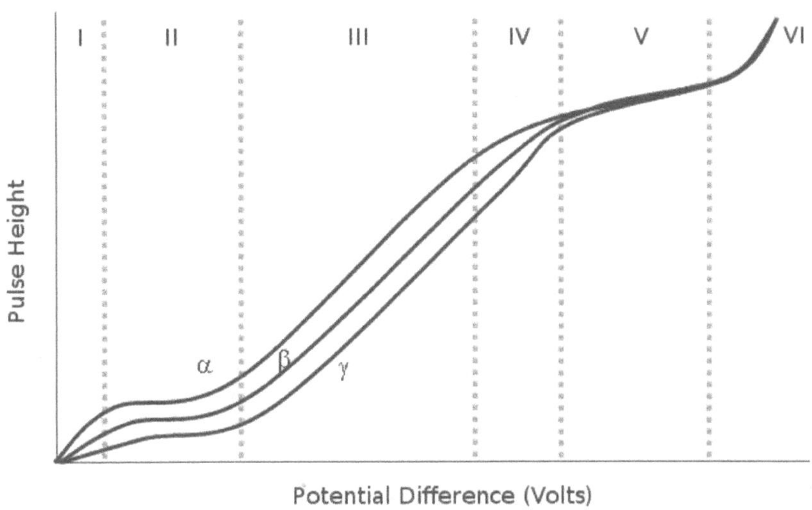

Figure 3. The different regions of a typical gas-filled radiation detector

2.1. IONIZATION CHAMBERS (ICs)

2.1.1. *Ion chamber detectors*

Gas filled detectors response depends on ion pairs produced as a result of the interaction between incoming radiation and the gas molecules within detector's chamber. The rate of ion pair migration towards anode and cathode of the detector depends on the applied voltage and the relationship between the applied voltage and the number of ions collected is well known. The size of the pulses produced when incoming radiation particles interact with the detectors chamber is related to the number of ions collected at the electrodes.

As the voltage to the detector is increased, a point is reached at which essentially all of the ions are collected before they can recombine. No secondary ionization or gas amplification occurs. At this point, the output current of the detector will be at a maximum for a given radiation intensity and will be proportional to that incident radiation intensity. Also, the output current will be relatively independent of small fluctuations in the power supply. The output of a gas-filled detector when 100% of the primary ion pairs are collected is called the saturation current.

Advantages

- Output current is independent of detector operating voltage. As a result, less regulated and thereby less expensive and more portable power supplies can be used with ion chamber instruments, and still offer a reasonably accurate response.
- Since the number of primary ion pairs is a function of the energy deposited in the detector by the incident radiation, the ion chamber response is directly proportional to the dose rate.
- Since exposure (x) is defined in terms of ionization of air by photons, an air-filled ion chamber, when used for photon radiation, yields the true exposure rate.

Disadvantages

- Since only primary ion pairs created by each radiation event are collected, the output currents are small. Independent current pulses large enough to measure are not formed by each ionizing event. Instead, the total current output created by many ionizing events is measured. Therefore, the sensitivity of a small ion chamber is very poor because a few ionizing events per minute do not create sufficient currents to be measured. A typical commercial portable ion chamber has a detector which produces a current of about $2e^{-14}$ amps per mR/h.
- Another consequence of the small output current is the effect humidity can have on the instrument response. The electronics associated with the detector must have a high impedance (approximately $1e^{15}$ Ω) to measure currents this small. The instrument incorporates insulators designed to maintain this high impedance. High humidity conditions can cause the formation of condensation on those insulators (The resistance of relatively pure water is approximately $1e^{7}$ Ω/cm). This condensation creates leakage paths which causes erroneous instrument response.

- Since anything which changes the density of the gas affects the response, changes in barometric pressure (or altitude) and/or ambient temperature can affect instrument response in some cases. This is particularly the case with thin-walled chambers, vented chambers, or chambers with windows. For instance, the response of a typical commercial portable ion chamber instrument decreases by 2% for each 10° increase in temperature, or decreases by 2.3% for each inch of mercury decrease in barometric pressure (4.6 % per psig).

2.1.2. *Proportional counters*

The proportional counter is a gas-filled detector introduced in the late 1940s to detect ionizing radiation. In this type of counter, proportional tubes are operated in pulse mode and rely on the phenomenon of gas multiplication to amplify the charge represented by the original ion pairs created within the gas. The interaction of ionizing radiation with the gas in the detector creates an ionization event that produces positive ions and free electrons. Gas multiplication is a consequence of increasing the electric field within the gas to a sufficiently high value. At low values of the field, the electrons and ions created by the incident radiation simply drift to their respective collecting electrodes. The mean number of pairs created is proportional to the energy deposited in the counter.

The pulse height produced by α-particles is greater than that produced by β-particles, therefore α-particles are counted at the lower voltages compared to β-particles. The operating voltage for both alphas and betas can be determined by operating the counter. If an alpha radioactive source, for instance ^{241}Am, is placed beneath the proportional counter in appropriate geometrical position and the voltage is slowly increased at a rate of 100 volts/run, a voltage will be reached at which the proportional counter begins to register counts in the scalar/counter. The counts registered are proportional to the applied voltage, however a stage will be reached in which the increasing voltage over a wide range would have little effect in the counting rate. This region is the alpha plateau, and the proper operating voltage should be selected relatively close to the threshold voltage within the lower 25–30% of the plateau in order to select optimal alpha voltage. Also, the operating voltage should be selected at a point where the plateau shows a minimum slope [5]. The beta operating voltage can also be determined by beta emitting source, such as ^{137}Cs, which is also placed beneath the proportional counter. However this region also detects α's, and correlation must be made.

Advantages

- They can detect α, β, γ or, if properly designed, neutrons.
- Based on the pulse height of the spectra, proportional counters can distinguish between different radiation types.
- The proportional counters have very little dead-time thus makes them very useful to count higher activity radiation sources.

Disadvantages

- They are expensive and demands operational expertise.
- Proportional counters are also sensitive to the environmental conditions such as temperature and humidity.
- They are potential fire hazard, for example P-10 gas is a mixture of methane and argon and methane can cause an explosion if exposed to a spark.
- Some proportional counters are relatively bulky, thus portability may become challenge in different circumstances.

2.1.3. *Geiger-Mueller (G-M) Counter*

A type of IC, was first introduced in 1928 by Hans Geiger and Walther Mueller to detect ionizing radiation. A typical G-M counter consists of a gas-filled G-M tube having a thin, mica end-window, a high voltage power supply for the tube, a counter to record the number of particles detected by the tube, and a timer which will stop the action of the counter at the end of the time interval. G-M counters employ gas multiplication to greatly increase the charge represented by the original ion pairs formed. Ion pairs, usually an ionized electron that migrate towards anode and the positively charged proton which travels towards cathode, are produced within the gas chamber and their migration rate depends on the applied electric field. The number of ions collected at the electrodes dictates the amplitude of the pulse produced as a result of interaction of radiation within the detector chamber. Typically 10^9–10^{10} ion pairs are formed during discharge phase of the G-M tube, producing a pulse of high amplitude therefore a pre-amplifier is not required with G-M tube but these counters produce a pulse for each entering particle and cannot differentiate between them (α, β, or γ).

Typically, the counting rate of G-M counter depends on the applied voltage. It is important to know about the operating voltage of a G-M counter to take measurements. Under threshold voltage, G-M counter is not sensitive enough to count radiation. Because of gas pressure, and number of times G-M tube has been used, every G-M tube has a characteristic response of counting rate versus voltage applied to the tube.

Over a certain range of voltages, called the plateau range, the counting rate is relatively critical to applied voltage. Above this plateau range, radiation counting is of no use because of continuous discharge region. The plateau curve of every tube that is to be used for the G-M tube should be drawn in order to determine the optimum operating voltage. The optimum operating voltage should be about one-third of the plateau [6].

Advantages

- They are simplest in principle among all gas-filled detectors.
- They can detect α, β, and γ with proper detector's design and calibration.
- After meeting geometric conditions of the chamber's cavity, one can use the following expression to measure absorbed dose D_m, with units in grays defined as joules per kilogram (J/kg), for different types of radiation as follows [2].

$$D_m = W S_m P \tag{8}$$

  Where: W = average energy loss per ion pair produced in the gas chamber, S_m = energy loss per unit density or relative mass stopping power of the material to that of the gas, P = number of ion pairs per unit mass produced in the gas.
- They can be used as radiation monitoring survey instruments as they are portable.

Disadvantages

- They have longer dead time typically ranging from 50 to 100 μs and slower response time.
- They are less sensitive to detect low levels of radiation.
- They were sensitive to environmental effects such as temperature, pressure, and humidity level and not anymore.

3. Typical Applications

Portable survey instruments used for measuring dose rates are typically ion chamber instruments. Ion chambers may also be used in several installed monitor systems such as the Area Radiation Monitor System (ARMS) and the various Process Radiation Monitors (PRM).

Acknowledgement: I extend my thanks to the NATO ATC organizing committee for the travel and acommadation award.

References

1. Radionuclide Transformations: Energy and Intensity of Emissions, International Commission on Radiological Protection Publication ICRP-38 (1983) Vol. 11–13, New York: Pergamon
2. Knoll GN (2000) Radiation Detection and Measurements, 3rd edition, New York: Wiley
3. Warren Davis, Ph.D. President, Davis Associates, Newton, MA
4. Leo WR (1994) Techniques for Nuclear and Particle Physics Experiments. 2nd revised edition, New York: Springer
5. Tsoulfanidis N (1995) Measurement and Detection of Radiation, 2nd edition, London: Taylor & Francis
6. Bevelacqua JJ (2004) Basic Health Physics, Weinheim: Wiley-Vch

Some important websites:

http://www.rstp.uwaterloo.ca/manual/detection/gas_filled/gas_filled_detectors.htm
http://en.wikibooks.org/wiki/Basic_Physics_of_Nuclear_Medicine/Gas-Filled_Radiation_Detectors
http://www.ortec-online.com/application-notes/an34/an34-content.htm
http://www.bookrags.com/research/particle-detectors-wop/
http://hss.energy.gov/NuclearSafety/techstds/standard/hdbk1122-04/Module_113_Study_Guide.pdf
http://www.osti.gov/energycitations/product.biblio.jsp?osti_id=6050613.
www.bu.edu/es/labsafety/ESMSDSs/MSP10.html

EFFICIENCY CALIBRATION OF A WELL-TYPE Ge DETECTOR FOR VOLUMINOUS SAMPLES IN CYLINDRICAL GEOMETRY

AYSE NUR SOLMAZ* AND DOGAN BOR
Institute of Nuclear Science of Ankara University
HALUK YÜCEL
Turkish Atomic Energy Authority (TAEK), Besevler Yerleskesi, 06100 Tandoğan, Ankara, Turkey

Abstract. To study environmental samples, a 44.8% relative efficient well-type Ge detector with an active volume of 218 cm^3 was calibrated in three different voluminous sample geometries. The models were tested by absolute full energy peak efficiency calibration with use of certified multinuclide standards. The efficiency results corrected for this experiment obtained self-absorption factors from the simple photon transmission experiments by using point-gamma sources for small cylindrical geometry and theoretically calculated self absorption factors based on the database related to mass attenuation coefficients for vial geometry. Tufa, soil, and marble were also measured at these counting geometries. The activity results obtained with these models were compared with each other.

Keywords: Radiation detection, well-type Germanium detectors

1. Introduction

The determination of radionuclides in various samples such as environmental, foodstuffs, etc., is often encountered with the difficulty in measuring low levels of radioactivities. A weak intensity peak observed in a gamma-ray spectrum of a sample can be masked by the intensity of the background in the peak area. Hence, the background has a direct impact on the statistical significance of the measured weak peak, and the level of background fluctuations is expressed by the uncertainty of the background. On the other hand, the low limit of the detection is proportional to square

* To whom correspondence should be addressed. e-mail: ansolmaz@ankara.edu.tr

root of background continuum counts under the peak region of interest, $\sqrt{B}$, where the proportionality factor varies with the confidence level chosen such as a k = 1.65 factor. The detection limit is a more useful quantity, which is generally expressed in counts, when introduced in the minimum detectable activity (MDA) parameter. The MDA (expressed in Bq) is defined as the smallest quantity of radionuclide which can be determined reliably, given the existing conditions of the spectral measurement. The MDA is inversely proportional to the absolute detection efficiency for a peak, and smaller MDA values can be obtained by lowering the background and increasing efficiency [1].

However, in practice, it is known that some parameters such as the detection efficiency, the background level in the detector assembly associated with passive shielding specifications limits to reduce the MDAs to the expected ones for a particular the counting system. Since accurate determination of the detection efficiency is a crucial requirement. The common method of calibration of the detector for measurement of small samples by assuming the sample to be equal geometrically to the commercially available standard radioactive point source, however, the situation is different for calibration of the detector for the measurement of voluminous samples [2]. In the present work, it is aimed that for close source-detector geometry, a well-type HPGe detector for high-resolution gamma-ray spectrometry is calibrated to measure the voluminous samples, which are small and relatively larger cylindrical beakers on the endcap and the vial placed in the well.

2. Materials and Methods

2.1. GAMMA-RAY SPECTROMETRY

The detector used in the radioactivity measurements was a well-type HPGe detector (Canberra Model GCW 4023) with the well diameter of 16 mm and the well depth of 40 mm. The HPGe crystal was a p-type and closed-end coaxial with an active volume of 218 cm^3, yielding to a measured relative efficiency of 44.8%. The HPGe well-type detector has also an energy resolution of 2.0 keV at 1,332.5 keV of ^{60}Co and of 1.16 keV at 122 keV of ^{57}Co, and peak-to-Compton ratio of 60.8:1 at 1,332.5 keV. A top-split opening 10 cm thick Pb shield (Canberra Model 747), jacketed by a 9.5 mm steel outer housing was used to reduce the room background. It also features a 1 mm thick Sn an 1.6 mm thick Cu graded liners prevents interference Pb X-rays, and the floor of shield has a 11.4 cm diameter hole blocked by an annular lead plug in which accommodate only the dipstick cryostat and detector cables, reducing thus the streaming path. To minimize scattered radiation from the shield, the detector was centered in it. The detector was interfaced to a 16 k ADC conversion/MCA channels spectral memory analyzer (Canberra Multiport II NIM Module) operating through a commercial gamma spectroscopy software (Genie-2000) including acquisition, peak searching, peak evaluation, energy/efficiency calculation mode, nuclide identification, etc. The coaxial well-type Ge detector was calibrated to collect 4,096 channel spectra with a gain of 0.75 keV/channel, thus covering up to 3,065 keV energy. The system dead time over all measurements were kept below 1%, and the measurement periods for the samples and standards varied between 3 h and 5 days to obtain good statistics of the spectrum counts.

2.2. CALIBRATION SOURCES

Two different types of radioactive materials were used as calibrators. First type was the certified reference materials (CRMs), namely: RGU-1 (U-ore), RGTh-1 (Th-ore) and RGK-1 (K_2SO_4), called IAEA/RG set which are certified and issued by IAEA–Analytical Quality Control Services (ACQS). The powdered CRMs have known concentration radioactivity values in ppm ($\mu g\ g^{-1}$) whose properties are given in Table 1.

TABLE 1. IAEA CRMs as calibrands

CRM code	Component/nuclide	Concentration	Confidence Interval[1]	Geographical origin/reference date
IAEA/RGU-1	Uranium	400 $\mu g\ g^{-1}$	$\pm 2\ \mu g\ g^{-1}$	Beaverlodge, Saskatchewan
IAEA/RGTh-1	Thorium	800 $\mu g\ g^{-1}$	$\pm 16\ \mu g\ g^{-1}$	Oka, Quebec, Canada
IAEA/RGK-1	Potassium	44.8%	±0.3%	–

[1] At 95% confidence interval

The second type calibrand was the mixed radionuclide gamma-ray reference standard (i.e., multinuclide standard source) emitting a series of gamma rays covering the energy range from 47 to 1,836 keV from the radionuclides ^{210}Pb, ^{109}Cd, ^{57}Co, ^{123m}Te, ^{51}Cr, ^{113}Sn, ^{85}Sr, ^{137}Cs, ^{88}Y, ^{60}Co, as given in Table 2. This source was purchased from Isotope Products Inc. traceable to PTB (Physikalisch-Technischen Bundesanstalt). The cylinder beakers and vials were filled with the sand matrix spiked with the above mentioned radionuclides (its density: $\rho = 1.7 \pm 0.1\ g\ cm^{-3}$).

TABLE 2. Certified multinuclide standard source[1]

Nuclide	E (keV)	f_γ (%)	Half life	Activity (kBq)	Total unc.[2] (%)
^{210}Pb	47	4.18	22.3 ± 0.2 years	56.1	11.4
^{109}Cd	88	3.63	462.6 ± 0.7 days	53.3	3.1
^{57}Co	122	85.6	271.79 ± 0.09 days	1.96	2.9
^{123m}Te	159	84.0	119.7 ± 0.1 days	2.54	2.9
^{51}Cr	320	9.86	27.706 ± 0.007 days	65.8	3.0
^{113}Sn	392	64.9	115.09 ± 0.04 days	10.4	2.9
^{85}Sr	514	98.4	64.849 ± 0.004 days	12.4	3.0
^{137}Cs	662	85.1	30.17 ± 0.16 years	8.68	2.9
^{88}Y	898	94.0	106.630 ± 0.025 days	19.7	2.9
^{60}Co	1,173	99.86	5.272 ± 0.001 years	10.3	2.9
^{60}Co	1,333	99.98	5.272 ± 0.001 years	10.3	2.9
^{88}Y	1,836	99.4	106.630 ± 0.025 days	19.7	2.9

[1]Reference date for decay correction: 01.10.2006
[2]At 95% confidence interval

2.3. SAMPLE PREPARATION

Two kinds of tufa samples were taken from Capadocia in the Center Anatolia, Turkey. Additionaly, a marble sample from Capital city, Ankara and soil sample from Elazig city, Turkey, were collected for investigating the activity concentrations. The samples were homogenized to a particle size less than 1 mm using the mechanical grinder and then were dried at 105°C to a constant weight. These homogenized samples were transferred into cylindrical beakers and vials and hermetically sealed. A set of samples cover an apparent density range of 0.97–1.37 g cm^{-3}. The samples were kept for a period of at least a month to ensure the radioactive equilibrium in ^{226}Ra with its daughter ^{222}Rn.

2.4. EFFICIENCY CALIBRATION METHODS USED FOR WELL TYPE HPGe DETECTOR

The efficiency calibrations of the well-type HPGe detector were performed for three different sized cylinders in the close geometry conditions, as shown in Figure 1. The used cylinder beakers named as (a) large cylindrical, (b) small cylindrical and (c) vial have the dimensions: 5.9, 3.6 and 3.5 cm in height, and 5, 4.3 and 1.4 cm in diameter, respectively. Two approaches were used for the efficiency calibrations.

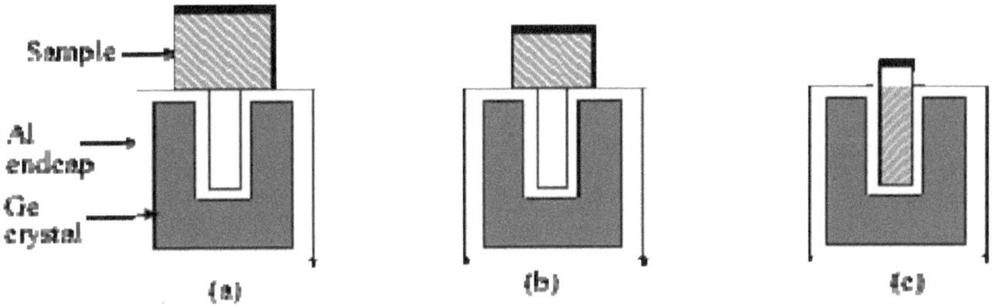

Figure 1. Schematic drawing of measurement setup: (**a**) large cylindrical beaker; (**b**) small cylindrical beaker; (**c**) vial

2.4.1. *Radionuclide specific efficiency calibration*

The first method is called the radionuclide specific photopeak efficiency (RSE) calibration that uses only the efficiency values of the specific radionuclides present both source and sample. Therefore, its applicability is limited only for the specific radionuclides exist in the used reference materials. In this work, the solid Certified Reference Materials (CRMs) such as IAEA-RG sets were chosen for this purpose because IAEA-RG sets as CRMs are commonly used and their properties are well known. From the measured gamma-ray spectra with IAEA-RG set, the full energy peak efficiency values have to be determined only for energies of gamma rays of the radionuclides present in the CRMs used. Then the determined efficiency values are used directly for the activity calculation of radionuclides contained in the samples counted in the same geometry as well as CRMs. Thus, this allows cancel out the true coincidence summing out/in effects in same magnitudes due to the specific gamma rays of some radionuclides measured in the close geometry conditions. Further, radionuclide specific efficiency calibration is a more simple way since it does not need to interpolation or extrapolation of the measured efficiency data. However, other correction factors such as self-absorption effects due to the differences in matrices and densities between samples and CRMs must be taken into account in the activity calculation [3].

2.4.2. *Absolute (full energy peak) efficiency calibration*

As is known, the main drawback of radionuclide specific efficiency (RSE) calibration is that it is not always possible to find appropriate CRMs, which are identical to both sample matrix and the radionuclides existing in the samples. Therefore, the more commonly used method is known as the absolute efficiency calibration or full-energy peak (photopeak) efficiency calibration (FEP) [4] with use of the mixed radionuclides or certified multinuclide standards. This second method also enables almost accurate efficiency results for the unknown gamma ray energies of radionuclides in the samples if suitable fitting procedure is applied to the measured efficiency data. In this procedure, it is emphasized that the interpolation or extrapolation of the measured efficiency data can lead to somehow serious errors in the fitted data especially for low energy region of the efficiency curve where the lack of sufficient measured points. The main limitation of absolute efficiency calibration method is that true and random coincidence losses cause substantial deviations in the experimental points especially in case of close geometry conditions, thus demanding the correction factors for the measured peak areas by exploiting the sophisticated procedures.

3. Results and Discussion

The photopeak efficiency at certain gamma-ray energy and sample geometry was calculated by:

$$\varepsilon(E) = \frac{[N_c / t_c - N_b / t_b]}{A \cdot f_\gamma(E)} \cdot F_i \qquad (1)$$

where $\varepsilon(E)$ is the efficiency; N_c , N_b and t_c, t_b are the gamma-ray peak areas and the counting times of the calibration source spectrum and background spectrum respectively; A is the activity of calibration source (Bq), $f_\gamma(E)$ is the gamma ray emission probability (gammas $s^{-1}Bq^{-1}$) and F_i are the correction factors for self-absorption, F_a and the factor for true coincidence losses (F_b), etc. The efficiency results for cylindrical geometries shown in Figure 1 were corrected for the experimentally obtained self-absorption factors from the simple photon transmission experiments by using point-gamma sources [4]. From the measured peak count rates for Figure 1a and b in uncollimated beam conditions, the self absorption factor, F_a for a given energy was calculated using the following equation

$$F_a = ln\left(\frac{N}{N_0}\right) \Big/ \left(\frac{N}{N_0} - 1\right) \qquad (2)$$

where N_0 is the count rate in the photopeak due to the source without attenuation and N is the count rate with attenuation.

Besides, when the exact composition of the sample was known, self-absorption factors for a vial in the detector well were expressed in terms of mass attenuation coefficient (cm^2 g^{-1}) of the sample and a parameter characterizing the well geometry [5, 6]. If L is the length of the sample and m is the sample mass, the emission ratio was calculated by

$$F_a = N / N_0 = 2e^{-k\mu m}\left\{\frac{1}{k\mu m}\sinh(k\mu m) - \frac{1}{(k\mu m)^2}(\cosh(k\mu m) - 1\right\} \quad (3)$$

where $k = 1/\pi aL$ is a geometric parameter and a is the radius of the sample. The calculated efficiency curve was expressed by the equation:

$$\varepsilon(E) = \sum_{i=0}^{n} a_i \cdot (\ln(E))^i \quad (4)$$

where a_i, are fitting constants of a fourth degree logarithmic polynomial to the weighted measured points from three replicate measurements.

According to the radionuclide specific efficiency calibration method described in Section 2.4.1, the experimental efficiency values for the gamma-ray energies used for activity measurement of the samples are presented in Table 3.

According to the absolute efficiency calibration method described in Section 2.4.2, the absolute efficiency curves, covering 40 to about 2,000 keV energy region for a well-type HPGe detector are illustrated in Figure 2 for the large cylindrical beaker, in Figure 3 for the small cylindrical beaker, and in Figure 4 for a vial measured in the well. The experimental uncertainties in the measured efficiency points are also indicated in the figures. The bottom figures show the percentage differences between the experimental and calculated efficiency values. When using a mixed radionuclide standard, several gamma lines are not included in the efficiency calibration curves, either because of their poor counting statistics (e.g. 320 keV of ^{51}Cr) or the interference with the annihilation line at 511 keV (e.g. 514 keV of ^{85}Sr). Eventhough they are already given in its certificate.

The activity results for ^{40}K, ^{226}Ra and ^{232}Th radionuclides were measured in three different sample geometries for several types of samples. The specific activity of ^{40}K was measured directly by its own gamma-ray at 1,460.8 keV (10.7%), while activities of ^{226}Ra and ^{232}Th were calculated based on the average activities of their respective decay products in equilibrium with parent. The specific activity of ^{226}Ra was measured using the 295.2 keV (18.2%), 351.9 keV (35.1%) keV gamma rays from ^{214}Pb and the 609.3 (44.6%), 1,120.3 (14.7%) keV from ^{214}Bi.

The specific activity of ^{232}Th was measured using the 338.4 (11.3%), the 911.2 (26.6%) keV from ^{228}Ac and 583.2 (corrected for branching: 30.6%) keV from ^{208}Tl. The activities were computed by Equation (1). The measured activities obtained from the peaks ^{214}Pb and ^{214}Bi were averaged to report as the equivalent activity for ^{226}Ra, and similarly the measured activities from ^{228}Ac and ^{208}Tl were averaged to report as the equivalent activity for ^{232}Th. The obtained activities for each sample were then divided by dry sample weight, m (kg) and corrected for self-absorption effects, F_a. Additionally, since 1,460.8 keV γ-ray (10.7%) of ^{40}K and 1,459.2 keV γ-ray (0.83%) of ^{228}Ac is formed a mixed peak in case sample contained thorium, the measured ^{40}K activity was corrected for this contribution [7] by the relation:

$$A_K(\text{corrected}) = A_K - 0.093 x A_{Th} \tag{5}$$

where A_K and A_{Th} are the measured activities for ^{40}K and ^{232}Th, respectively.

At 95% confidence level, the MDA was calculated by the known Currie equation [8]:

$$MDA = \frac{2.71 + 3.29\sqrt{B(1 + n/2m)}}{\varepsilon(E)\cdot f_{\gamma}(E)\cdot t\cdot w} \tag{6}$$

where B is the is the background area under the peak; n is the number of channels in the peak region of interest; m is the number of background channels on each side of peak; $\varepsilon(E)$ is the photopeak efficiency of the peak; $f_\gamma(E)$ is the gamma-ray emission probability; t is the counting time in seconds and w is the dried sample weight expressed in kg.

Thus, the experimentally determined activity concentrations (Bq kg^{-1}) of the nuclides for several samples in three different counting geometries mentioned above are presented in Table 4. The mean of measured activity concentrations were obtained by using the efficiency values calculated from the FEP efficiency curves with uncorrected and corrected for self-absorption and obtained by using the experimental efficiency values uncorrected and corrected for self-absorption from the radionuclide specific efficiency method.

The MDAs in large cylindrical beaker are 1 Bq kg^{-1} for ^{226}Ra, 1 Bq kg^{-1} for ^{232}Th and 7 Bq kg^{-1} for ^{40}K, and those in a small cylindrical beaker are 2 Bq kg^{-1} for ^{226}Ra, 1 Bq $kg^{-1}$for ^{232}Th and 9 Bq kg^{-1} for ^{40}K and those in a vial measured the well are 3 Bq kg^{-1} for ^{226}Ra, 2 Bq $kg^{-1}$for ^{232}Th and 22 Bq kg^{-1} for ^{40}K, respectively.

TABLE 3. The radionuclide specific efficiency results

				Radionuclide specific efficiency (%)					
CRM code	Nuclide	E (keV)	f_γ (%)	Large cylindrical beaker		Small cylindrical beaker		Vial	
				$\varepsilon_e \pm u(\varepsilon_e)^1$	$\varepsilon_e \pm u(\varepsilon_e)^2$	$\varepsilon_e \pm u(\varepsilon_e)^1$	$\varepsilon_e \pm u(\varepsilon_e)^2$	$\varepsilon_e \pm u(\varepsilon_e)^1$	$\varepsilon_e \pm u(\varepsilon_e)^2$
RGU-1	^{214}Pb	295.2	18.2	3.70 ± 1.72	2.52 ± 0.28	4.79 ± 1.65	3.82 ± 0.28	32.49 ± 0.29	30.11 ± 0.29
RGTh-1	^{228}Ac	338.4	11.3	3.27 ± 0.89	2.28 ± 0.70	4.12 ± 1.07	3.27 ± 0.67	29.41 ± 0.68	25.86 ± 0.68
RGU-1	^{214}Pb	351.9	35.1	3.23 ± 1.72	2.25 ± 0.26	4.19 ± 1.65	3.39 ± 0.30	27.27 ± 0.31	25.39 ± 0.31
RGTh-1	^{208}Tl	583.2	30.6	1.80 ± 0.85	1.33 ± 0.65	2.25 ± 1.05	1.85 ± 0.64	8.08 ± 0.67	7.48 ± 0.67
RGU-1	^{214}Bi	609.3	44.6	1.80 ± 1.72	1.34 ± 0.29	2.29 ± 1.66	1.93 ± 0.35	8.51 ± 0.39	8.04 ± 0.39
RGTh-1	^{228}Ac	911.1	26.6	1.35 ± 0.95	1.05 ± 0.77	1.70 ± 1.12	1.45 ± 0.75	8.30 ± 0.77	7.84 ± 0.77
RGU-1	^{214}Bi	1,120.2	14.7	1.12 ± 1.75	0.88 ± 0.41	1.40 ± 1.67	1.23 ± 0.37	4.38 ± 0.61	4.19 ± 0.61
RGK-1	^{40}K	1,460.8	10.7	0.97 ± 2.82	0.75 ± 0.42	1.18 ± 1.01	1.06 ± 0.38	6.67 ± 0.32	6.37 ± 0.32

[1]Measured efficiency results corrected for self-absorption
[2]Measured efficiency results uncorrected for self-absorption

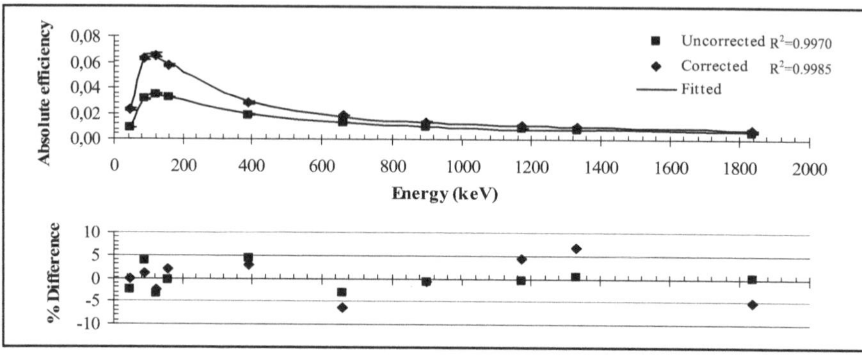

Figure 2. Absolute efficiency calibration of the HPGe detector for large cylindrical beaker, uncorrected and corrected for self-absorption

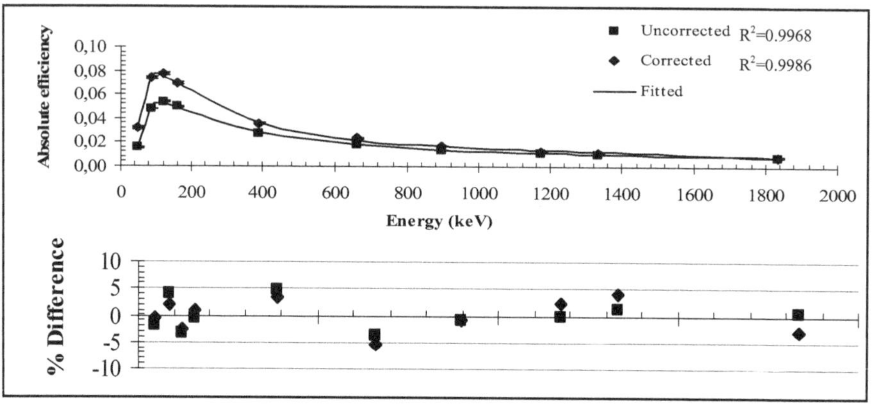

Figure 3. Absolute efficiency calibration of the HPGe detector for small cylindrical beaker, uncorrected and corrected for self-absorption

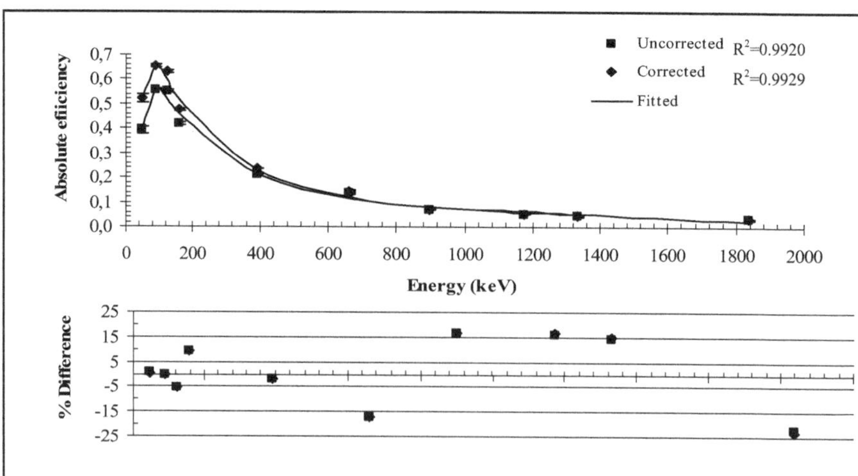

Figure 4. Absolute efficiency calibration of the HPGe detector for a vial in detector well, uncorrected and corrected for self-absorption

TABLE 4. Activity concentrations of the samples

Sample	Method	Large cylindrical beaker A ± u(A)[1] (Bq kg^{-1})			Small cylindrical beaker A ± u(A)* (Bq kg^{-1})			Vial A ± u(A)* (Bq kg^{-1})		
		^{226}Ra	^{232}Th	^{40}K	^{226}Ra	^{232}Th	^{40}K	^{226}Ra	^{232}Th	^{40}K
Tufa 1	FEP	78 ± 2	81 ± 2	1,107 ± 22	79 ± 4	79 ± 2	1,127 ± 20	75 ± 4	66 ± 5	1,631 ± 28
	RSE	77 ± 3	82 ± 2	1,057 ± 36	76 ± 4	81 ± 2	1,035 ± 17	70 ± 4	61 ± 4	1,131 ± 18
Tufa 2	FEP	72 ± 2	76 ± 2	904 ± 17	74 ± 3	73 ± 2	908 ± 20	39 ± 6	34 ± 6	773 ± 28
	RSE	72 ± 2	77 ± 2	864 ± 21	72 ± 3	76 ± 2	837 ± 17	37 ± 6	32 ± 6	545 ± 19
Soil	FEP	6 ± 1	<MDA	<MDA	6 ± 3	<MDA	<MDA	16 ± 3	<MDA	<MDA
	RSE	7 ± 1	<MDA	<MDA	6 ± 3	<MDA	<MDA	14 ± 2	<MDA	<MDA
Marble	FEP	6 ± 1	<MDA	<MDA	8 ± 2	<MDA	<MDA	15 ± 3	<MDA	<MDA
	RSE	6 ± 1	<MDA	<MDA	7 ± 2	<MDA	<MDA	14 ± 3	<MDA	<MDA

[1]Estimated uncertainty, based on one standard deviation (±1σ)
FEP: Full energy peak efficiency (absolute) method
RSE: Radionuclide specific efficiency method

4. Conclusion

A general empirical fits to full energy peak efficiency calibrations for the extended sources placed on the endcap and a vial measured in the well. The fitted results yielded the accuracies of less than 5% on average for the cylindrical geometries, which make it suitable for routine measurements of environmental samples. The chosen efficiency function $\varepsilon(E) = \sum_{i=0}^{n} a_i \cdot (\ln(E))^i$ gives the best fitted efficiency values for a 40–2,000 keV energy range for the three counting geometries. However, the differences between experimental and calculated efficiency values for the vial measured in the well-type of HPGe detector are very large ranging from 23% to 16% especially for some radionuclides due to true coincidence effects. This was attributed to mainly the true coincidence summing effects because significant deviations were observed for gamma-ray energies of ^{60}Co and ^{88}Y which have well known coincidence effects, in the vial measured in the detector well.

The self-absorption effect due to varying matrices and densities were seen from the efficiency results. Particularly for low energy range (<200 keV) measured efficiency values increased regarding to self-absorption correction. Therefore the self absorption correction factors were determined experimentally in the cylinder beakers measured on the endcap and theoretically calculated from the mass attenuation coefficients. Additionally, an efficiency model called radionuclide specific efficiency calibrations were tested for three different close counting geometries by use of certified reference materials (IAEA/RG set). This model gave also very accurate results for the cylindrical geometries placed on the endcap. However, the obtained activity results from the vial geometry measured in the detector well are generally erroneous for some radionuclides when the ratios of the measured/certified activities are compared. Since this model seems to be useful for specific radionuclides in environmental measurements when the same radionuclides are found both in sample and certified reference standard. Then, few environmental samples such as tufa, soil and marble were measured at these three different counting geometries. The activity results from cylindrical counting geometries for these samples are close each other within experimental uncertainty limits. This study shows that the efficiency calibration of a detector is still a laborious and tedious work and there are serious problems in the close counting geometries due to mainly true coincidence summing effects and low counting statistics.

Acknowledgement: Authors would like to thank the Directorate of Institute of Nuclear Sciences of Ankara University for his support to this study. NATO ATC organizing committee provided financial support to present this paper.

References

1. Nir El Y (1998) Application of reference materials in the accurate calibration of the detection efficiency of a low-level gamma-ray spectrometry assembly for environmental samples. J Radioanal Nucl Chem 227:67–74
2. Alfassi ZB (2000) In-house absolute calibration of gamma-emitting radioactive voluminous samples: detector setups using natural radionuclides. J Radioanal Nucl Chem 245:561–565
3. Debertin K et al. (1988) Gamma- and X-ray Spectrometry with Semiconductor Detectors. Amsterdam: North-Holland, 395 pp.
4. Cutshall NH et al. (1983) Direct analysis of ^{210}Pb in sediment samples: Self-absorption corrections. Nucl Instrum Meth 206:309–312
5. Appleby PG et al. (1992) Self-absorption corrections for well-type germanium detectors. Nucl Instrum Meth B 71:228–233
6. Yucel H et al. (1998) Use of the 1001 keV peak of ^{234m}Pa daughter of ^{238}U in measurement of uranium concentration by HPGe gamma-ray spectrometry. Nucl Instrum Meth A 413:74–82
7. Lavi N et al. (2004) On the measurement of ^{40}K in natural and synthetic materials by the method of high-resolution gamma-ray spectrometry. Radiat Meas 38:139–143
8. Gilmore G et al. (1995) Practical Gamma-Ray Spectrometry. Chichester: Wiley, 322 pp.

ENVIRONMENTAL MONITORING AT KFKI CAMPUS

LÁSZLÓ SÁGI AND ATTILA NAGY*
Atomic Energy Research Institute, Budapest, Hungary

Abstract. In the paper the instrumentation of Environmental Protection Service are summarized. The equipments used in on line and off-line monitoring are presented. The calibration facilities and system of personnel dosimeter also are demonstrated.

Keywords: Environmental monitoring, emergency preparedness

1. Introduction

At Central Research Institute of Physics (KFKI) campus, the Budapest Research reactor and the Institute of Isotope Co. Ltd. dealing with production of wide variety of radioactive isotopes is in operation. Controlling the radioactive contamination environmental monitoring stations has been installed. The stations are equipped to monitor the dose rate, the air contamination. For controlling the internal exposure the whole body counting instrument is in operation. The Environmental Protection Service is also responsible for personal dosimeter. For the emergency situations a well equipped mobile laboratory is in the alert.

* To whom correspondence should be addressed. e-mail: sagi@aeki.kfki.hu

2. Methods

2.1. ON LINE DATA COLLECTION

The reconstructed Data Collecting System processes every minute and then visualizes:

- Dosimetrical parameters in the field (dose rate from 17 GM monitors on the KFKI Campus)
- Release parameters from the Budapest Research Reactor (stack release monitoring)
- Meteorological data (wind speed, wind direction, and precipitation, etc.)

2.2. OFF-LINE DATA COLLECTION

The radioactive contamination is monitored by:

- Air sampling
  (at four stations by gross beta measuring)
- Dust water sampling
  (by gross beta measuring)
- Fallout sampling
  (by gross beta measuring)

2.3. IRRADIATION LABORATORY

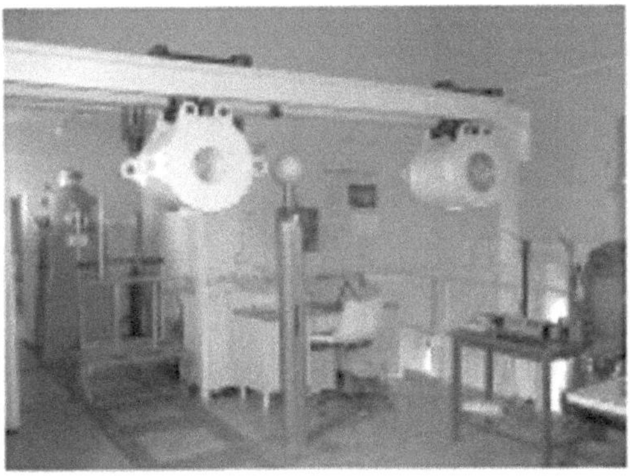

For the calibration of different types of detectors (gamma, beta and neutron), five irradiation instrument are utilized. Their control electronics were modernized this year. The main users are various groups from AEKI's Health Physics Department.

2.4. EXTERNAL DOSIMETRY

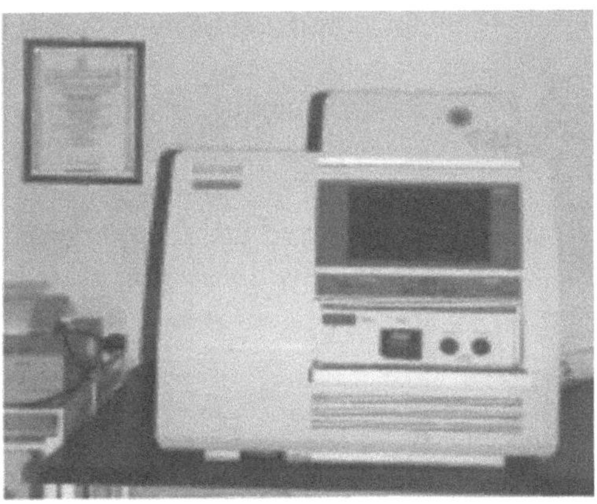

People dealing with or exposed to ionizing radiation are equipped with film and TL dosimeters. Evaluation of TL dosimeter is done with the instrument above.

2.5. INTERNAL DOSIMETRY
(Whole body counting)

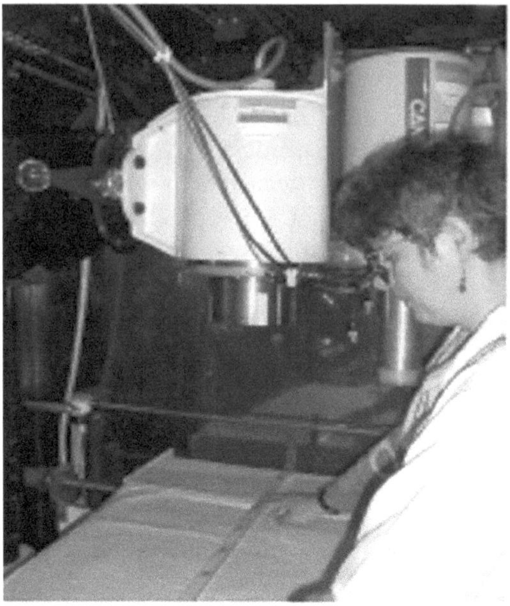

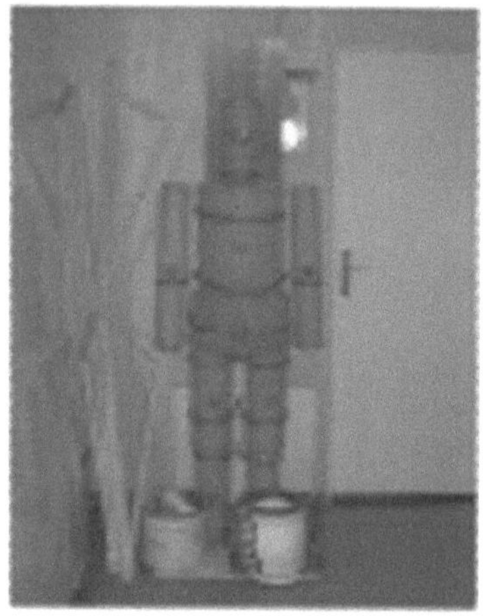

Incorporation of radionuclides involves measuring by whole body counting instrument at very low background. The whole body doses are evaluated and are then archived in at Individual's Report.

2.6. MOBILE LABORATORY

In emergency situations the mobile laboratory is of special importance. With a view to evaluating radioactive contamination, this vehicle equipped with an in situ gamma spectrometer, a thyroid monitor, a survey meter, an air sampling unit and a GPS monitor.

2.7. NEW DEVELOPMENTS

Parallelly of upgrading the environmental system of Paks NPP, a new so called reference station has been installed in KFKI campus in last year. The reference station equipped by:

Aerosol sampling

Continuous monitoring

- By plastic scintillator: for elementary, aerosol and organic form of iodine
- By NaI scintillator: Continuous gamma spectra

Meteorological station

- Wind parameters and precipitation

3. Results

3.1. ONLINE MEASURING

The dose rates measured by the reconstructed Data Collecting System are recorded in every 10 min. The processed data for the year of 2007 are presented in Table 1.

3.2. OFF-LINE MEASURING

Gross beta measuring: The radioactivity concentration monitored by air sampling at the second station by gross beta measuring were found not to exceed the value of 1.0E+00 Bq/m^3, from January to June and from July to December 2007.

TABLE 1. Dosimetrical data for the year of 2007 in the field (dose rate from 17 GM monitors on the KFKI Campus)

No. of station	Number of data (for 10 min)	Number of valid data		Out of operation		Failure		Average	Standard deviation
		Number	(%)	Number	(%)	Number	(%)	(nGy/h)	σ
1	52,705	52,538	99.7	158	0.3	9	0.017	103.95	4.74
2	52,705	52,546	99.7	158	0.3	1	0.0018	109.19	5.01
3	52,705	52,531	99.66	158	0.3	16	0.03	119.7	4.44
4	52,705	52,547	99.7	158	0.3	0	0	135.7	4.65
5	52,705	52,206	97.2	1,499	2.8	0	0	104.12	4.49
6	52,705	52,547	99.7	158	0.3	0	0	120.54	29.4
7	52,705	52,547	99.7	158	0.3	0	0	124.44	5.88
8	52,705	52,942	43.5	29,762	56.5	1	0.0018	142.8	67.2
9	52,705	52,546	99.7	159	0.3	0	0	116.7	7.8
10	52,705	52,546	99.7	159	0.3	0	0	125.6	22.1
11	52,705	52,546	99.7	159	0.3	0	0	125.1	6.75
12	52,705	52,546	99.7	159	0.3	0	0	113.17	4.7
13	52,705	52,545	99.7	159	0.3	1	0.0018	117.6	55.2
14	52,705	52,546	99.7	159	0.3	0	0	533.8	49.8
15	52,705	52,544	99.7	159	0.3	2	0.0037	114.8	5.3
16	52,705	52,545	99.7	159	0.3	1	0.0018	112.2	15.6
17	52,705	52,544	99.7	159	0.3	2	0.0037	109.74	7.35

Gamma spectrometry:

- Fall out sampling: The Be-7 radioactivity concentration monitored by fall out sampling at the second station by gamma spectrometry were found between 4.0E+00 and 7.0E+01 Bq/m^2.
- High volume air sampling: The ^{125}I and ^{131}I (elementary end aerosol forms) radioactive concentration monitored by air sampling at the first station by gamma spectrometry. The radioactivity concentrations for ^{125}I were found to be between 2.0E-05 and 1.5E-02 Bq/m^3. The radioactivity concentrations were found to be between 2.0E-05 and 3.0E-02 Bq/m^3 for ^{131}I. For aerosol form, these values were found as; 2.0E-05 and 2.0E-02 Bq/m^3 for ^{125}I and 1.0E-05 and 5.0E-04 Bq/m^3 for ^{131}I.

4. Conclusions

All of part of environmental monitoring system and personal dosimetry at the KFKI Campus are well elaborated. By this the internal and external control of exposure of personnel and the population dependably unbound and is a good practice for a nuclear installation.

Acknowledgement: It is grateful acknowledge to Professor Gul Asiye AYCIK (Director from NATO country) for organizing the Advanced Training Course in the frame of "The NATO Science for Peace and Security Programme". It is to thank for NATO for supporting the participation on the course arranged on excellent place in Turkey.

THE PROBLEM OF VULNERABLE IONIZING RADIATION SOURCES IN REPUBLIC OF MOLDOVA

ELENA MURSA*
National Agency for Regulation of Nuclear and Radiological Activities (NARNRA), 1, A. Russo Street, Chisinau, MD2068, Republic of Moldova

Abstract. The reviews of radiological accidents showed, that in many cases, the root cause was the weak control of the source for a certain period of time before it became orphaned. In the Republic of Moldova basically, physical protection and safety of powerful radioactive sources are provided with passive protection measures. Because of financial crisis, some organizations are not capable of providing the appropriate safety of radioactive sources at the fullest extent. Such, have been reported some worrying cases of vulnerable powerful radioactive sources in Moldova which easily could become "orphans", also some incidents spend with radioactive samples and radioactively contaminated materials. The instruments used in Moldova for detecting illicit movement of radioactive materials are the portable and stationary detectors in Laboratory System of Radiological Monitoring.

Keywords: Radioactive sources, illicit trafficking, monitoring and survey instruments

1. Introduction

The work of managing radioactive sources program in the Republic of Moldova is provided under financial support of international organizations. Moldova is a signatory to six International Conventions and has arrangements for the exchange of information relating to radiological emergencies with several neighboring countries. The practice shows that a significant benefit in detection and prevention of occurrence of "orphan" radioactive sources can render only from international cooperation, through exchanging

* To whom correspondence should be addressed. e-mail: elmour_l@mail.ru

G.A. Aycik (ed.), *New Techniques for the Detection of Nuclear and Radioactive Agents*,

information about radiation source, nuclear materials and device characteristics, suppliers, and import/export records.

Despite potential danger of radioactive sources to those who enters close contact to them, they are safely used in a daily medical practice on care of patients and treatment alongside with other applications in the field of the industry, agriculture and a science.

A significant number of serious radiological accidents in the world have involved uncontrolled, so-called "orphan" radioactive sources, which has never been under regulatory control or has been abandoned, lost, misplaced, stolen or otherwise transferred without proper authorization. The reviews of radiological accidents showed, that in many cases, the root cause was the weak control of the source for a certain period of time before it became orphaned. The world's metal recycling industries have been particularly vulnerable to the Orphan sources can find their way into metal scrap destined for recycling. People who find them, attracted by the prospect of economic gain, sometimes sell the source for its metallic value to scrap dealers who usually are not aware of the radioactive content. Thus, the source enters into the worldwide scrap inventory which, because of the latest global opening of the markets, has become essentially uncontrollable. More than 2,300 reports of sources found in scrap metal are stored in the NRC's database. Sometimes, it is known that radiation sources were melted after detection of radiation contamination in imported commodities. The NRC has detected a number of such cases [1]. Orphan radioactive sources have caused fatal or serious injuries to the unknowing individuals who have found them. Many factors can lead to loss of control of radioactive sources, including: ineffective regulations and regulatory oversight; the lack of management commitment or worker training; poor source design; poor physical protection of sources during storage, transport and use; abandonment due to economic factors; as well as theft or other malicious acts.

The main institutions in Republic of Moldova that own and maintain powerful radioactive sources are: Republican Oncology Institute of the Ministry of Health installations for teletherapy, brachtherapy and diagnostic used for medical purposes; Scientific Research Institutes of the Academy of Sciences; the radioactive waste storage.

Basically, physical protection and safety of powerful radioactive sources are provided with passive protection measures, like restricted access to places with installations with radioactive sources. Almost all radioactive sources are placed at specially designed premises, which are supplied with security signaling systems, and in some use security guards (Storage of radioactive waste products, Moldavian – Joint Enterprise CEMA).

Due to sharp financial crisis, some organizations are not capable of providing the appropriate safety of radioactive sources at the fullest extent. This especially concerns the enterprises of processing industry and scientific research institutes which inherited production and experimental lines and installations with radioactive sources and also objects on conditioning and storing of radioactive waste products which in the majority cases, are not used. In particular, during 2001–2002 there have been reports of lost low activity sources (up to 1 mCi) of Cs-137. Search of these lost sources have not led to any success.

2. Experimental

The politics of regulation of radiological activities in every country must ensure: the protection of individuals and the population as a whole against the radiation exposure that they are expected to incur as a result of the normal uses of radiation sources; the safety of the radiation sources in order to prevent the occurrence of accidents and, should they nevertheless happen, to mitigate their consequences; and, the security of the radioactive materials in order to prevent any relinquishing of control over their use [2].

The effective detection range depends on the amount and type of radiation emitted by the source and also on the possible presence of shielding materials that may reduce the amount of radiation that reaches the detector. Fortunately, the most intense and dangerous sources normally are the most susceptible to detection [3]. Several types of instruments already are in use in Laboratory System of Radiological Monitoring from Moldova for detecting illicit movement of radioactive materials:

- Portable detectors "Fieldspeak", "IdentiFinder" using an inorganic scintillator as detector. They include a multi-channel analyzer for gamma spectroscopy to identify radioactive materials emitting gamma rays; γ-pagers – 20 units for Border Crossing Stations, NaI multichannel analyzer "EasySpec" (CANBERRA), Survey meter "Ludlum"-2241-2 (α, β, γ-detector), portable contamination monitor "CONTAMAT FHT 111M" (α, $\beta+\gamma$ surface contamination), Manual Sample Changer "FHT 770K/57" (low α, β activity), Roentgen meter – radiometer DP-5B (β, γ low resolution), γ-survey meter DP64 (car mounted).
- Two stationary gates for car monitoring designed for border control points, Figure 1. They typically include alarms and display instrumentation and are automated to enable screening of people, luggage, or vehicles.

Figure 1. Stationary gate for car monitoring

In 2007, a new TACIS program, under which the Customs Point will be provided by other new advanced detectors, has been started, in Moldova.

A worrying case of vulnerable powerful radioactive sources in Moldova which easily could become "orphans":

The large Cesium-137 sources "Gamma-KOLOS" used in agricultural activities. Originally, these sources were mounted on trucks, Figure 2.

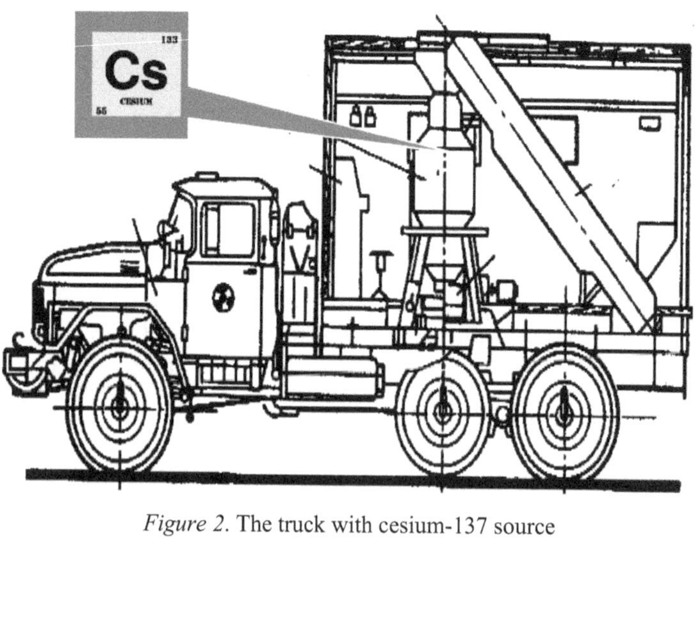

Figure 2. The truck with cesium-137 source

Many were stored in precarious conditions such the case with five KOLOS irradiators, each of them with 13 canals. Every canal contained five Cs-137 RS (65 radioactive sources) with total activity ^{650}Ci stood at open-air on storage, Figure 2.

In 2004 under the "Tripartite Initiative" between the International Atomic Energy Agency, the Russian Federation and the USA on securing and managing radioactive sources, in the Republic of Moldova 92 sealed radioactive sources (inclusive "Gamma-KOLOS") from 11 sites were collected, transported, stored and placed into the new storage facility, Figure 3.

Figure 3. Collecting and transportation of radioactive sources

Another case of the vulnerable source were Gamma-field Complex with two high power Co-60 sources – "OGUP" with activity ^{240}Ci and "Flora-M" with activity ^{643}Ci, which ware without of any protection at Agrarian University, Figure 4.

Figure 4. Gamma-field complex with two Co-60 sources

In 2002, with financial support of IAEA, the problem of dismantled and transported at storage facility ware solved. For improvement of radiation safety condition at waste storage, according to the contract with Department of Energy of the USA, on the basis of the Base Standard Contract 5493-A-D2, have been performed following works:

- The new elevated warehouse has been constructed of a monolith for storage of the high activity radiation sources, such as the equipment "Gamma Field", "KOLOS" and others.
- Repair of an office building is made and is attached an additional building for the personnel of protection.
- The signal system and video-monitoring equipment in a warehouse of storage with radioactive sources are mounted.
- A power plant (diesel engine) to supply electricity.

Some incidents spend in Moldova with radioactive samples or radioactively contaminated materials such as contaminated scrap metal also have been reported to the illicit trafficking database and are included in the statistics:

- 1999 RW Repository ^{60}Co source found (activity < 13 Ci).
- 2000 CCP "Otaci" vehicles caring mineral fertilizers (40 K) was stopped (August, September, November).
- On 2.01.2001 CCP "Sculeni" empty vehicle from Belts coming back from Romania had ^{137}Cs pollution (deactivated by a CP team).
- On 3.05.01 missing of two ^{137}Cs sources from Falesti sugar factory was detected (1 mCi).
- On 20.06.01 missing of two ^{137}Cs sources from Balti ironed concrete factory detected.
- On 05.09.02 contaminated metal with ^{60}Co (181 t) on the metal factory. The source has been included in technological process an alloy of metal. This metal has been sent abroad. As a result of it have returned back at the enterprise where it is stored in a special protected place.
- September 2004 CCP "Tudora" ceramic furniture for baths from China.

Notwithstanding what the happened cases in Tudora and Otaci do not represent illicit trafficking or pollution in consequence of the lost source, they show that at customs the radiation control of the goods is conducted.

The detected trafficking incidents appear to involve opportunists or unsophisticated criminals, motivated by the hope of profit.

In 2005, the IAEA carried out the mission to appraise the effectiveness of national regulatory infrastructures governing both the safety and security of radioactive sources and to promote the adoption of information systems to manage source inventories and control systems.

Moldova is a signatory to six International Conventions; Early Notification of a Nuclear Accident, Physical Protection of Nuclear Material, Assistance in the Case of a Nuclear Accident or Radiological Emergency, Convention on Nuclear Safety, Civil Liability for Nuclear Damage and the Nuclear Weapons Non-Proliferation Treaty. In 2007, Moldova has joined the Illicit Trafficking data base maintained by IAEA. Moldova has arrangements for the exchange of information relating to radiological emergencies with Romania, Ukraine and Belarus, exchange of safety-related information with Romania and a quadrilateral agreement on transport of nuclear material with Ukraine, Russia and Bulgaria.

3. Conclusion

Moldova has proper legislation and regulations in place to deal with the safety of radioactive sources and the security of radioactive materials, and they have established independent regulatory authorities able to license sources and enforce requirements. A significant benefit in detection and prevention of occurrence of "orphan" radioactive sources can render only from international cooperation, through exchanging information about, radiation source, nuclear materials and device characteristics, suppliers, and import/export records.

Acknowledgement: I am thankful to NATO ATC organizing committee for the financial support to make this presentation possible.

References

1. Gonzalez AJ (1999) Strengthening the safety of radiation sources and the security of radioactive materials: timely action, IAEA Bulletin
2. International Basic Safety Standards for Protection Against Ionizing Radiation and for the Safety of Radiation Sources (1996) IAEA, SS-115
3. Ortiz P, Friedrich V, Wheatley J, Oresegun M (1999) Lost and found dangers orphan radiation sources raise global concern, IAEA Bulletin

MEASUREMENT OF NATURALLY OCCURRING RADIONUCLIDES WITH SEVERAL DETECTORS: ADVANTAGES AND DISADVANTAGES

CONSTANTIN PAPASTEFANOU*
Aristotle University of Thessaloniki, Atomic and Nuclear Physics Laboratory, Thessaloniki 54124, Greece

Abstract. This section lecture will review the interaction of radiation with matter to explain spectral features and their interpretation, including peak identification and energy determination, backscattering peaks, single and double escaping peaks and proper use of control charts of radionuclides. It is designed to provide a practical introduction to gamma-ray spectroscopy for those new to the field of gamma-ray spectroscopy, but also provide practical applications to those who are currently performing gamma-ray spectroscopy. It is intended for anybody who will be doing routine and specialized gamma-ray spectroscopy, as well as quality assurance officers and data validators who may have a need to understand gamma-ray spectroscopy measurements.

Keywords: NORM, TENORM, scintillation detectors, solid state detectors, semi conductor detectors, gamma-ray spectroscopy, radiation measurements in-situ

1. Introduction

Measurement of naturally occurring radionuclides in fields (in-situ) or in Lab (in-vitro) is made by sodium iodide thallium activated, NaI(Tl) detectors and/or solid state (semiconductor diode) HPGe detectors coaxial, planar or well type by using various spectroscopic systems linked with accumulation data processing units. For spectral analysis of gamma-ray photon peaks, various libraries with softwares have been developed and established and are available to the users. Those gamma-ray spectrometers are classified according to their resolution and the efficiency in detecting

* To whom correspondence should be addressed. e-mail: papastefanou@physics.auth.gr

G.A. Aycik (ed.), *New Techniques for the Detection of Nuclear and Radioactive Agents*,

and measuring the radionuclide concentrations through the gamma-ray peaks of the gamma-ray photons emitted by the naturally occurring and/or man-made produced radionuclides in various kinds of environmental samples.

2. Scintillation Detectors

The more usual way of detection and measurement of gamma-rays for a long time period in the decades of 1950s–1970s was by using a scintillation detector [1–5]. The scintillation detector spectroscopy system is consisted of a scintillation crystal of sodium iodide thallium activated, NaI(Tl) of various sizes, a photomultiplier, a series of nuclear electronic instruments, such as high voltage bias supply, a linear spectroscopic amplifier, a pulse counting unit which could be a single channel analyzer, SCA or multichannel analyzer, MCA . In the decades of 1980s and 1990s, a PCA card started using for analyzing the gamma-ray spectra in PC computing facilities [6] and in nowadays picospec digital multichannel analyzer units, dMCA-pro-card are used [7]. These units include a spectroscopy amplifier, a digital multichannel analyzer and a high voltage bias supply altogether in one unit.

The main advantages of the scintillation detectors, NaI(Tl) against the other type detectors, e.g. of Ge detectors used for detection and measurements in the gamma-ray spectroscopy are the following:

- High efficiency
- Possibility of counting of a high number of events of gamma-ray decays
- Possibility of measurement of gamma-ray photon energy with a satisfying accuracy
- Low cost of scintillation detectors against the solid state (semiconductor diode) Ge detectors

This of course, implies bad resolution which compensates for the high efficiency.

When a gamma-ray photon falls onto the scintillation crystal NaI(Tl) and inserts in its volume, it causes the excitation effect. That means, the photon interacts with the matter of crystal by

- The photoelectric effect
- The Compton effect and
- The pair production effect

The final result of excitation of the scintillation crystal is the emitting of photons (fluorescent light photons) by the crystal which is lead to the photocathode of a photomultiplier, PM by contributing of the reflector surrounding the scintillation crystal except one side, i.e. the side of connection of crystal with the PM. The fluorescent light photons have a wavelength ranging from 3,900 to 5,100 Å. The number of the emitted fluorescent light photons is proportional to the energy of the gamma-ray photon which fell onto the crystal and absorbed in the crystal matter by the electrons of the previously mentioned three effects. These fluorescent light photons after their production fall onto the photocathode of PM, which has a small work function and extract or produce photoelectrons which are then multiplied by the dynodes of PM. Finally, in the anode of PM arrive 10^6 to 10^8 electrons which consist an electronic pulse of about 50 mV. The electronic pulses due to the gamma-ray photons fallen onto the scintillation detector are linearly amplified and by energy discrimination and analysis are then counted.

The energy response of NaI(Tl) crystal is shown in Figure 1. In fact in this figure, the cross section of absorption of gamma-ray photons, μ as a function of their energy is shown. The gamma-ray photon absorption coefficient, μ is the sum of three coefficients, that is σ_f (photoelectric effect), σ_c (Compton effect) and σ_p (pair production effect).

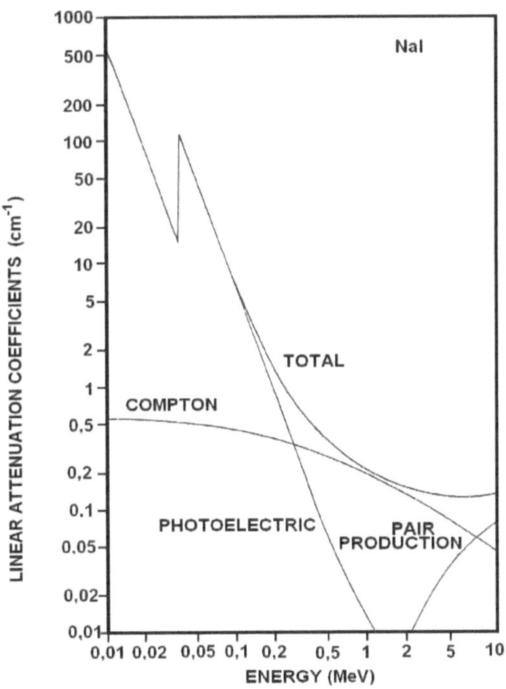

Figure 1. Mass absorption coefficient of gamma-ray photons for a scintillation detector, NaI(Tl) as a function of their energy

For gamma-ray photons with energy higher than 300 keV, absorption of gamma-ray photons is mainly due to the Compton effect, while for gamma-ray photons of low energies their absorption is due to the photoelectric effect. For the gamma-ray photons with energies above 1.02 MeV (2 × 0.511 MeV, the rest mass of the electron and positron), their absorption is due to the pair production effect.

In the case that a gamma-ray photon with energy E_o falls onto the scintillation crystal, NaI(Tl) if photoelectric effect is taking place, then photoelectrons are emitted with energy $E_p = E_o - b$, where b is the binding energy of electrons in the K, L, M., … shells in the atoms of Na and I. b_K is 1.072 keV for Na, 33.170 keV for I and 85.531 keV for Tl. That energy is attributed in the form of Roentgen radiation photons during the cascades and the filling of shells of atoms. These Roentgen photons might be considered as prompt with the photoelectric effect. These Roentgen photons have two alternative possibilities. The first one and mainly dominant is to be absorbed in the crystal. The second is to escape. In the case of absorption, the fluorescent light photons that are emitted in the crystal matter and excite the PM photocathode have energy proportional to E_o. If they escape, the fluorescent light photons have energy proportional to E_{es} and

$$E_{es} = E_o - \Delta b \qquad (1)$$

In the case of NaI (crystal), it is

$$E_{es} = E_o - 28 \text{ keV} \qquad (2)$$

The peak due to the gamma-ray photons with energy E_o is called photopeak. The peak due to the escape of Roentgen photons with energy E_{es} is called escaped peak of the photoelectric effect. This escape peak of the photoelectric effect is too weak and not easily observed in the ordinary crystal.

The electrons produced by the Compton effect have continuous distribution for their energy in the gamma-ray spectrum. The energy of these electrons is given by

$$T_e = E_o\,[1 - 1 / (1 + (1 - \cos\theta)\, E_o/0.511)] \qquad (3)$$

where θ is the angle of scattered photons. So, the energy of electrons, T_e is varying between 0 and $(T_e)_{max}$ which is

$$(T_e)_{max} = E_o / (1 + 0.511/2\, E_o) \qquad (4)$$

where $\theta = 0^\circ$ in Equation (3).

The theoretical distribution of electron energy due to Compton effect is shown in Figure 2. The distribution of electron energy is appeared in the gamma-ray spectrum as "Compton edge".

As the Compton effect is also taking place in the materials surrounding the scintillation crystal NaI(Tl), while the scattered photons in these interactions have "supplementary" energy distributions of those of Compton electrons, it is concluded that in the gamma-ray spectrum the peak of the backscattered photons due to Compton effect by the surrounding materials of the crystal will be appeared with energy $(E_{bs})_{min}$ as

$$(E_{bs})_{min} = E_o / (1 + 2 E_o/0.511) \quad (5)$$

where $\theta = 180°$ in Equation (3).

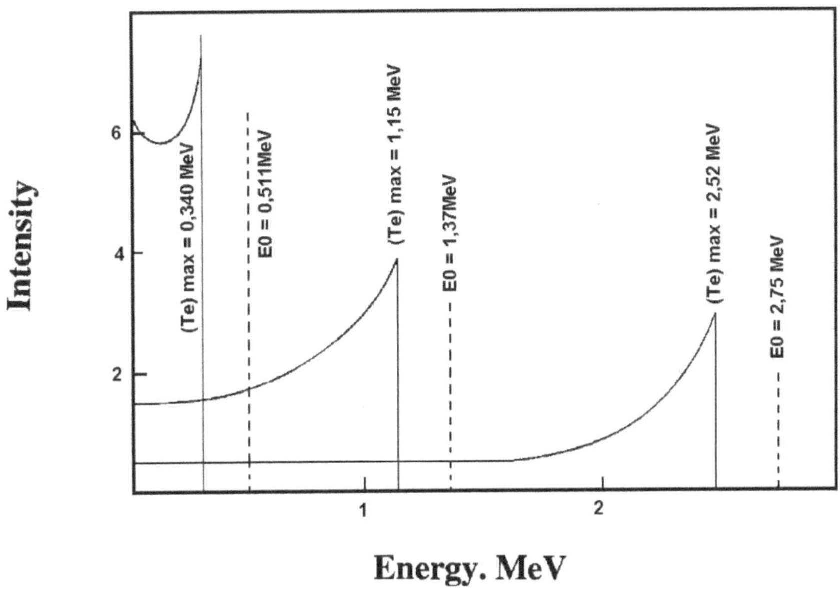

Figure 2. Distribution of Compton electrons for the marked energies of the gamma-ray photons

The gamma-ray spectrum of ^{137}Cs is shown in Figure 3. The photopeak due to E_o= 0.662 MeV, the Compton edge $(T_e)_{max}$ = 0.477 MeV and the backscattering Compton peak $(E_{bs})_{min}$ = 0.184 MeV. In the case that the gamma-ray photon energy E_o is greater than 1.02 MeV, it is possible for pair production to be taken place. The two charged particles, electron and positron of pair production have total kinetic energy E_o – 1.02 MeV.

Positron of pair production is thermalized in the NaI(Tl) crystal in about 10^{-11} s. Finally, it annihilates by an electron by producing two photons of 0.511 MeV in energy. These two photons could be absorbed in the crystal material and the total energy deposited in the crystal is equal to E_o. It is also possible for one of the photons to be absorbed and the other to be escaped, or both the photons to be escaped.

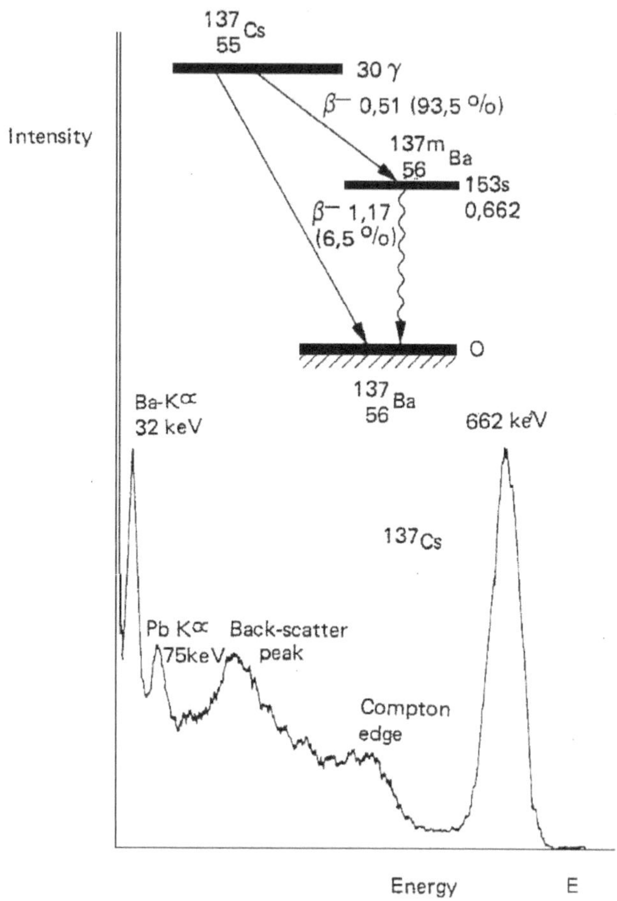

Figure 3. Gamma-ray spectrum of ^{137}Cs obtained by a scintillation detector, NaI(Tl) and showing the photopeak of 662 keV, the Compton edge of 477 keV and the backscattering Compton peak of 184 keV

As the pair production effect is possible to take place in the materials surrounding the scintillation crystal NaI(Tl), two photons of 0.511 MeV to be escaped. So, in the case that Eo > 1.02 MeV, the escaping peak of 0.511

MeV and the double escaping peak of 1.02 MeV will be appeared in the gamma-ray spectrum. In the case that Eo >> 1.02 MeV, except the Compton edge and the backscattering Compton peak, three other peaks will be appeared in the gamma-ray spectrum:

- Eo
- Eo – 0.511 MeV single escaping
- Eo – 1.02 MeV double escaping

The fact that the photons will escape or not, and consequently the above peaks will be appeared strong or not, depends on the crystal size.

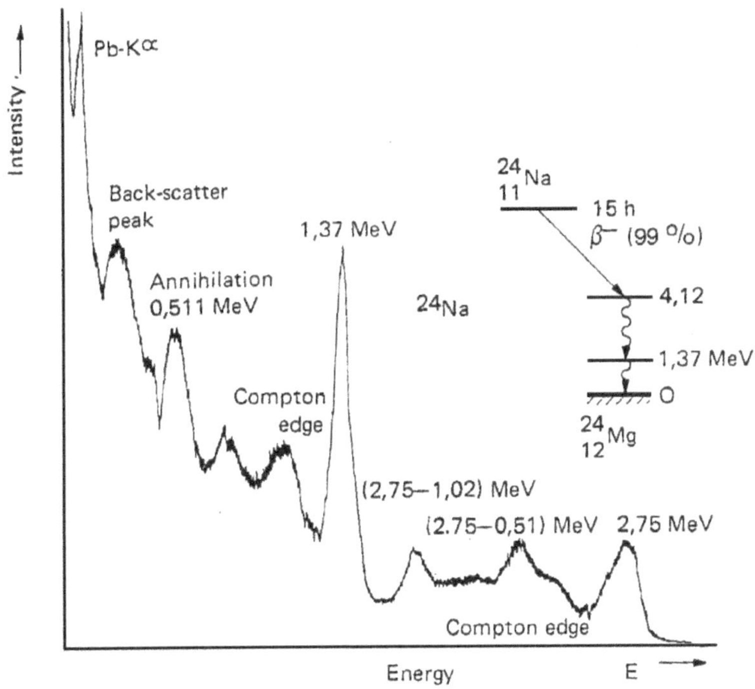

Figure 4. Gamma-ray spectrum of ^{24}Na obtained by a scintillation detector, NaI(Tl) and showing the 511 keV annihilation peak , the 1.37 and 2.75 MeV gamma-ray photon peaks and the single and double escaping peaks

Figure 4 shows the gamma-ray spectrum of ^{24}Na which emits in a decay two gamma-ray photons with energies 2.75 MeV and 1.37 MeV, respectively. So, except the photopeaks and other peaks, the single escaping of 0.511 MeV and the double escaping of 1.02 MeV as well as the peak of gamma-ray photons of 0.511 MeV are shown in the gamma-ray spectrum of Figure 4. Additionally, in the gamma-ray spectra there are the following peaks:

- Peaks due to Roentgen radiation of the material surrounding the crystal, e.g. Pb, Figures 3 and 4
- Peaks due to Roentgen radiation of the decay product of a nucleus if this is an alpha- or beta-emitter, e.g. Ba for the case of ^{137}Cs, Figure 3
- Sum peaks

These peaks are due to the possibility of two accidental photons or two prompt photons to be absorbed simultaneously in the crystal or not. The latter case is rather due to the appropriate geometry of crystal – source of radiation. For example, if a source of ^{60}Co is set in a well type NaI(Tl) crystal, then the two different gamma-ray photons with energies of 1.17 and 1.33 MeV, respectively, is possible to be absorbed simultaneously in the crystal and a sum peak of 2.5 MeV will be appeared in the gamma-ray spectrum as shown in Figure 5, together with the gamma-ray photon peaks of 1.17 and 1.33 MeV.

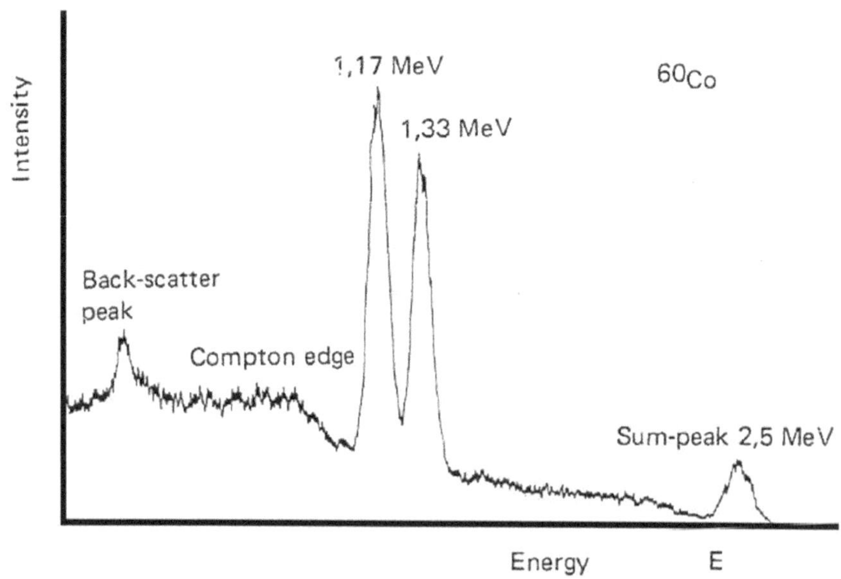

Figure 5. Gamma-ray spectrum of ^{60}Co obtained with a well type scintillation detector, NaI(Tl) and showing the 1.17 and 1.33 MeV gamma-ray photon peaks and the sum peak of 2.5 MeV

The scintillation detector is subject in various statistical fluctuations resulting that the total efficiency and its background to be depended on the measuring conditions. The high voltage bias supply and the pulse height of the low discriminator to be selected appropriately, so that, the measuring system to have high efficiency, E and low background, B, that means, the figure-of-merit (FOM), E^2/B to be maximum, where E is the efficiency counting rate, i.e. the ratio of net count rate to the real activity of source of radiation.

The output electronic pulses of the spectroscopy linear amplifier due to the detection of gamma-ray photons are having different heights from 0 to 10 V and producing corresponding peaks in the gamma-ray spectrum. The width of a gamma-ray photon peak is due to:

- The energy dispersion in the detector crystal.
- The natural width of the gamma-ray peak which is of the order of a keV. This is due to the uncertainty of the excited state of the nucleus when decays according to the Heisenberg principle, $\Delta E \, \Delta t >= h$.
- The noise introducing by the electronic devices of the gamma-ray spectroscopy system.

The resolution of the scintillation detector, δ is determined by the equation

$$\delta = \Delta E/E \; 100 \,\% \qquad (6)$$

where ΔE is the full width at the half maximum of the gamma-ray peak (photopeak), FWHM . It is evident that the best resolution means low value of δ. The factors that affect the resolution of the scintillation detector NaI(Tl) are due to:

- Light dispersion in the NaI(Tl) crystal
- Light dispersion in the photocathode of photomultiplier, PM
- Electron production in the photocathode of PM
- Electron collection by the first dynode
- Electron multiplication successively by the dynodes

Therefore, the resolution of the scintillation detector, δ depends on:

- The quality of the scintillation crystal and the photomultiplier, PM.
- The energy of the gamma-ray photons. The dependency of the resolution, δ on the energy of the gamma-ray photon there is an empirical equation

  $$\delta^2 = a + b/E \qquad (7)$$

  where a, b are constants, i.e. $a = 1.11 \times 10^{-3}$ and $b = 4.06 \times 10^{-3}$.
- The high voltage supply in the PM through the spectroscopy preamplifier.

A best resolution for a scintillation detector NaI(Tl) is that of 6%, while the resolution of 8% is considered as reference for the gamma-ray photons of $E_o = 0.662$ MeV of ^{137}Cs.

In the gamma-ray spectroscopy system for analyzing the gamma-ray photon peaks in the spectra, a single channel analyzer, SCA or a multichannel analyzer, MCA with a PCA card or a picospec digital multichannel analyzer, DMCA-pro-card with 1,024 or 4,096 channels [7] is used.

3. Solid State (Semiconductor Diode) Detectors

3.1. Ge DETECTORS

Germanium, Ge detectors are semiconductor diodes having a p-i-n structure in which the intrinsic (i) region is sensitive to ionizing radiation particularly X-ray and gamma-rays. Under reverse bias, an electric field extends across the intrinsic or depleted region. When, for example, gamma-ray photons interact with the material within the depleted volume of the detector, charge carriers (holes (+) and electrons (−) are produced and are swept by the electric field to the p and n electrodes. This charge, which is in proportion to the energy deposited in the detector by the incoming gamma-ray photon, is converted into a voltage pulse by an integral charge sensitive preamplifier.

Because Germanium has relatively low band gap, these detectors must be cooled in order to reduce the thermal generation of charge carriers (thus reverse leakage current) to an acceptable level. Otherwise, leakage current of about 0.01 nA induced noise destroys the energy resolution of the detector. Liquid nitrogen, which has a temperature of 77K is the common cooling medium for such detectors. The detector is mounted in a vacuum chamber which is attached to or inserted into an LN_2 Dewar. The sensitive detector surfaces are thus protected from moisture and condensable contaminants. By use of both p-type and n-type Germanium, there are used diffused, implanted, and barrier contacts.

The main advantage of Ge detectors against the scintillation detectors, NaI(Tl) used for detection and measurement in the gamma ray spectroscopy is the resolution, Figure 6. The resolution of Ge detectors is one order of magnitude higher than that of the scintillation detectors, NaI(Tl). This is due to the fact that the energy needed to produce a pair of a hole (+) and an electron (−) is just about 2 eV. For example, the full width at half maximum, FWHM of the 0.662 MeV photopeak of ^{137}Cs is about 0.002 MeV (2 keV) or less for Ge detectors. This equivalent with a resolution value of $0.002/0.662 = 0.3\%$.

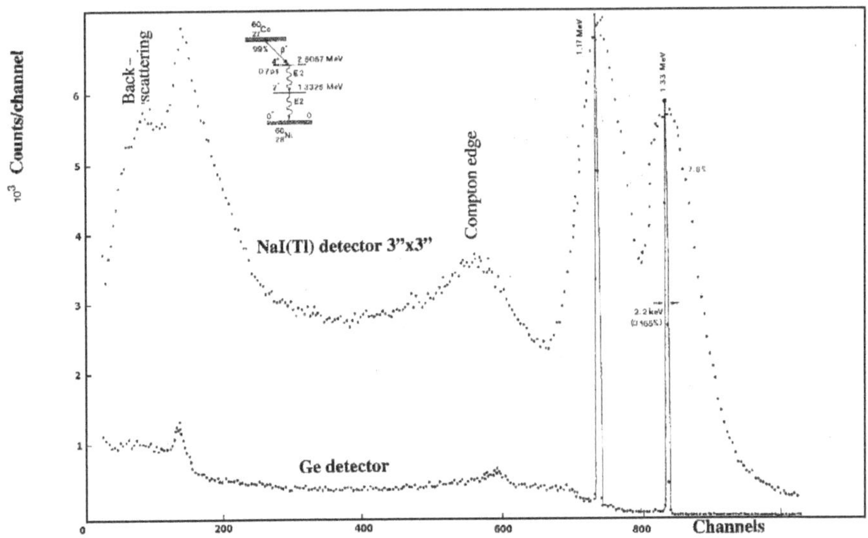

Figure 6. Gamma-ray spectra of ^{60}Co obtained by a Ge detector and a scintillation detector, NaI(Tl), respectively, for comparison and distinguished the resolution of the detectors

The specific characteristics of Ge detectors are the following:

- The resolution which is as low as 1.8 keV at 1.33 MeV gamma-ray photons of ^{60}Co the peak-to-Compton ratio, P/C which is as high as 70:1. The peak-to-Compton ratio, P/C of Ge detector is the mean height of Compton edge in the gamma-ray spectrum of ^{60}Co (1.040–1.096 MeV) to the height of photopeak of 1.33 MeV of ^{60}Co.
- The efficiency which is as high as 45%. The efficiency of Ge detector is the relative efficiency of a photopeak, i.e. the area under the photopeak (counting rate) of Ge detector compared with the respective area of the photopeak of a scintillation detector, NaI(Tl) 3" × 3" in size of crystal for the gamma-ray photons of 1.33 MeV of ^{60}Co with the source of radiation set in a 25 cm distance from the end cap of the detector. Figure 7 shows a comparison in the efficiency between two typical Ge detectors of p-type and n-type.

To increase the efficiency of gamma spectroscopy system linked by a Ge detector, the system is surrounded by an Anti-Compton, suppression system which operates as a mantle of shielding and consisted of a series of scintillation detectors, NaI(Tl) with large volumes appropriately mounted to reject the background counts.

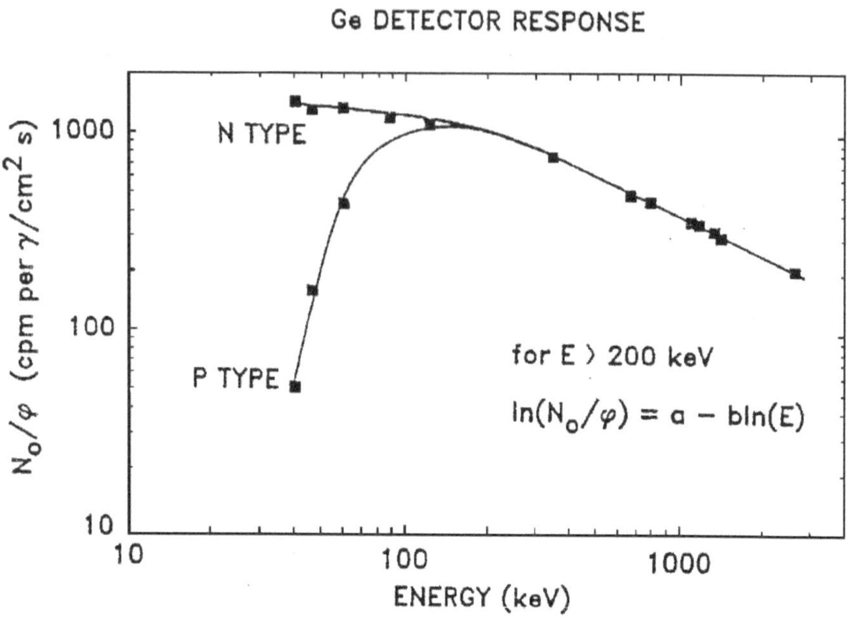

Figure 7. Comparison in the efficiency between the p-type and n-type Ge detectors

3.2. SILICON DETECTORS

Si(Li) detectors cover the energy range mainly of X-ray photons and of course the low energy gamma-ray photons from a few hundred eV to 50 keV or so. These are surface barrier lithium drifted detectors.
The specific characteristics of the silicon detectors are:

- The sensitive area ranging from 12.5 up to 200 mm^2.
- The depletion depth ranging from 2 to 5 mm.
- The Be window about 25 μm thick.
- The resolution (full width at half maximum, FWHM). This is considered for α-rays of ^{241}Am with energy 5.486 MeV (85%), ranging from 12 to 42 keV, averaging 20 keV, for β-rays of ^{137}Cs with energy maximum 0.514 MeV (94%) averaging 18 keV, for γ-rays of 0.122 MeV (85.6%) of ^{57}Co ranging from 150 to 200 eV and for X-rays of 5.9 keV of ^{55}Fe averaging to 250 eV.

4. In Situ-Gamma-Ray Spectrometry

The term *in situ* is taken from Latin and translates to “in the original place”. Thus, given a site where radioactivity and radiation levels are under investigation, the term "in situ gamma-ray spectrometry" implies that a spectrum of the ambient gamma-ray flux would be collected at the site and analyzed, principally to identify and quantify the radionuclides present. Sometimes, a less academic term is used, i.e. "field gamma-ray spectrometry", which implying that one is in the outdoor environment.

The technique of *in situ* gamma-ray spectrometry had its origin during the time of atmospheric nuclear weapons testing where it was found to provide quick, reliable information on the components of the outdoor radiation environment. It provided a means to separate natural background radiation from man-made sources of radiation and give quantitative results. Over the years, it has been employed by various groups for assessing sources of radiation in the environment not only through group based detectors, but with aircraft systems as well. It proved particularly useful following the Chernobyl accident and was employed by a number of Laboratories in Europe. It should prove adaptable to site assessments in the current era of environmental restoration.

The power in the technique of *in situ* gamma-ray spectrometry lies in the fact that a detector placed over a ground surface measures gamma radiation from radioactive sources over an area of several hundred square meters. As an example of the effective ground area being measured by a detector at 1 m above the ground, Figure 8 shows the relative contribution to the fluence from different rings of ground area about the detector for a typical source of fallout ^{137}Cs (gamma rays energy of 0.662 MeV) in the environment. The "field of view" for the detector would be larger for higher energy sources closer to the soil surface. In contrast, a soil sample would represent an area of but a few tens or hundreds of square centimeters. In practice, an effective characterization of a site would involve *in situ* gamma-ray spectrometry in conjunction with soil sampling. As part of an overall program, *in situ* gamma-ray spectrometry provides a means to assess the degree of contamination in areas during the course of operations in the field, thus guiding the investigator on where to collect samples. It can also substantially reduce the number of samples that need to be collected and subsequently analyzed.

Some of the limitations of *in situ* gamma-ray spectrometry need to be pointed out from the start. Due to the nature of radiation transport through matter (the soil and air), it is for the most part limited to the measurement of gamma emitters and, to some extent, X-ray emitters. Even so, the attenuation properties of soil are such that buried sources are not likely to be detected with measurements performed above ground. At 0.662 MeV, 30 cm of soil will cut out about the 97% of the primary flux from a buried point source. Of course, a detector can be lowered into a bore hole for measuring a buried source and techniques applied for interpreting a collected gamma-ray spectrum. While this constitutes an *in situ* measurement in the broad sense of the term, it will not be addressed in this tutorial. We will instead limit the subject area to measurements performed near ground level, where the distribution of the source of radiation is expected to be spread out over a fairly large area and where the source of radiation is at or near the surface of the soil.

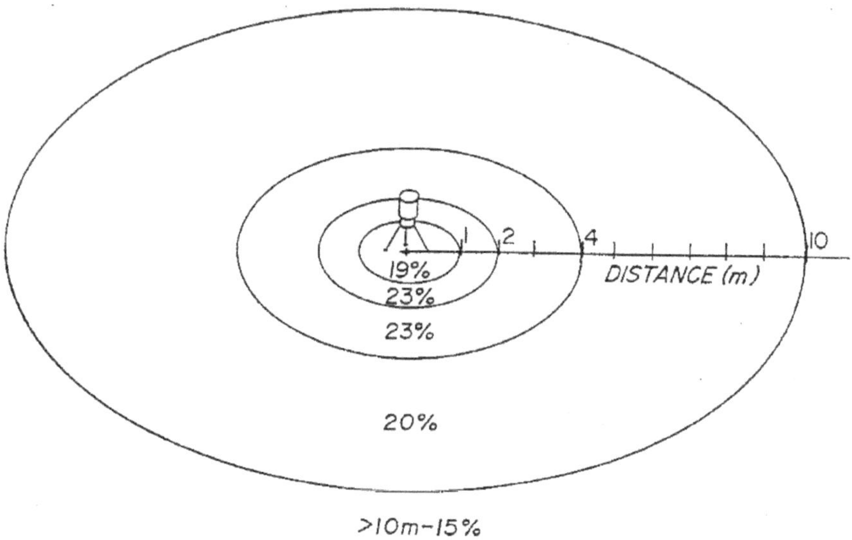

Figure 8. Contribution to total 0.662 MeV primary gamma-ray photon flux at 1 m above ground for a typical ^{137}Cs source distribution

Although measurements can be conducted with scintillation detectors NaI(Tl), as they were in the 1960s, the energy resolution of solid state Ge detectors and the fact that they are available with efficiencies as great as that of a 3" × 3" NaI(Tl) detector make them the detectors of choice. As with any counting system, the size of the detector that is needed is related to the source strength, the counting time, and the desired statistical counting error. For typical environmental radiation fields, a detector with a quoted 25% efficiency would be large enough to give a 5% (1σ) counting error for natural gamma emitters given a 1 h counting time. A quick 10 min counting time would be sufficient to provide lower limits of detection on the order of 100 Bq m^{-2} for many common fission product radionuclides residing at the surface of the soil. Higher sensitivity and/or reduced counting times can be achieved by using larger detectors. Depending upon the application, a smaller detector might actually be a better choice in order to reduce counting time when making measurements in highly contaminated areas.

Another consideration is the choice between a p-type and n-type Ge detector regarding the crystal. For applications that involve the measurement of low energy gamma rays, such as from ^{241}Am (59.5 keV), the n-type detector has better sensitivity, Figure 7. Older lithium drifted Ge detectors can function perfectly well, however, the fact that intrinsic or high-purity Ge detectors can warm up without damage makes them the most suitable for field measurements. Quality Ge detectors can be expected to have resolutions of 2 keV or better at 1.332 MeV. Better energy resolution allows a greater separation of two peaks that are close in energy. Also, each individual peak is narrower and therefore lower statistical counting errors are achieved since there is less continuum counts under the peak. Modern Ge detectors are equipped with built-in preamplifiers. For field measurements where battery power is used, it is important to specify a low-power preamplifier when ordering a detector. This will extend the operational time in the filed since the preamplifier is a principal draw on power.

Although measurements in the field can be performed with a Ge detector in almost any type cryostat-dewar configuration, performance and ease of handling is best achieved with a small dewar (1–2 L) that can be tripod mounted with the detector facing down, For convenience, a 24 h liquid nitrogen holding time is desirable as this requires filling only once a day, although it may be safer to maintain a twice a day schedule. Ge detectors can also be cooled with electrically powered apparatus, however, this may not be as convenient for field measurements with battery-powered equipment. It is possible to mate small dewars to automatic filling apparatus in the laboratory or to larger gravity-feed storage dewars. As for orientation, a detector facing sideways (the axis of symmetry parallel to the ground) should be avoided because it introduces complicated angular corrections.

A Ge detector can be connected to a full laboratory instrumentation package that is carried in a van and powered with a motor generator of battery bank. This was the norm in the early days of field gamma-ray spectrometry. Today, it is far more convenient to make use of portable battery-powered analyzers that are specifically designed for field measurements. These units not only serve as multichannel pulse height analyzers, but they also provide preamplifier power and high voltage to the detector. This type of analyzer with the Ge detector and a set of connecting cables is all that is needed for a complete gamma-ray spectrometry system.

One additional component needed for practical application of *in situ* gamma-ray spectrometry is a method of spectrum storage since it is likely that many spectra will be collected during the course of a site investing-ation. Some portable analyzers have built-in compact disk (CD) or mini cassette data storage capability, while others rely on an external portable audio digital video (DVD) or audio cassette recorder. The PC-based systems have the advantage of being able to store numerous gamma-ray spectra directly on the internal disk drive.

The ideal site for collecting a gamma-ray spectrum would be a large (20 m diameter or more) flat, open area with little or no natural or man-made obstructions. The area to be measured can be scanned first with a suitably sensitive survey meter to insure that there is rough uniformity in background dose rate. It is also possible to move the Ge detector about and obtain quick gamma-ray spectra (1–5 min), observing that a full absorption peak counting rate does not change substantially for a radionuclide under study. For measuring fallout radionuclides that were deposited in the past, the land should not have been disturbed by plowing or by wind or water erosion. For standard measurements, the Ge detector should be at height of 1 m above the ground, although a variation of as much as 50 cm in either direction will not introduce a large error. While collecting a gamma-ray spectrum, personnel should stand away from the detector. Since the operator may wish to examine the gamma-ray spectrum during collection, it is best to position the analyzer away from the detector using cable lengths of a few meters.

As with any gamma-ray spectrometry system, the amplifier gain and analyzer conversion rate must be adjusted to provide a gamma-ray spectrum in the energy region of interest (ROI). For environmental gamma radiation, this would be from 50 keV out to 2.615 MeV, normally the highest energy gamma line seen. For a 4,096 channel analyzer, a conversion rate of 1 keV per channel will suffice in most cases, although 0.5 keV per channel may be desirable in certain situations to take advantage of the higher energy resolution of the detector at low gamma-ray photon energies.

5. Inferred Quantities

5.1. CONCENTRATION IN SOIL

The fundamental quantities used for *in situ* gamma-ray spectrometry include full absorption peak counting rate (N), fluence rate (Φ), and radiation source activity (A). In practice, one would like a single factor to convert from the measured peak counting rate, N in a gamma-ray spectrum to the radiation source activity level in the soil or the dose rate in air. This factor can be calculated from three separately determined terms as follows:

$$N_f / A = (N_f / N_o)(N_o / \Phi)(\Phi / A) \quad (8)$$

where N_f/A is the full absorption peak counting rate at some energy, E, from a gamma-ray transition for a particular radioisotope per unit activity of that radioisotope in the soil; N_o/Φ is the full absorption peak counting rate per unit fluence rate for a plane parallel beam of gamma-ray photons at energy, E, that is normal to the detector face; N_f/N_o is the correction factor for the detector response at energy, E, to account for the fact that the fluence from an extended radiation source in the environment will not be normal to the detector face but rather distributed across some range in angles; and Φ/A is the fluence rate at energy, E, from gamma-ray photons arriving at the detector unscattered due to a gamma-ray transition for a parallel radioisotope per unit activity of that radioisotope in the soil.

The fluence rate, $\Phi(E)$ at the detector is given by the expression

$$\Phi(E) = R(E) / 4\pi x^2 \quad (9)$$

where R(E) is the gamma-ray emission rate at that energy and x is the source to detector distance. The attenuation effect of the source encapsulation should be taken into account along with that of the air between the source and detector, particularly for low energy gamma rays and large values of x.

The term N_o/Φ is purely detector dependent, while the term N_f/N_o is dependent on both the detector characteristics and the radiation source geometry. The term Φ/A is not dependent on the detector characteristics but rather on the radiation source distribution in the soil.

Having determined the three separate quantities, N_o/Φ, N_f/N_o, and Φ/A, their product yields the desired conversion factor, N_f/A. for radionuclides uniformly distributed with depth in the soil ($\alpha/\rho = 0$, where α is the inverse of the relaxation length (cm), ρ is the soil density (g cm^{-3}), the term A is in units of activity per unit mass. As such, there is no need to determine the soil density.

Although the assumption of a uniform profile in the soil for natural gamma emitters is generally safe, unusual situations where there is markedly different soil strata of varying radionuclide concentration may produce anomalous results. This situation been used. Also, evaluations of the ^{238}U series must be done with the awareness that ^{222}Rn escapes from the soil and that the important gamma-emitting decay products, ^{214}Pb and ^{214}Bi, may not be in equilibrium with ^{226}Ra in the soil. In fact, there may be a measurable contribution to the fluence rate at 1 m above the soil from the decay products in the air, particularly under atmospheric inversion conditions. Disequilibrium is also possible for the ^{232}Th series due to the exhalation of ^{220}Rn (thoron), although this is less likely to be as severe due to its relatively short half-life.

Another effect that may interfere with the interpretation of a gamma-ray spectrum is that of ^{222}Rn decay products scavenging during precipitation. In this situation the ^{214}Pb and ^{214}Bi assume a surface radiation source distribution that can considerably alter the flux and dose rate. For this reason (and to keep people and equipment dry!), it is best to avoid measurements during and for about 2–3 h following rain.

It is possible to consider a fallout radionuclide product as having a uniform profile if it is deeply distributed or has been mixed through soil cultivation. Depending upon the radiation source gamma-ray photon energy, plowing to depth of 15–30 cm essentially accomplishes this. Although the distribution does not extend to infinity in a situation such as this, in terms of the total gamma-ray photon flux seen above ground, it is effectively infinite in depth. For in situ applications such as this, the concentration that is measured can be considered as representative of the surface soil.

5.2. DEPOSITION/INVENTORY

For radionuclides that are exponentially distributed with depth ($\alpha/\rho > 0$), the ∞ term A is in units of activity per unit area. Although the results of analyses of environmental samples are frequently reported in terms of concentration, the fundamental quantity that is of most use for assessing fallout radionuclide products is the deposition (sometimes referred to as deposition density or inventory). Whereas the deposition remains a constant, the concentration of a fallout radionuclide product will vary depending upon the depth distribution. To illustrate this point, consider a radionuclide such as ^{137}Cs that was deposited in an area 45 years ago from atmospheric nuclear weapons testing. Where the surface soil has retained it, a sample down to 5 cm will yield some concentration, x. On an adjacent strip of land that was plowed deeply, the same sampling protocol will yield a concentration of perhaps only 0.2×. Obviously, this would be a flawed scheme for investigating a potential local radiation source of contamination. Instead, consider a soil core that was taken down to 30 cm. The measured concentration of an aliquot of this sample should be multiplied by the entire sample mass to give the total activity in the core and then divided by the sample area to give activity per unit area. This would yield the same result for both sites. The only precaution is to sample to a great enough depth to collect essentially all of the deposited activity.

In order to make an accurate assessment of deposited activity with *in situ* gamma-ray spectrometry, an estimate or actual measurement of α/ρ

must be made. As such, the time of deposition must be taken into account and assurances that no erosional or human activities such as plowing have disturbed the site. For fresh fallout that is dry deposited, the assumption of a surface source of radiation ($\alpha/\rho = \infty$) is generally not justified due to the effects of soil surface roughness which effectively buries the source and lowers the fluence at the detector. Wet deposition processes will also tend to distribute the fallout within the surface soil layer such that the assumption of a surface source of radiation would not be correct. Experience has shown that a more realistic assumption of α/ρ would be on the order of 1–10 $cm^2\ g^{-1}$. Depending upon the degree of uncertainty that is acceptable, experimental determination of the profile may be required through soil sampling. For deposition that occurred in the past, soil sampling is generally required to obtain an accurate value of α/ρ.

In making measurements of deposition of radionuclides in the environment, one must be aware of the sensitivity of the inferred inventory to the value of α/ρ. Figure 9 shows an example of the results of a calibration for a 22% efficient Ge detector. The conversion factor, N_f/A, is plotted as a function of the source depth constant, α/ρ, for the commonly encountered fission product radionuclide ^{137}Cs. The conversion factor is seen to change

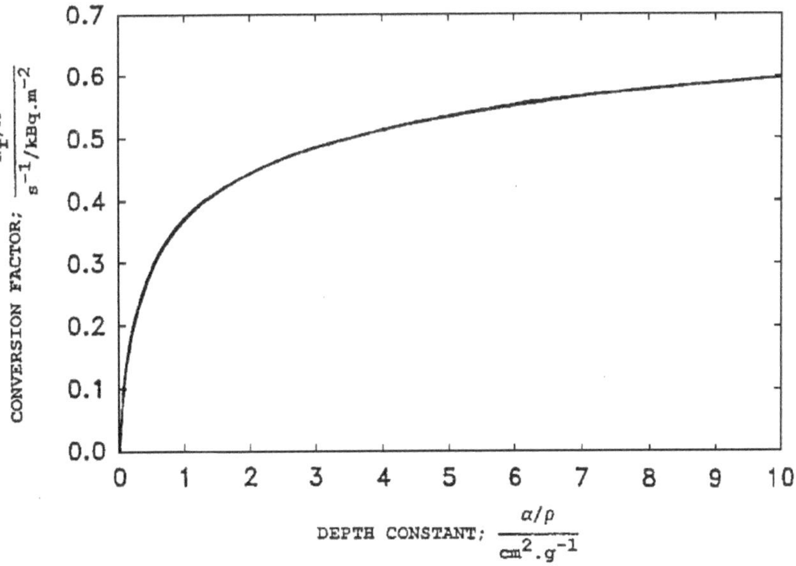

Figure 9. Conversion factor N_f/A as a function of the source depth parameter α/ρ

relatively little for values of α/ρ >1 cm^2 g^{-1} (shallow source depth distribution) as compared to values of $\alpha/\rho < 1$ cm^2 g^{-1} (deep source depth distribution). In effect, the error made in inferring the source activity will not be large for a fresh deposition event even if the profile is not precisely known. Conversely, if a measurement of aged fallout is made, accurate results will only be obtained if the profile is determined by some independent means, i.e. soil sampling.

5.3. DOSE RATE IN AIR

One of the most useful quantities that can be determined with *in situ* gamma-ray spectrometry is the dose rate in air (or the exposure rate) for the individual radionuclides present at a site. To do this, the results of transport calculations are used for the infinite half space geometry and the exponential source distribution. The conversion factors, I/A, exposure rate per unit activity in the soil, can be found in [8–10]. One can incorporate these factors directly into the detector calibration using the relationship

$$N_f / I = (N_f / A) / (I / A) \qquad (10)$$

where N_f/I is the full absorption gamma-ray photon peak count rate per unit exposure rate for that radionuclide. The factor I/A takes into account all of the gamma rays emitted in the decay of that radionuclide. Therefore, one does not have to analyze every gamma-ray photon peak for that radionuclide. In practice, however, it is best to analyze more than just one peak, especially if they are well separated in energy to check agreement.

What is not obvious in this analysis is the fact that the derived quantity, N_f/I, is less sensitive to α/ρ than is N_f/A. This result from the fact that as the source distribution in the soil gets deeper, the primary flux decreases relatively rapidly compared to the scattered component. This scattered component still contributes to the dose rate. To illustrate this, Figure 10 compares these to the two calibration factors N_f/I and N_f/A as a function of the relaxation depth, α^{-1}, where the soil density = 1.6 g cm^{-3}. This range in depth profiles extends from a fresh deposit to one that is perhaps 30 years old. It can be seen that the exposure rate factor varies by only 50% or so whereas the inventory factor varies by about a factor of 7. Thus, only a rough estimate of the depth profile is needed to predict the dose rate. At the same time, substantial errors can be made in the inventory estimate if the wrong depth profile is used.

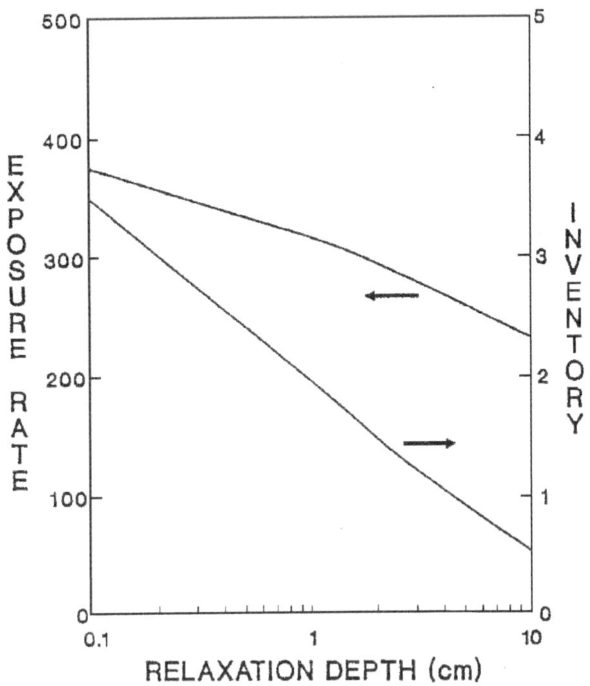

Figure 10. Exposure rate and inventory conversion factors as a function of the relaxation length for ^{137}Cs

6. Radiation Sources in the Environment

6.1. NATURAL EMITTERS

Virtually any gamma-ray spectrum collected over soil will reveal the presence of the three primordial natural radionuclides, ^{238}U, ^{232}Th and ^{40}K. In the case of ^{238}U, detection and measurement is made through the analysis of its decay products, principally ^{214}Pb and ^{214}Bi. For ^{232}Th, the decay products ^{228}Ac, ^{228}Th and ^{208}Tl are commonly used. As mentioned previously, these radionuclides are generally distributed uniformly with depth in the soil. As such, the appropriate quantity to report is the concentration, i.e. the specific activity (Bq kg^{-1}). Since these natural radionuclides are likely to contribute substantially to the total gamma-ray flux, the exposure rate or dose rate in air is a useful quantity to report as well. The summation of all contributions to the dose rate should be made and compared to a reading from an instrument such. Table 1 lists some of the more prominent gamma-ray peaks that are seen in the gamma-ray spectrum and which are the best to analyze. As a standard practice, the conversion factors N_f/A and/or N_f/I should be computed for these gamma-ray peaks, as they almost always be used.

One characteristic of an *in situ* gamma-ray spectrum is that the continuum rises substantially at low gamma-ray energies from the absorption of scattered radiation in the air by the crystal of Ge detector. This makes it difficult to detect and analyze gamma-ray peaks below about 200 keV. For instance, the rather weak 186 keV gamma-ray peak of ^{226}Ra superimposed on this large continuum does not usually give highly precise results due to the counting error.

One cosmogenically produced radioisotope that can sometimes be seen in the gamma-ray spectra is ^{7}Be (477 keV, 53 day half-life). Since it is produced in the atmosphere and deposited on the earth's surface, it can be expected to have an exponential profile like that of a typical fission product radionuclide in fallout. Due to its short half-life, it can be expected to lie close to the soil surface and thus have a high value of α/ρ.

6.2. FALLOUT EMITTERS

Due to nuclear weapons testing in the atmosphere, measurable amounts of the fission product radionuclide ^{137}Cs can be seen in surface soils around the world. Also, many areas especially in Europe, showed the activation product, ^{134}Cs, along with additional amounts of ^{137}Cs from Chernobyl fallout. Other, less intense, and shorter-lived radioisotopes from Chernobyl, such as ^{125}Sb and ^{106}Ru, can be sometimes seen as well. For common fallout product radionuclides, such as ^{125}Sb and ^{106}Ru and for other radio-isotopes that one would expect to encounter, it is useful to determine the conversion factor N_f/A and plot it for several different values of α/ρ. A smooth curve can be drawn through the points or a fit can be applied, as shown in Figure 9.

For in situ applications where there is potential for inhomogeneity in the horizontal distribution of deposited activity due to sparse ground cover, accurate measurements can still be performed providing that the scale of these inhomogeneities is small in comparison to the field of view of the detector. As an example, fallout in semi-arid regions may tend to clump under scattered plants from the effects of wind blown soil. If the depth distribution of the radionuclides is approximately the same for bare ground, as well as under the plants, no correction is needed as the application of the appropriate conversion factor for that depth distribution will yield the average inventory for that site. However, it is possible that

TABLE 1. Principal gamma-ray photon energies of natural radionuclides

Energy (keV)	Nuclide	Parent series	Comments
186	^{226}Ra	^{238}U	Low intensity, high continuum, cannot be resolved from ^{235}U peak at 185 keV
239	^{212}Pb	^{232}Th	Strong peak, contribution from ^{224}Ra peak at 241 keV, interference from ^{214}Pb peak at 242 keV
295	^{214}Pb	^{238}U	Generally clean peak, fairly strong
352	^{214}Pb	^{238}U	Generally clean, strong peak
583	^{208}Tl	^{232}Th	Generally clean, strong peak
609	^{214}Bi	^{238}U	Strong peak, interference from 605 keV peak if ^{134}Cs is present
911	^{228}Ac	^{232}Th	Generally clean, strong peak
965 + 969	^{228}Ac	^{232}Th	Doublet, not as strong as 911 peak
1,120	^{214}Bi	^{238}U	Reasonably strong, continuum relatively low
1,461	^{40}K	–	Clean, strong, only peak for this nuclide
1,765	^{214}Bi	^{238}U	Reasonably strong, continuum low
2,615	^{208}Tl	^{232}Th	Clean, strong, continuum very low

there may be two or more distinct depth profiles associated with the various ground covers in which case separate determinations must be made. The infinite half space in this circumstance can be considered a collection of subspaces, each with its own characteristic radionuclide inventory and depth profile. The conversion factor for field spectrometry is then computed as an average, weighted by the fraction of the total deposited activity associated with each ground cover. An estimate of this can be made through selected soil sampling to determine the inventory and by measuring the fraction of the half space for each ground cover. In a strict sense, the *in situ* gamma-ray spectrum in this situation does not provide an independent measure of the deposited activity in that there is a reliance on the data provided by the soil samples. However, the average conversion factor is bounded by the range in respective values for each type of ground cover. This range may be small compared to the variation in deposition density so that the in situ gamma-ray spectrum provides a reasonably accurate average without resorting to far more extensive soil sampling. For more details on this subject, the reader is referred to [11–14].

6.3. COSMIC RADIATION

A portion of the continuum seen in a gamma-ray spectrum obtained by a Ge detector is due to the interaction of cosmic-ray secondary radiation in the crystal of the detector. The degree of this contribution can be estimated from the count rate above the 2.615-MeV line from ^{208}T. Generally, it is a small fraction of the count rate due to terrestrial gamma radiation. The overall effect is to increase somewhat the error associated with the analysis of a peak in the gamma-ray spectrum. In that the continuum under that peak is slightly higher.

It is important to realize, however, that a measurement of the external dose rate will included a contribution from the cosmic-ray component. Many survey instruments have some response to cosmic radiation. If a comparison is made between a survey instrument reading and the sum of the dose rates inferred from peak analysis with a Ge detector, it must be remembered that the latter provides only the terrestrial gamma-ray component.

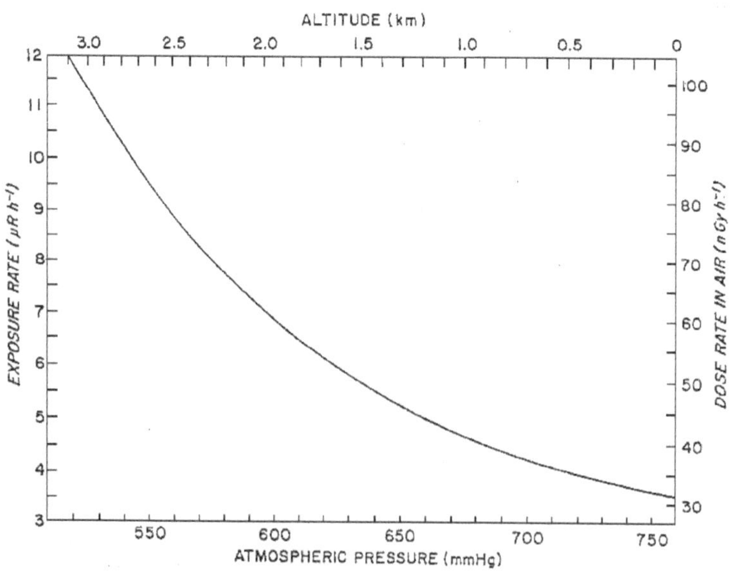

Figure 11. Mid-latitude conversion factor for cosmic-ray dose and exposure rate as a function of altitude and atmospheric pressure

In general, the dose rate from cosmic radiation increases towards the earth's poles and decreases toward the equator (latitudinal effect) [15]. For mid-latitudes, Figure 11 provides a useful conversion from altitude/ pressure to cosmic-ray dose rate. A reading with a pressure meter would be the preferred method with which to infer the cosmic-ray component. In place of this, a geological survey map can be used to find one's altitude. In using this chart, a limitation on its accuracy must be recognized. There are variations of a few percent with the 11-year solar cycle and somewhat smaller variations with season. During periods of maximum solar activity (as measured by sunspots by scientists [16, 17], the cosmic component tends to be lower, while during period of a "quit" sun it is higher. The overall uncertainty given both these spatial and temporal variations is estimated to be on the order of 10%.

7. Epilogue

Most of us who deal with aspects of ionizing radiation in the environment are familiar with basic dose rate and/or radioactive sample measurements using survey instruments. Perhaps we can recall those times where we have walked about a site with a meter in our hand and measured external or internal radiation levels. This constitutes an *in situ* measurement in its most basic form, one which deals with a single parameter such as the exposure rate. For more information, one can take a sample from this same site, perhaps soil, and return it top the laboratory for analysis. Gamma-ray spectrometry by a Ge detector might then be employed to determine the specific radionuclides present in the samples. This could be done for strictly qualitative purposes or perhaps to convert measured concentrations of radionuclides in the samples to the exposure rate at the original site using conversion factors. The technique of *in situ* gamma-ray spectrometry combines radioelements of both of these methods for characterizing the external radiation field. By using high resolution De spectrometers placed over the ground, a spectrum of gamma radiation collected in the field can be used to identify radionuclides present in a qualitative manner by simply looking for the presence of peaks at characteristic energies of gamma-ray photons. At a higher level of sophistication, one can convert the measured peak count rate into some more meaningful quantity such as the concentration of these radionuclides in the soil or, in the case of deposited fallout, the activity per unit area. It is also possible to infer the contribution of each individual radionuclide to the dose rate in air.

This tutorial will introduce you to these techniques the individual will be able to adapt the techniques to unique situations. To this end, a basic grounding in the theory is given, however short-cut methods are also presented for those who may use the techniques for approximate measurements. For more details on certain aspects, appropriate references are given.

References

1. Heath RL (1962) IRE Trans Nucl Sci NS-9 No. 3. 294
2. Heath RL (1964) Scintillation spectrometry gamma-ray spectrum catalogue IDO-16880 Volumes 1 and 2
3. Siegbahn K (1968) Alpha- beta- and gamma-ray spectroscopy. Vol. 1. North-Holland, Amsterdam
4. Knoll GF (1979) Radiation detection and measurement. Wiley, New York
5. Tsulfanidis N (1983) Measurement and detection of radiation. McGraw-Hill, New York
6. Tennelec (1988) Nucleus PC Analyzer. Tennelec/Nucleus Inc 601 Oak Ridge Turnpike Oak Ridge TN 37831-2560
7. Target (2000) PicoSPEC digital Multichannel-Analyzer System/dMCA-pro Target Systemelectronic GmbH Kollner strasse 99 D-42651 Solingen, Germany
8. Beck HL (1972) The physics of environmental gamma radiation fields. In: The Natural Radiation Environment II. Eds. Adams JAS, Lowder WM, and Gesell TF. CONF-720805-P2 pp. 101–134. Technical Information Center/US Department of Commerce Springfield Virginia 22161
9. Beck HL, De Campo J and Gogolak CV (1972) In situ Ge(Li) and NaI(Tl) gamma-ray spectrometry. US DOE Report HASL-258
10. Beck HL (1980) Exposure rate conversion factors for radionuclides on the ground. US DOE Report EML-378
11. Miller KM (1984) A spectral stripping method for a Ge spectrometer used for indoor gamma exposure rate measurements. US DOE Report EML-419
12. Miller KM, Helfer IK (1985) In situ measurements of ^{137}Cs inventory in natural terrain. In: Environmental Radiation 1985 Proceedings of the 18th Midyear Topical Symposium of the Health Physics Society, pp. 243–251
13. Helfer IK, Miller KM (1988) Calibration factors for Ge detectors used for field spectrometry. Health Phys 55:15–24
14. Miller KM, Shebell P (1993) In situ gamma-ray spectrometry. US DOE Report EML-557
15. Papastefanou C (2007) Chlorine-39 in rainfall at temperate latitude (40° N). J Environ Radioact 92:175–182
16. Ioannidou A, Papastefanou C (1994) Atmospheric Beryllium-7 concentrations and sun spots. Nucl Geophys 8:539–543
17. Papastefanou C, Ioannidou A (2004) Beryllium-7 and solar activity. Appl Radiat Isotopes 61:1493–1495

PREPARATION OF RADIONUCLIDES AND THEIR MEASUREMENT BY HIGH RESOLUTION γ-SPECTROMETRY, β-SPECTROMETRY AND HIGH RESOLUTION α-SPECTROMETRY

FLAVIA GROPPI*, MAURO L. BONARDI AND LUIGI GINI
Università degli Studi di Milano and INFN-Milano, LASA, Radiochemistry Laboratory, via F.lli Cervi 201, I-20090 Segrate, Milano, Italy
ZEEV B. ALFASSI
Department of Nuclear Engineering, Ben Gurion University of Negev, Beer-Sheva, Il-84105, Israel

Abstract. The use of High Specific Activity Radionuclides HSARNs, obtained by either proton, deuteron or alpha cyclotron irradiation, followed by selective radiochemical separation from the irradiated target in No Carrier Added (NCA) form, is a powerful analytical tool. The main applications of these radionuclides are to medical dignosis and therapy in addition to toxicological, environmental and industrial studies. These NCA radionuclides and their Specific Activity (A_S) can reach values close to the theoretical "carrier-free" one A_S(CF). The "real" A_S(NCA) must by measured by either analytical or/and radioanalytical techniques. The "accurate" knowledge of behavior of cross-sections of each reaction, as a function of beam energy (excitation functions) is mandatory. The minimization of Isotopic Dilution Factor is achieved also. Some examples of production and radioanalytical quality control methods for HSARNs are presented.

Keywords: γ-spectrometry, α-spectrometry, β-spectrometry, radionuclide measurements, HSARNs

* To whom correspondence should be addressed. e-mail: flavia.groppi@mi.infn.it

G.A. Aycik (ed.), *New Techniques for the Detection of Nuclear and Radioactive Agents*,

1. Introduction

The applications of the radionuclides play a fundamental role in life sciences; their greater user is the nuclear medicine, which, nowadays, has become an indispensable branch of medical science and, as the production of radionuclides and labelled compounds for medical applications, is becoming an important branch of nuclear and radiochemistry [1, 2].

The labelling of monoclonal antibodies (or fragments of peptides) that is "specific" for the over expressed receptors of the neoplastic cells, or the correspondents antigen, is an extremely suitable strategy for dealing with such pathologies. Obviously these labelled compounds, specific for the target, have also to be selective, in order to minimize the dose to the surrounding tissue.

The production of innovative medical radionuclides, both for PET and SPECT diagnostics and for metabolic radiotherapy required the selection of an appropriate radionuclide depending on a number of criteria and parameters, affecting usefulness and feasibility. The first is directly related to the radiological performance of the ionising radiation in relation to tissue and its morphology, with a major distinction between the effects of alpha and beta-particles; usefulness is also directly related to the proper choice of RN half-life. The second depends also on convenience and safety aspects in the preparation and the handling of the RN's. In general, an ideal radiotracer must be characterised by the following requirements:

- High specific activity (A_S) (i.e.: activity/mass of isotopic carrier)
- High activity concentration (C_A) (i.e.: activity/volume or mass of substrate)
- High radionuclidic, radiochemical and chemical purities

Furthermore, in order to allow the use of these labelled compounds on living organisms, it is necessary to guarantee their biological compatibility, that means physiological pH, sterility, physiological values of both ionic conductivity and osmotic strength.

We define *specific activity* (A_S) of a RN the ratio between the activity $A(t) = N(t)\,\lambda$ of RN (Bq) and the mass of "*isotopic carrier*" present in the sample and in the SI units this quantity is given in $Bq \cdot kg^{-1}$, or in most practical cases $GBq \cdot \mu g^{-1}$. With the term "isotopic carrier", we mean the overall *mass of both stable and radioactive nuclides* m(t) of same element or mass of isotopic carrier (kg), at time t (s) of application envisaged, where λ (s^{-1}) is the decay constant of RN concerned and N(t) is the number of atoms of RN at time t (s):

$$A_S(t) = \frac{N(t) \cdot \lambda}{m(t)} \qquad (\mathrm{Bq} \cdot \mathrm{kg}^{-1}) \qquad (1)$$

This quantity "must" not be misunderstood with the *activity concentration* (C_A) of a RN defined as:

$$C_A(t) = \frac{N(t) \cdot \lambda}{m(substrate)} \qquad (\mathrm{Bq} \cdot \mathrm{kg}^{-1}) \qquad (2)$$

where m(substrate) is the mass ($\mathrm{Bq{\cdot}kg^{-1}}$) or volume ($\mathrm{Bq{\cdot}\ m^{-3}}$) of either substrate or solvent where RN is diluted. A_S and C_A present the same units in SI and are confused each other very often, even their significance (and definition) is completely different.

The A_S of a RN, if no other nuclides of the same Z are present, is named Carrier Free, A_S(CF), that is the maximum "theoretical value" of A_S, reached by a RN "absolutely free" of isotopic carrier (both stable and radioactive). This quantity is an *intensive parameter* that depends on the physical properties of RN only, in accordance to the definition:

$$A_S(CF) = \frac{N_{AV} \cdot \lambda}{M} \qquad (\mathrm{Bq} \cdot \mathrm{kg}^{-1}) \qquad (3)$$

where: N_{Av} is the Avogadro's constant and M is the atomic mass of the RN.

This parameter is a *physical constant*, not depending on either decay time or chemical or physical environment. The higher is M the lower is A_S(CF); conversely, the shorter is the half-life of RN, the higher is A_S. Actually, the atomic mass of different RNs of Periodic Table from medium to high Z varies by less than one order of magnitude, while the half-life is the parameter that determines the order of magnitude of A_S(CF). In practice, a CF condition can be achieved in a few selected cases. In most cases the RN is diluted in a variable amount of Isotopic Carrier, which is constituted by both stable and radioactive isotopes of same Z.

For practical purposes, it is very useful introducing another quantitative parameter: the Isotopic Dilution Factor (IDF) of the RN (or labeled compound) defined as the total number of stable and radioactive nuclides of same element to the number of atoms of RN (or labeled compound) of interest. From the definition of A_S and A_S(CF):

$$IDF(t) = \frac{A_S(CF)}{A_S(t)} \qquad \geq \quad 1 \qquad (4)$$

Thus, IDF(t) is a *non dimensional quantity* that is always greater than the *unity*. Just in case when the RN is in a real "carrier free" form, the IDF is equal to 1. In common cases of NCA radiotracers, the IDF(t) presents values varying from some tens up to several thousands. In order to optimize radionuclidic purity increasing at the same time A_S of light ion produced radionuclides, two main approaches must be considered. A. An accurate knowledge of "experimental" excitation functions of nuclear reactions involved, as a function of projectile energy, as well as the knowledge of the excitation functions of both stable and radioactive nuclides produced by side reactions, that are isotopic with the radionuclide of interest; B. the optimization of very selective radiochemical separations of the radionuclide produced, without "intentional" addition of either isotopic or even isomorphous/iso-dimorphous carrier. The radionuclides and labeled compounds obtained in this way are named No Carrier Added (NCA). Figure 1 reports the different definitions of specific activity.

Even without the "intentional" addition of "isotopic carrier", the contamination by isotopic carrier is unavoidable and it is due to chemical impurities present in both target materials, target holders, chemicals used during radiochemical processing and even by contact of a HSARN with the polluted environment. Whenever the radiotracer is used to label any chemical compound or radiopharmaceutical, the "isotopic carrier" is easily introduced into the sample by the labeling procedure. Nevertheless, it is evident that, if, in order to administer a radiopharmaceutical or labeled compound use is made of a physiological medium or metallic injection needles, the sample is contaminated in principle by any element present in the Periodic Table, excluded a few "artificial" ones.

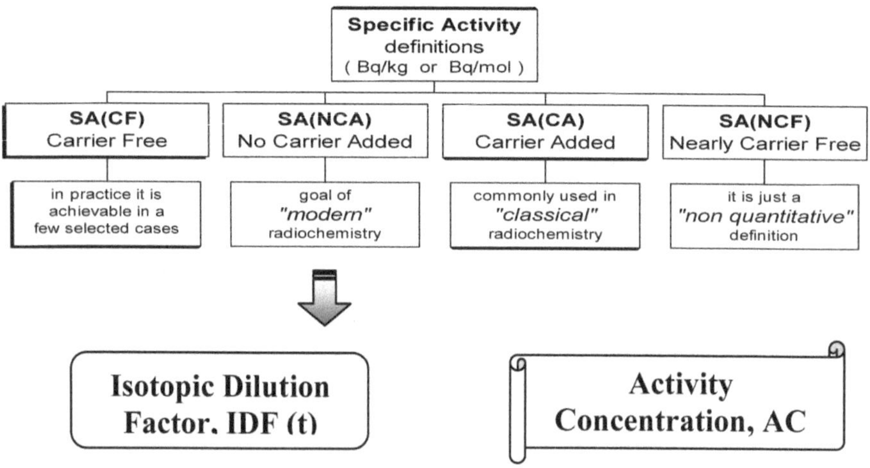

Figure 1. Different definitions of A_S for RNs and labeled compounds. The concept of IDF is related to definition of A_S, while C_A has a completely different meaning (see text)

1.1. THIN-TARGET EXCITATION FUNCTIONS AND NUCLEAR REACTION CROSS-SECTIONS

An accurate knowledge of *thin-target excitation functions* for production of radionuclides and radionuclidic (isotopic in particular) impurities coming from side nuclear reactions, is of primary importance for optimising both yield, radionuclidic purity and even specific activity of radionuclides obtained by charged particle irradiation. This concept was stressed several times and recommended by the International Atomic Energy Agency [3–9, and all refs included].

In accelerator production of radionuclides, thin-target yield $y(E)$ is defined as a function of projectile energy E (MeV), at the End Of an Instantaneous Bombardment (EOIB), as the *slope at the origin of the growing curve of the activity per unit beam current* (A/I) of a specific radionuclide versus irradiation time τ, for a target in which the energy loss ΔE is negligible (i.e.: a few percent in practical cases) in respect to the projectile energy E itself. In practice y(E) is defined as the *second derivative* of (A/I) in respect to particle energy and irradiation time, calculated when the irradiation time tends to zero (EOIB), Equation (5). From experimental point of view, y(E) is measured by activation of thin-target at different energies, followed counting by either off-line gamma spectrometry or other radiometric/radioanalytical techniques, by use of

$$y(E,\tau) = y(E)_{EIOB} = y(E,0) = \left(\frac{\partial\, (\partial\, [A/I]}{\partial E\ \partial \tau} \right) \quad \text{with } \tau \to 0 \tag{5}$$

$$y(E) = y(E)_{EIOB} = \frac{C_\gamma}{(\varepsilon_\gamma \alpha_\gamma LT)\ Q\ \Delta E}\ D(RT)\ \ G(\tau)\ \ e^{\lambda WT} \tag{6}$$

Equation (6). The correction factors G(τ) and D(*RT*) defined in Equations (7) and (8) take into account the decay of radionuclide during irradiation and counting times respectively.

In Table 1 are summarised and defined the various parameters used in these equations and in the text.

The thin-target yields of different RNs, produced by direct and side reactions, are numerically fitted, taking into account the overall statistical errors as weights.

$$D(RT) = \frac{\lambda\, RT}{1 - e^{-\lambda\, RT}} \tag{7}$$

$$G(\tau) = \frac{\lambda \tau}{1 - e^{-\lambda \tau}} \tag{8}$$

The "effective" cross-section σ*(E) of Equation (9) as a function of projectile energy E, is proportional to thin-target yield, Equation (10). Nevertheless, the physical meaning of this parameter is poor, being only a raw summation of the several cross-sections $\sigma_i(E)$ of all *reaction channels* concerned, weighed with weight w_i on target isotopic composition (with $\Sigma w_i = 1$):

$$\sigma^*(E) = \sum_i w_i \, \sigma_i(E) \tag{9}$$

$$y(E) = \frac{\sigma^*(E)\, N_a\, \lambda}{a.m.\ \gamma z\ e \left(\dfrac{dE}{dx}(E) \right)} \tag{10}$$

The term γz (non dimensional) is the effective projectile charge (1 for protons or deuterons) and e is the elementary charge (i.e.: $1.602\ 10^{-19}$ C):

1.2. THICK-TARGET YIELD

Thick-Target Yield Y(E,ΔE) is defined as a *two parameter function* of incident particle energy E(MeV) *incoming onto* the target and energy loss ΔE(MeV) into the target itself, obtained by *integration* of thin-target excitation function y(E), Equation (11). This approach holds in the strict approximation of a monochromatic beam of energy E, not affected by either intrinsic energy spread or straggling:

$$Y(E, \Delta E) = \int_{E-\Delta E}^{E} y(x)\, dx \tag{11}$$

Both energy loss and energy straggling were computed by Monte Carlo codes, like SRIM 2008, that are based on the former Bethe-Bloch theory. In case of *total particle energy absorption* in the target, for a nuclear reaction of physical energy threshold E_{th}, the function Y(E,ΔE) reaches a value Y(E, E-E_{th}), for ΔE = E-E_{th}, that represents mathematically the *envelope* of the Y(E,ΔE) family of curves. This envelope is a *monotonically increasing curve*, never reaching either a maximum or a saturation value, even if its slope becomes negligible for high particle energies and energy losses. Some *loci of the maxima* of Y(E,ΔE) curves are present in most cases.

As a relevant conclusion, target thickness larger than the "effective" one, must be avoided, because it leads to higher *power density* Pd ($W \cdot g^{-1}$) deposited by beam itself (with charge γz, intensity I [μA] and energy loss ΔE [MeV]), into target material of mass m (g), instead of target cooling system, Equation (12). From technological point of view, the total energy absorption into the target does not present any advantage, due to the high stopping-power $dE(E)/dX = S(E)$ (in $MeV \cdot g^{-1} \cdot cm^2$) of fast ions in materials, while on the contrary production yield $Y(E, E-E_{th})$ does not increase further:

$$Pd = \frac{I[\mu A] \cdot \Delta E[MeV]}{\gamma z \cdot m[g]} \qquad \left[\frac{W}{g}\right] \tag{12}$$

TABLE 1. Physical quantities and symbols used in text and equations

RT = real counting time (s)	Na = Avogadro constant
LT = live counting time = RT – DT (s)	I = beam current (A)
DT = counting dead time (s)	Q = integrated beam charge (C)
WT = t = waiting time after the EOB (s)	E = beam energy in the thin-target (MeV)
EOB = end of bombardment	ΔE = energy loss in thin/thick target (MeV)
EOIB = end of instantaneous bombardment	S(E) = stopping power ($MeV \cdot g^{-1} \cdot cm^2$)
A = N λ = activity (Bq)	y(E) = thin-target yield ($Bq \cdot C^{-1} \cdot MeV^{-1}$)
N = number of radioactive atoms (atoms)	$Y(E,\Delta E;\tau,t)$ = thick-target yield ($Bq \cdot C^{-1}$)
$\lambda = \ln(2) \cdot (t_{1/2})^{-1}$ = decay constant (s^{-1})	RNP = radionuclidic purity (%)
$t_{1/2}$ = radionuclide half-life (s)	EOP = end of radiochemical processing
$1 \cdot \lambda^{-1}$ = mean life (s)	$C_A(t)$ = activity concentration ($Bq \cdot kg^{-1}$)
C_γ = net counts of photo-peak chosen	$A_S(t)$ = specific activity ($Bq \cdot kg^{-1}$)
α_γ = emission intensity of photo-peak chosen	CF = carrier-free
ε_γ = efficiency of photo-peak chosen	$A_S(CF)$ = CF specific activity ($Bq \cdot kg^{-1}$)
τ = irradiation time (s)	NCA = no carrier added
EOSB = End Of Saturation Bombardment	$A_S(NCA)$ = NCA specific activity ($Bq \cdot kg^{-1}$)
$(A/I)^{sat}$ = saturation activity ($Bq \cdot A^{-1}$)	CA = carrier added
$\sigma^*(E)$ = effective cross-section (cm^2)	$A_S(CA)$ = CA specific activity ($Bq \cdot kg^{-1}$)
$\sigma_i(E)$ = reaction channel cross-section (cm^2)	(NCF = nearly carrier-free)
w_i = isotopic fraction of target	($A_S(NCF)$ = NCF specific activity) ($Bq \cdot kg^{-1}$)
γz = projectile charge	IDF = isotopic dilution factor (≥1)
e = elementary charge (Coulomb)	MA_S = molar specific activity ($Bq \cdot mol^{-1}$)

Finally, this set of Thick-Target Yields $Y(E, \Delta E)$ and *maxima* permits calculating *a priori* the optimal irradiation conditions for producing radionuclides with higher as possible yield, radionuclidic purity and specific activity. In many cases, the best irradiation conditions *do not correspond to the maximum projectile energy* available. In order to join the advantages of the accurate knowledge of thin-target excitation functions and cross-sections of radionuclide of interest and its radioisotopic impurities, very selective *radiochemical separations* were optimised for separating the radionuclide concerned from irradiated target without addition of isotopic carrier or in other words in NCA form.

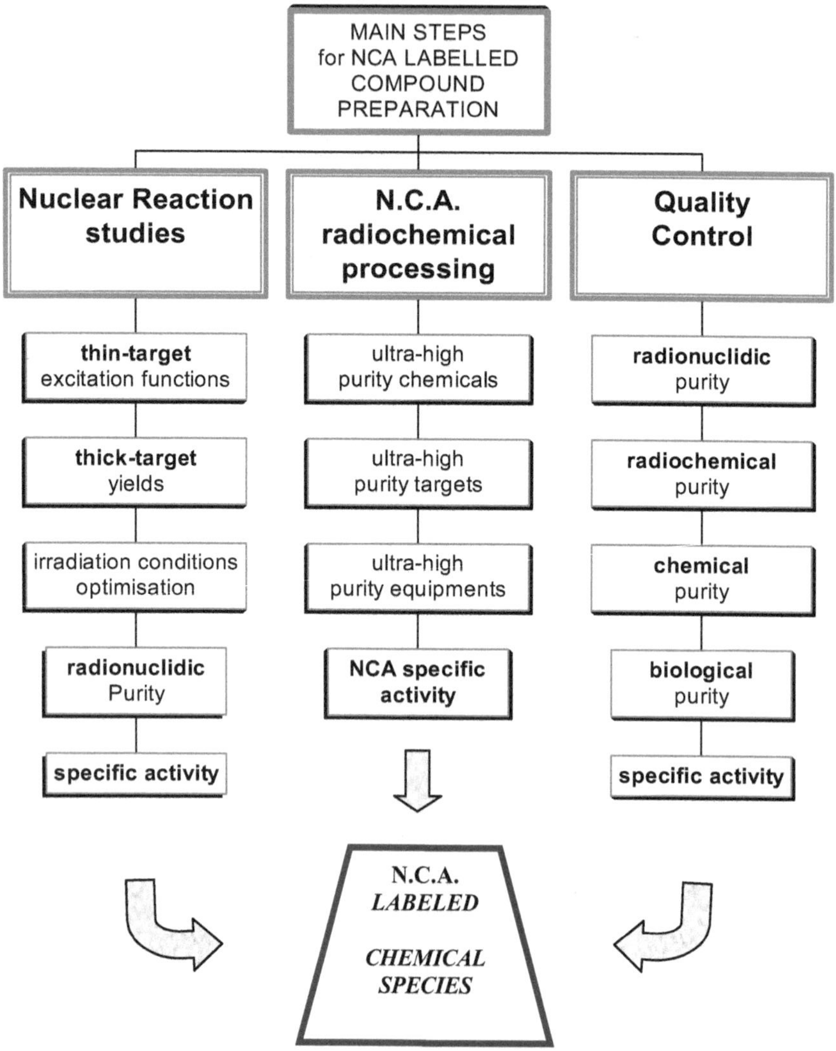

Figure 2. Main steps that must be performed for the optimisation of purity and specific activity of accelerator produced RNs and labelled compounds for uses in the life sciences (cell cultures, laboratory animals, humans)

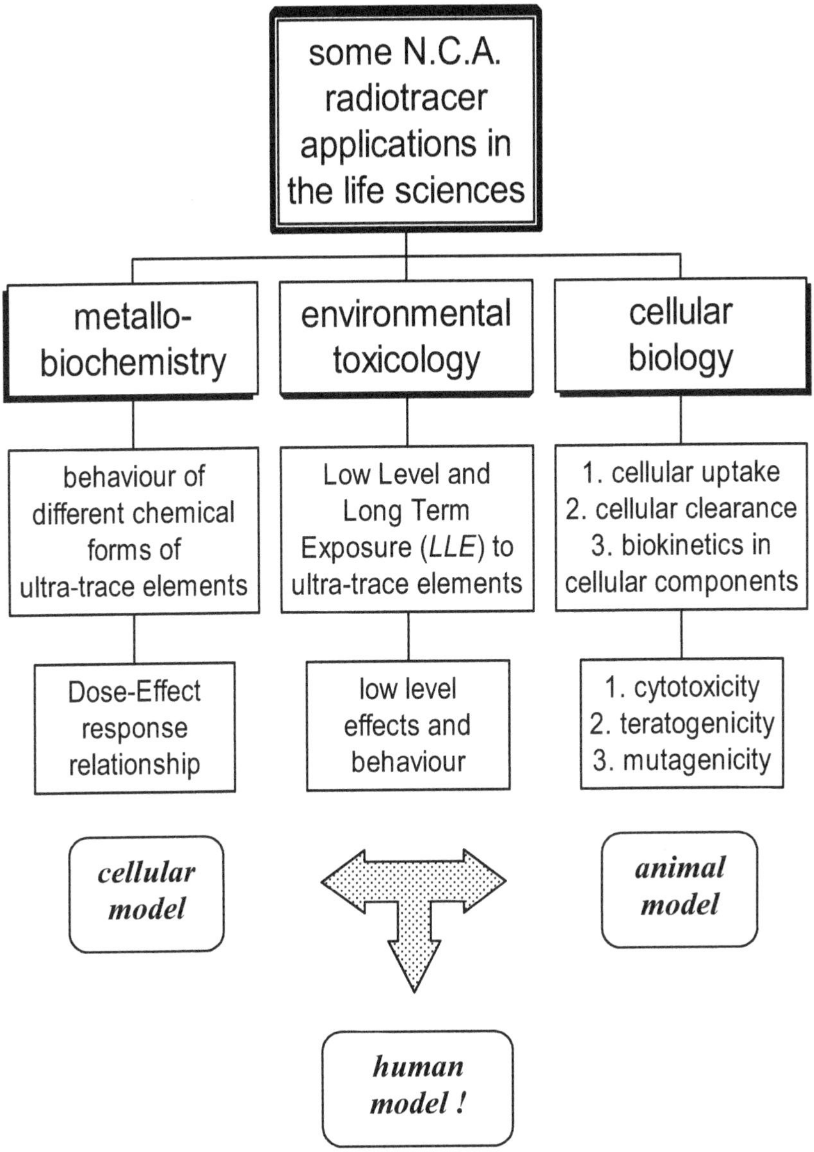

Figure 3. Main modern applications of HSARNs in NCA form (in laboratory, in-vitro, in-vivo)

A complete scheme showing the different steps that must be followed for reaching these results and achieve very high purity RNs and labeled compounds is reported in Figures 2 and 3 are reported the main uses of HSARNs.

In order to determine the *real* experimental value of A_S (NCA) of short-lived HSARNs we used in our Laboratory a wide range of very sensitive analytical, radioanalytical and radiometric techniques, that are summarised in Table 2.

TABLE 2. Main analytical, radioanalytical and radiometric techniques used at LASA (Segrate, Milano) for determination of A_S of RNs presented in this work. In many cases the analysis was carried out after speciation by radiochromatographic techniques

Groups of techniques	Technique	A few selected applications
Nuclear activation	Neutron activation analysis	Re, Os, Sc, Cu, many others
	Proton activation analysis	Pb, Tl, Ge, Bi, others
Electroanalytical	Anodic and cathodic	Se, As, others
	Stripping voltammetry	Zn, Ga, Cd, In, others
Mass spectrometry	Dipole mass spectrometry	Tc
	Quadrupole mass spectrometry	Light elements
	Inductively coupled plasma-MS	Ti
Atomic spectrometry	Graphite-furnace atomic absorption	Pb, Cd, Cu, Fe
	ICP-optical emission spectrometry	Ti
Colorimetry	UV–VIS	Tl(I)
Gamma emitters	High resolution HPGe spectrometry	All gamma emitters
	Low resolution Na(Tl) spectrometry (by decay fitting)	A few pure gamma emitters
Beta and alpha Emitters	High resolution Si (PIPS) spectrometry	Low gamma and X emitters, U, Th, Np, Pu, Am
	Liquid scintillation counting (by decay fitting)	^{64}Cu, ^{66}Ga, ^{67}Ga
Miscellaneous	Radio-release of $^{123}I_2$	Speciation of $^{202}Tl(III)$
	Radio-release of $^{14}CO_2$	Mineralization yield

At the Radiochemistry Laboratory of LASA, a wide range of high specific activity accelerator-produced radionuclides have been produced for a long time in NCA form, for both basic research and application purposes and in particular for uses in metabolic radiotherapy and in related PET and SPECT diagnoses. In particular in the more recent years [10–21] our studies have been devoted to the following RNs production:

1. NCA ^{211}At/^{211g}Po, produced by ^{209}Bi(α, 2n) reaction, with internal spike of gamma emitter ^{210}At from ^{209}Bi(α, 3n) reaction (and small amount of ^{210}Po as radiotoxic long-lived impurity), for high-LET radio- and immuno-radiotherapy.
2. NCA ^{64}Cu, produced by ^{nat}Zn(d,αxn) and ^{nat}Zn(d, 2pxn) reactions for simultaneous β^+/β^- metabolic radiotherapy with intrinsic PET imaging, including the short-lived radionuclide ^{61}Cu, and also ^{66,67}Ga as highly relevant radionuclides.
3. ^{186g}Re, produced by ^{186}W(p, n) reaction for bone metastases pain palliation by negatron (1.1 MeV) metabolic radiotherapy including SPECT imaging.
4. NCA ^{177g}Lu, produced by the alternative indirect ^{176}Yb (d, p) and direct ^{176}Yb (d, n) reaction route for radiotherapy and for SPECT.
5. ^{103}Pd, as a commonly used radionuclide for therapeutic treatments, produced by the alternative ^{103}Rh(d, 2n) reaction route.
6. Studies intended to the production of therapeutic ^{230}Pa/^{230}U alpha-emitters by the reaction ^{232}Th(p, 3n).

2. Experimental

Nuclear activations: Since the beginning of the 1970s, a large range of HSARNs, have been produced at former Cyclotron Laboratory of the UNIMI. Several nuclear data for radionuclide production, measured in our Laboratory, are presently recommended by the Nuclear Data Section of IAEA, Vienna. Recently, nuclear activations are carried out at the cyclotron of JRC-Ispra of EC (Scanditronix MC40, Uppsala, Sweden), as well as at the TRIGA MARK II (General Atomic, USA) nuclear reactor of Pavia. The accelerator ($K = 38$) has proton and alpha beams, with energy variable up to 38 MeV. Deuteron beams of energy up to 19 MeV are also available.

The nuclear reactor is a conventional 250-kW research reactor, with 30% enriched ^{235}U fuel rods. A thermal nuclear flux varying from 10^{12} to $9 \cdot 10^{12}$ $n \cdot cm^{-2} \cdot s^{-1}$ was used in different cases (INAA and isotope production). The epi-thermal component of the neutron spectrum was experimentally measured and was less than 10%.

Nuclear reaction choice, energy threshold, Coulomb barriers: The choice of more suitable nuclear reaction for production of each specific RN is based on a preliminary study. *Q* values, energy thresholds for a charged particle on a target and incoming particle Coulomb Barrier are calculated by the mass defects with kinematics in the laboratory system.

High-resolution gamma spectrometry: It was performed by four 50 cm^3 coaxial HPGe detectors, with intrinsic efficiency of 15% (EG&G, USA), with a peak to Compton ratio of 30:1 at 1,332.50 keV, FWHM 2.3 keV (^{60}Co point source), connected to spectroscopy amplifiers (EG&G, mod. 572, 672) and A/D 2 × 4,096 multichannel analysers, MCAs (EG&G, mod. 918A, 919). The amplifiers were set to gain a channel/energy conversion factor of 2:1. Gamma spectra in the energy range up to 2,000 keV were acquired and analysed by advanced s/w packages. The efficiency data, obtained by decay corrected certified point source of ^{152}Eu ^{226}Ra and ^{133}Ba have been fitted by both Gamma Vision and s/w package Table Curve 1.10 for Windows (Jandel Scientific, AISN Software, Germany).

Beta spectrometry:

1. Liquid scintillation counting, LSC, was performed by a 1,000 channel Liquid Scintillation Counter (Beckman, mod. LS5000TD, USA), with: three energy window capability, random coincidence monitor and Horrocks Number quenching correction method. The quenching correction curves were experimentally determined by addition of increasing amounts of CH_3NO_2 and CCl_4 to scintillation cocktail (Packard, Ultima Gold, Ionic Fluor, Pico Aqua, USA).
2. A higher-resolution liquid scintillation portable spectrometer with α/β pulse shape analysis (PSA) discriminator (Hidex, Finland, mod. Triathler).

Larger activities were measured by well ionisation chamber (Capintec, mod. CRC-15R, USA).

Alpha-spectrometry:

1. A Si surface barrier detector (EG&G, Ortec, 600 mm^2), with resolution of 27 keV (FWHM). In this case, to improve the resolution of the measurements, plastic collimators of 5 mm in diameter were put between the sample and the detector, in spite of the overall decrease of the count efficiency. The detector was calibrated with certified sources of ^{241}Am and ^{233}U (2%, CEA, France).
2. The two liquid scintillation counting, LSC.

Metal determinations were performed with GFAAS (Varian SpectrAA30, LASA), ICP-OES (Perkin Elmer 5500, JRC-Ispra), for most elements (e.g. Pb, Tl, Cd, Ti, V, Cr, Fe, Pd, Zn, Cu, etc.).

For some particular metals (i.e. Os, Au, Ir, Ru, Sc, Sn) use was made of instrumental nuclear activation analysis at the TRIGA nuclear reactor.

Some metal trace determinations (As, Se, Sb, Cu, Zn, Ga) were performed by cathodic and anodic stripping voltammetry also (ASV). A polarography voltampermeter, with Hg dropping electrode (Metrohm Heriseau, mod. Polarecord E506, E608, 648, Switzerland) was used. For most elements, an intercomparison between the different analytical techniques was carried out. Some irradiated targets were mineralised by a microwave digestion system (Milestone, MLS 1200 MEGA, USA), before being chemically analysed.

Specific activity determination: In most cases, in order to determine the experimental value of A_S(NCA), we use radiometric techniques, like β and γ spectrometry *hyphenated* with elemental analysis techniques, like GF-AAS, ICP-OES, NAA and ASV.

3. Results and Discussion

In order to apply the previous definitions and concepts to a practical case, we present in this section some examples related to few selected HSARNs production.

Astatine-211 is an α-emitter that has gained considerable interest for cancer treatment. Its half-life of 7.2 hours matches with the biological half-life of most carrier molecules envisaged for Radionuclide Therapy (RNT) and Radioimmunotherapy (RIT) and would therefore be suitable for applications that require long times for uptake. Its decay scheme exhibits practically 100% yield for the emission of α-particles, with very low intensity gamma emissions [8, 9]: in fact ^{211}At, with its main α to ^{207}Bi at 5.870 MeV 41.80%, is at secular equilibrium by EC (58.20%) with its ultra-short-lived α-emitting daughter Po-211g ($t_{1/2}$ = 516 ms, nuclear spin $9/2^+$, 7.450 MeV 98.89%). The system ^{211}At/^{211g}Po combines a high linear energy transfer LET_∞ of ≈ 100–130 eV·nm^{-1} with a short range of around 70 μm. The emission of Auger electrons contributes significantly to the dose to tissue in the nanometer range, adding to the desired therapeutic effect in targeting single cancer cells. However, the production of ^{211}At via the nuclear reaction ^{209}Bi(α, 2n)^{211}At requires compact cyclotrons, larger than the common baby cyclotrons used for production of PET radionuclides, that can accelerate alpha particles to an energy in the region of 28–29 MeV. This production route is nevertheless preferable to ^{7}Li ion bombardment of ^{209}Bi, producing ^{211}Rn via the (^{7}Li, 5n) reaction, that decays into ^{211}At, or other heavy ion beam methods, or proton induced spallation, that require even more expensive or rare equipment. The direct method based on the nuclear reactions ^{209}Bi(α, 2n)^{211}At seems in any case the most satisfactory, because it leads to a high yield and low contamination by the only radioisotopic impurity ^{210}At, that can be kept either at levels lower than a few percents assuming it as internal spike, or not produced if the α beam energy is kept less than 28.61 MeV because the Q value of the interfering nuclear reaction $^{209}Bi(\alpha, 3n)^{210}At$ is equal to 28.1 MeV, as confirmed from the experimental cross section curves for the two nuclear reactions as shown in Figure 4, that gives only the fit lines for the excitation functions.

Furthermore, as can be easily calculated by Q values, the production of ^{210}Po is allowed in the energy range from 15 to 29 MeV through the direct reactions $^{209}Bi(\alpha, t)$, (α, dn) and (α, p2n); regarding these direct reactions, Table 3.

Figures 5 and 6 show gamma spectra from Bi targets irradiated with α-particles with energies of 28.8 and 32.8 MeV, respectively. As it was expected, the amounts of ^{210}At are quite different: 0.001% and 7.4% of activity respectively.

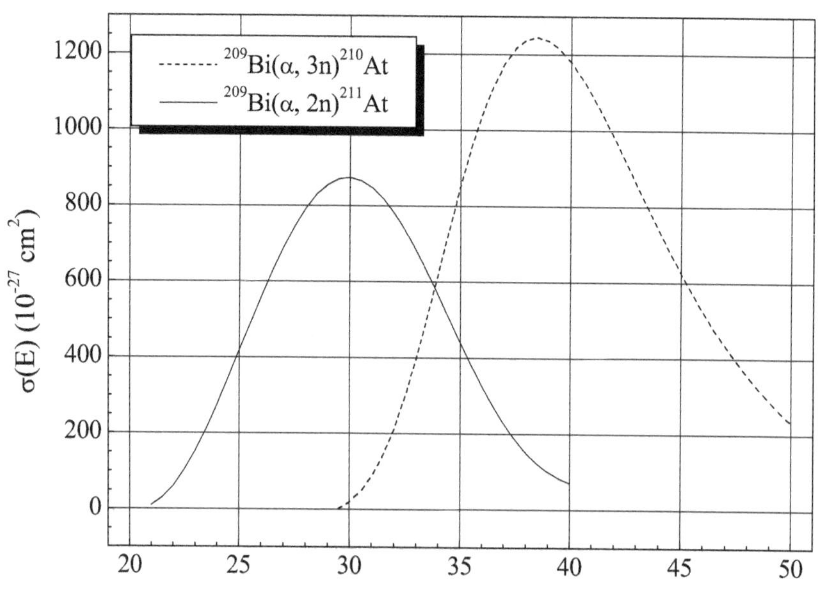

Figure 4. Fit of the experimental cross-section data for the $^{209}Bi(a,2n)^{211}At$ and $^{209}Bi(a,3n)^{210}At$

TABLE 3. Nuclear reactions for production of At RNs of interest with the energy threshold and the corresponding half lives

Reaction	E_S(MeV)	$t_{1/2}$
$^{209}Bi(\alpha, 2n)^{211}At$	20.7	7.214 h
$^{209}Bi(\alpha, 3n)^{210}At$	28.6	8.1 h
$^{209}Bi(\alpha, p\ 2n)^{210}Po$	23.8	
$^{209}Bi(\alpha, t)^{210}Po$	15.1	138.376 days
$^{209}Bi(\alpha, d\ n)^{210}Po$	21.5	

The low count-rate of the 1,064 keV line, that is present only in the spectrum of Figure 5, a spectrum that was measured for a very long time of 50,000 s, is due to the long-lived ^{207}Bi isotope, confirming that the short-lived ^{211m}Po is not formed in the EC of ^{211}At. The very small contribution of ^{210}At in the irradiation with 28.8 MeV α particles, confirms the right choice of the irradiation energy to obtain an extremely pure production of ^{211}At isotope and the correct energy calibration of the beam

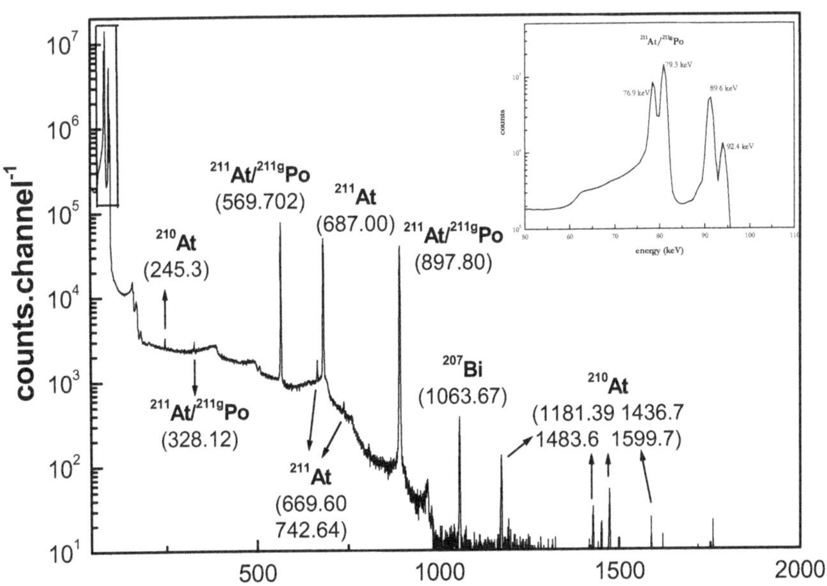

Figure 5. Gamma spectrum obtained with a HPGe detector from a Bi target irradiated with 28.8 MeV alpha-particles. In the zoom region, the Po X-ray emissions (see text) is clearly seen

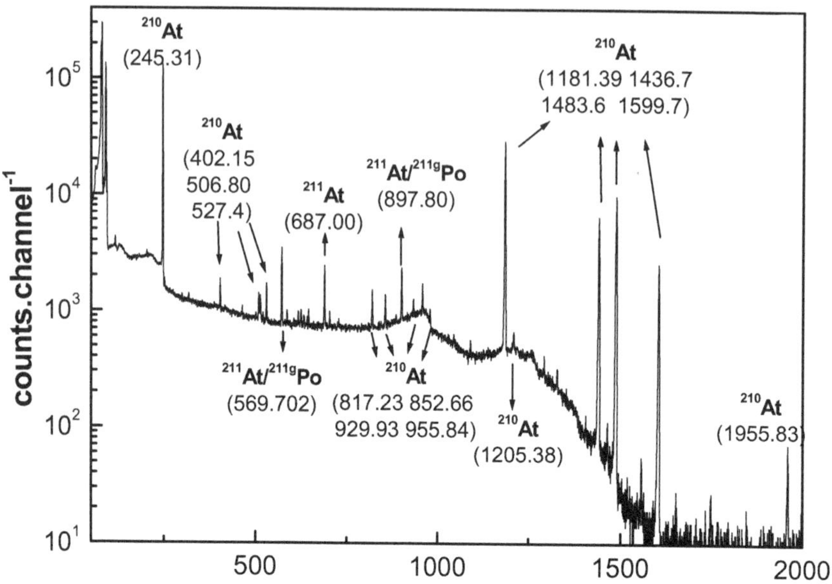

Figure 6. Gamma spectrum obtained with a HPGe detector from a Bi target irradiated with 32.8 MeV alpha-particles

line of the cyclotron used. In spite of large amount of auto-absorption, it is important to realize that the highest count rates are for the 77 and 88 keV X-rays of polonium. Thus they were be used to measure low activities of ^{211}At/^{211g}Po. For the 28.8 MeV irradiation, our experimental thick-target yield, calculated from the α spectrum of irradiated Bi, measured before any chemical treatment, was 8,085 $\pm$ 176 MBq·C^{-1}, calculated at the End Of an Instantaneous Bombardment (EOIB). This value agrees very well with the value 8,341 MBq·C^{-1}, obtained by integration of the literature data for the excitation functions of this reaction, Figure 4, from 28.8 down to 20 MeV: a discrepancy of ~ 3%, is considered well within the experimental error of both measurements.

In order to produce NCA ^{211}At/^{211g}Po for metabolic radiotherapy, a suitable radiochemical separation of At radionuclides from Po by-products and from the Bi target is mandatory, without voluntary addition of an isomorphous carrier. The first radiochemical separation adopted was a classical "wet" method based on liquid/liquid extraction, carried out in transparent Teflon PFA funnels to prevent the adsorption of radioactive species on surfaces. The radionuclidic purity of the different radio-chemistry fractions produced along the procedure were measured with a HPGe γspectrometry, that allows to distinguish between ^{211}At and ^{210}At and is effective to determine the contamination of the ^{211}At preparation with ^{210}At. For solutions with low activity concentration the amount of ^{211}At was also determined by $K_{\alpha1}$, $K_{\alpha2}$, $K_{\beta1}$, and $K_{\beta2}$ X-rays measurement due to their much larger intensities, while due to the larger intensity of the α-lines of ^{210}At, even small percentages of ^{210}At can be determined. In addition the different fractions were measured by:

1. α spectrometry using a *Si surface barrier detector*. In this case the different liquid fractions obtained were deposited on Ag foils (5 × 5 mm^2), by stirring the Ag foil inside the solution of ^{211}At and counted under an oil-free rotational pumping vacuum system (KNF). Although α–spectra measurements has high resolution leading very well resolved peaks, they require time consuming procedures during the deposition stage accompanied by a low and unreliable overall radiochemical recovery yield, yielding a practical counting efficiency much lower then the theoretical value of 50% as a consequence of a 2π counting geometry. The α spectrum of the deposition on a Ag metal foil from the original dissolution solution before the first L/L extraction of the 32.8 MeV irradiation is shown in Figure 7. The peaks of ^{211}At, ^{211g}Po and ^{210}Po are clearly resolved. The α spectrum of the aqueous phase after the solvent–solvent extraction is given in Figure 8, which clearly shows only the presence of the ^{210}Po peak, indicating that only Po remained in the aqueous phase while all the At was removed almost quantitatively. Conversely the organic phase contains only the At radionuclides and there is no presence of ^{210}Po as demonstrated by the αspectrum reported in Figure 9.
2. Two different *liquid scintillation counting and spectrometry systems*. In particular, the Triathler "multilabel tester" with PSA discrimination, permitted α activity measurements to be made in a very short time with very high count efficiency (about 100% and 4π counting geometry). In this way it was possible to reduce the measurement times, especially mandatory for short-lived radionuclides and to reduce doses to personnel, although on account of less well resolved spectrum. However this method is not able to separate directly the α line of ^{211}At at 5,870 keV from its possible contaminant ^{210}Po (daughter of ^{210}At) with α energy of 5,304 keV, whose content was determined after complete decay of At nuclides. The same method was adopted for determination of ^{210}Po in the aqueous phase, showing that this radionuclide was produced by direct $^{209}Bi(\alpha, xpyn)$ reactions, other than by decay of ^{210}At, Table 3. Figures 10a and b give a typical spectrum of $^{211}At/^{211g}Po$

obtained with the liquid scintillation in 3D and 2D. Using the counts related to the windows indicated in Figure 10b the typical decay curves (log counts vs. decay time) for the α peaks of both ^{211}At and ^{211g}Po are given in Figure 11. Both peaks give the half-life of ^{211}At, as ^{211g}Po is at secular equilibrium with ^{211}At. The discrepancy between the measured half-life of ^{211}At and the literature value is not larger than 1.2%.

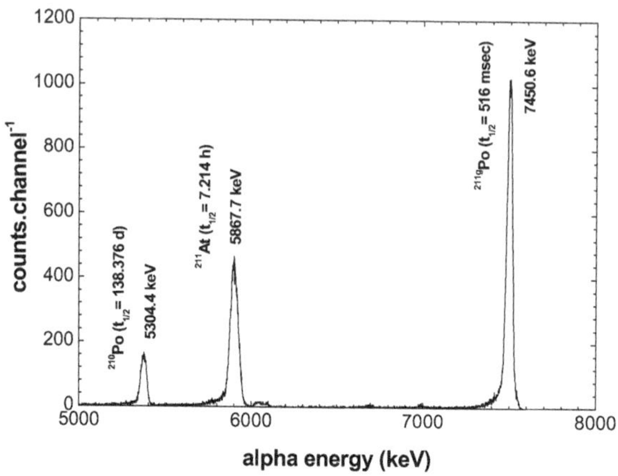

Figure 7. Alpha spectrum from a deposit on Ag of the solution extracted from Bi irradiated with 32.8 MeV alpha-particles after dissolution by 8 M HCl (before the L/L extraction), measured with a Si surface barrier detector and a 5 mm collimator, 2 days after the EOB

With this procedure we found that the ^{211}At is separated into the organic phase with a yield greater than 99% and is quantitatively separated from the Bi and ^{210}Po impurities. On the other hand for the next step it is necessary to bring it back to the aqueous phase (stripping or back-extraction), which can be done at either basic pH or with a reducing agent. The best results were found with both basic pH and a reducing agent such as sodium sulphite or sodium thiosulphate. However, the reducing agent should be avoided due to the requirements of the next labelling procedure. In order to overcome this problem a thermocromatographic radiochemical separation based on the fusion of the target at high temperature was developed. In this method only At is distilled and it is decontaminated from the target and from the Po impurities. It consists of the sublimation of

At radionuclides and Po radio-impurities in air-flux at 1.2 atm, the interaction of these elements with a chemical filter of CaO for the purification and the trapping of At radionuclides in a cryotrap. This method allowed to obtain a radiochemical yield larger than 64% with a radio-impurities concentration smaller than $3 \cdot 10^{-3}$%; in these separations we obtained the first Italian thermochromatogram separation.

Copper-64 ($t_{1/2}$ = 12.701 h, β^- 579 keV End Point, 39%; β^+, 653 keV End Point, 18%, annihilation radiation 511 keV, 43%, γ at 1,345.84 keV, 0.473%), is a radionuclide suitable for labelling of a wide range of radiopharmaceuticals, for both PET imaging (maximum range of β^+ in soft tissue ≈2.7 mm and "average" range ≈1 mm), as well as systemic and radionuclide immunotherapy of tumours. Among the several methods for production of NCA ^{64}Cu (together with the short-lived positron emitter ^{61}Cu, $t_{1/2}$ = 12.70 h, β^+ , 1,300 keV End Point, 61%), we investigated the deuteron irradiation via (d,αxn) plus (d, 2pxn) nuclear reactions on natural Zn targets in the energy range up to 19 MeV with *simultaneous* production of ^{66}Ga, ^{67}Ga and ^{65}Zn, ^{69m}Zn. As a relevant by-product of ^{64}Cu production. ^{67}Ga applications in nuclear radiodiagnostic medicine are known since a long time.

The thin-target excitation functions for ^{61}Cu, ^{64}Cu, ^{66}Ga, ^{67}Ga, ^{65}Zn and ^{69m}Zn RNs, were experimentally measured after irradiation of "thin" zinc targets of regular thickness, in which the particle energy loss was of the order of 100 keV at high energy and about 400 keV at the lower energies. The stacked-foil technique was adopted, with catchers and Al/Ti monitor foils to control the reliability of the charge integration system. As beam intensity monitor the following nuclear reactions were chosen: the ^{27}Al(d, X)^{24}Na in the energy region from 10 to 19 MeV and ^{nat}Ti(d, xn)^{48}V in the energy region below 10 MeV, by using the cross-section data recommended by IAEA [9]. Three stacks of six thin Zn foils each were irradiated at 19, 14 and 10 ± 0.2 MeV deuteron energy, with a deuteron beam current of 100 nA.

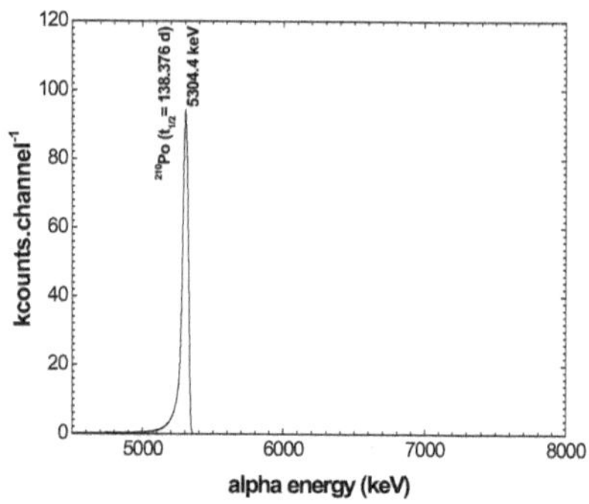

Figure 8. Alpha spectrum of the deposit on Ag from the aqueous fraction after L/L extraction, obtained by a Si surface barrier detector

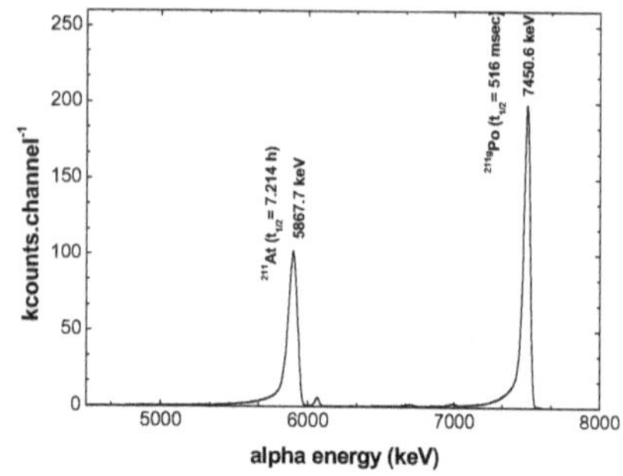

Figure 9. Alpha spectrum from a deposit on Ag of the organic solvent fraction after L/L extraction, obtained by a Si surface barrier detector

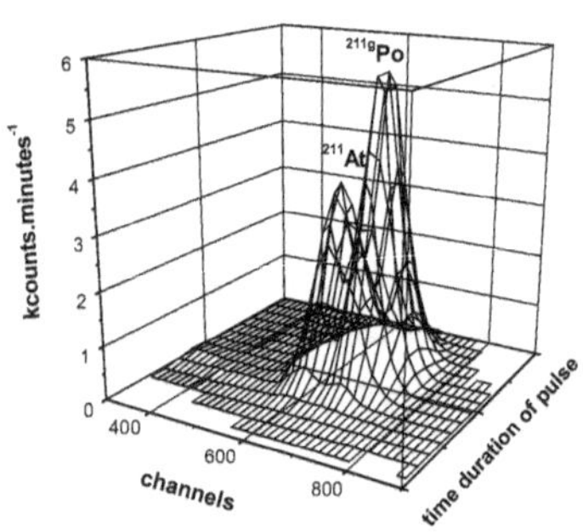

Figure 10a. 3D Alpha spectrum of the deposit on Ag from the aqueous fraction after L/L extraction, obtained by the LSC detector with pulse shape discriminator

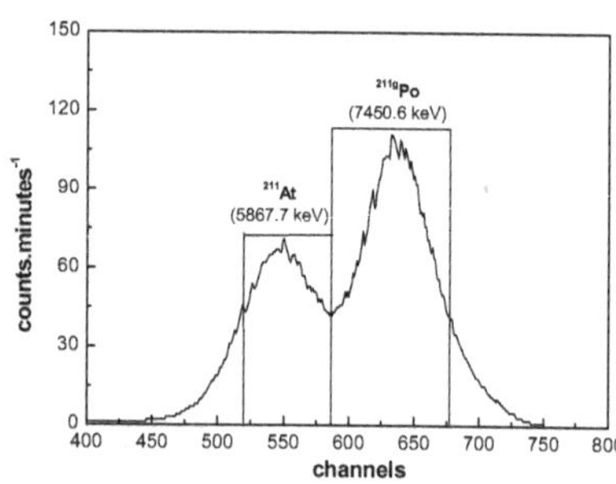

Figure 10b. 2D Alpha spectrum from a deposit on Ag of the organic solvent fraction after L/L extraction, obtained by the LSC detector with pulse shape discriminator

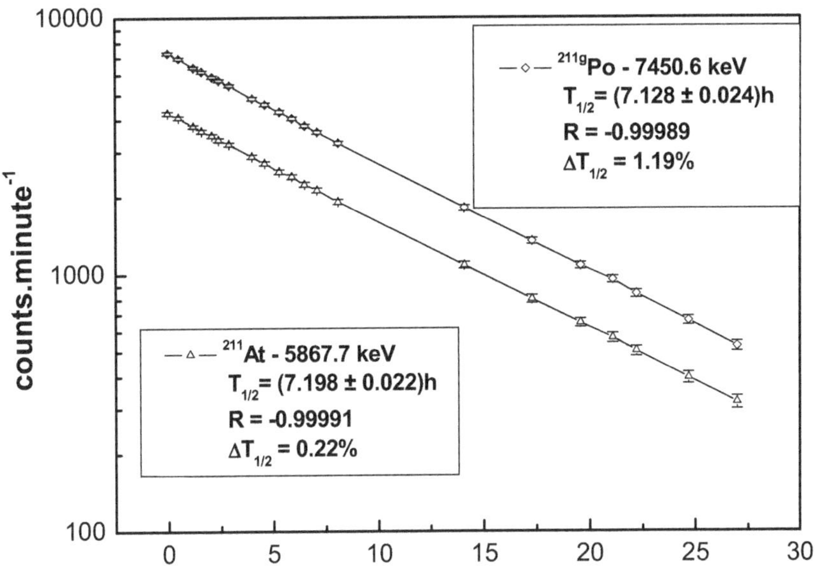

Figure 11. Typical decay curves of alpha peaks of ^{211}At and ^{211g}Po at 5,870 and 7,451 keV, respectively, obtained by liquid scintillation counting with PSA from the data of the two windows indicated in Figure 10b

In Figure 12a is reported a typical gamma spectrum of deuteron irradiated natural Zn target. Figure 12b is the zoom of the energy range corresponding to the only gamma emission at 1,345.84 keV of ^{64}Cu, the gamma line (α = 0.473%) for ^{64}Cu, even if very close to the two lines of ^{66}Ga, is very well resolved and allows an accurate determination of the ^{64}Cu activity. The reliability of such measurements is confirmed by the χ^2 value of the decay curve, as shown in Figure 13.

The experimental thin target yields from the complete experiment are shown in the Figure 14. In particular from this figure the ^{64}Cu and ^{61}Cu excitation function crossing suggested the possibility to optimise the irradiation conditions in order to obtain a ^{64}Cu with high yield and radionuclidic purity. The yield of ^{61}Cu is very much higher at higher deuteron energies, thus the optimum irradiation conditions do not correspond to the maximum energy available.

The thin-target excitation functions for the two radionuclides were integrated, obtaining at 19 MeV a very good agreement with the thick-target yield experimentally measured as reported in Figure 15a and b.

For producing simultaneously NCA ^{64}Cu and ^{66,67}Ga, a suitable radiochemical separation of Cu RNs both from Ga and Zn target is required.

Among possible methods to recover ^{64}Cu, we developed a wet radiochemical separation. The overall radiochemical yield for Cu was > 80% at EOP. Nevertheless, variable amounts of Cu carrier are present in any ultra-pure material and chemical; thus, we determined the real A_S (NCA) by GF-AAS. An IDF° = 200 was achieved, under the "mild" irradiation conditions adopted (4 h irradiation with a 1 μA deuteron beam current), with an experimental value of A_S (NCA) > 700 MBq μg^{-1}. This value is very high under the irradiation conditions adopted and could be increased sharply using a long irradiation, with the maximum beam current available at Ispra Cyclotron Laboratory. HPGe γ-spectrometry was used for RNP determination and to follow the radiochemical separation steps A gamma spectrum of the high pure Cu fraction, that after the radiochemistry contains less than 0.01% of the initial activity of Ga RNs, is reported in Figure 16. Moreover, the decay of the high purity ^{64}Cu and the fractions containing Ga RNs were followed by beta spectrometry also, by liquid scintillation counting. In Figure 17 is reported an example spectrum of Cu fraction. The radiochemical yields for ^{66,67}Ga were >98%.

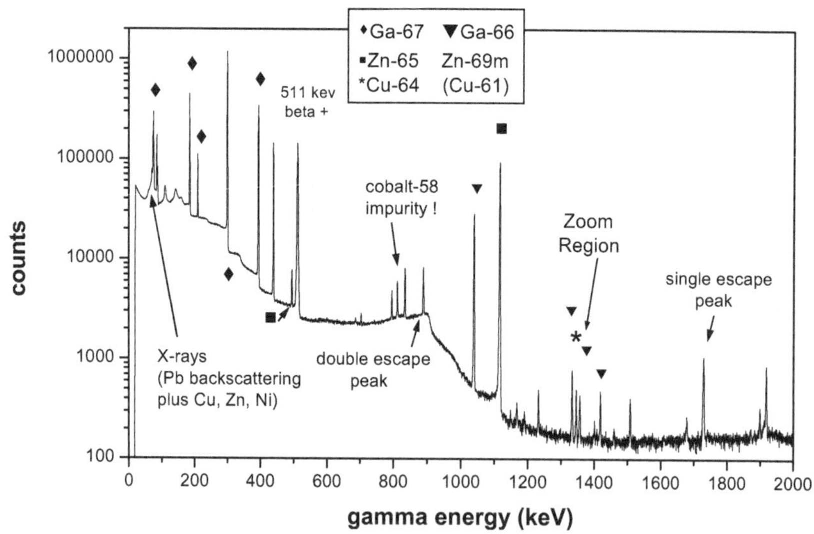

Figure 12a. Typical HPGe spectrum of deuteron irradiated Zn target. The spectrum was collected by shielding the target with a 7 mm thick high energy positron annihilator and 3 mm Pb absorber, in order to increase S/N ratio in high energy region (zoom region of interest). Both single and double escape peaks are detectable

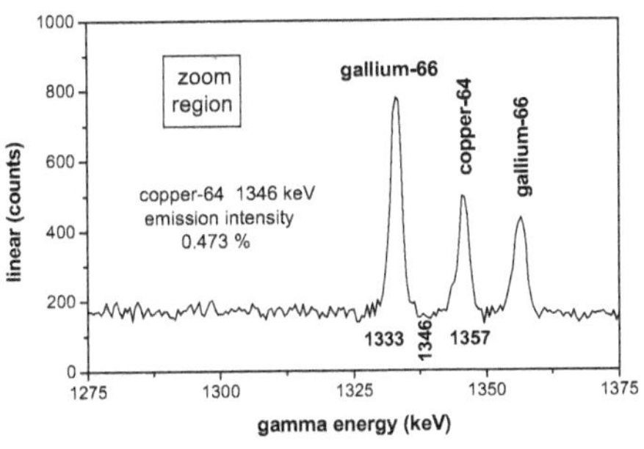

Figure 12b. Significant γ line spectroscopic data put in evidence in the energy interval of interest

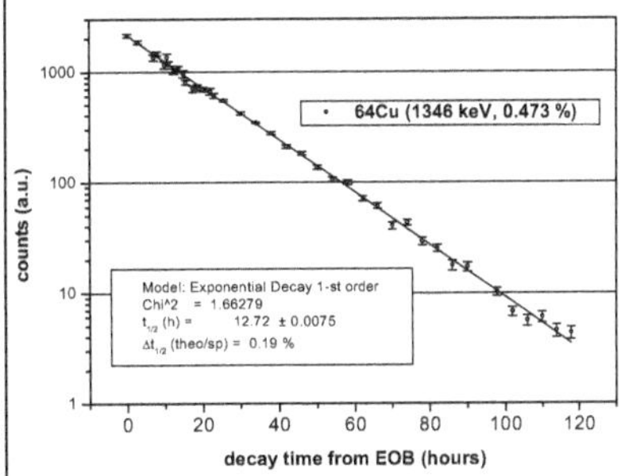

Figure 13. Decay regression fitting of first order for ^{64}Cu gamma emission at 1,346 keV (0.473%), with evidence for the relevant χ^2 value

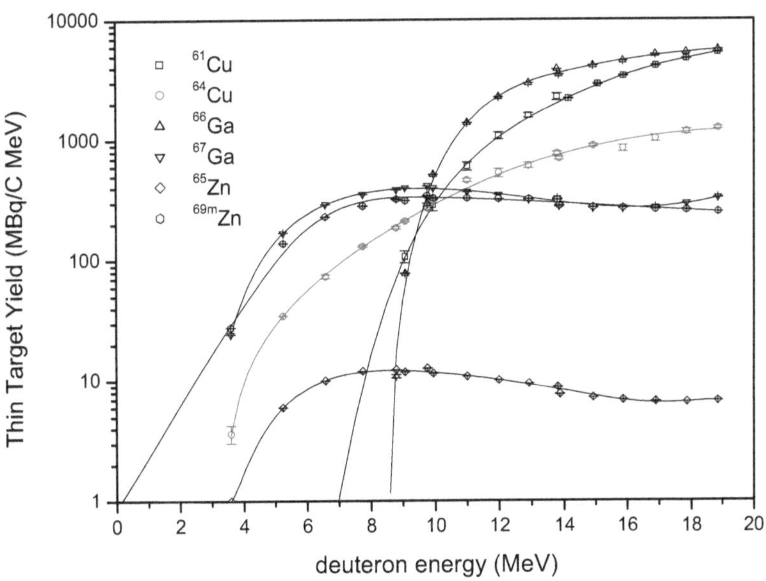

Figure 14. Experimental thin target yields obtained for the different $^{nat}Zn(d, X)$ nuclear reactions

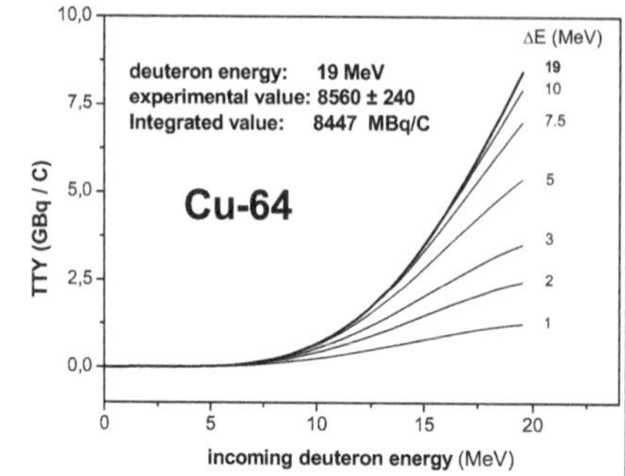

Figure 15a. Integrated thick-target yield for ^{64}Cu as a function of energy loss into the target. The data at 19 MeV deuteron energy are compared with the experimental value

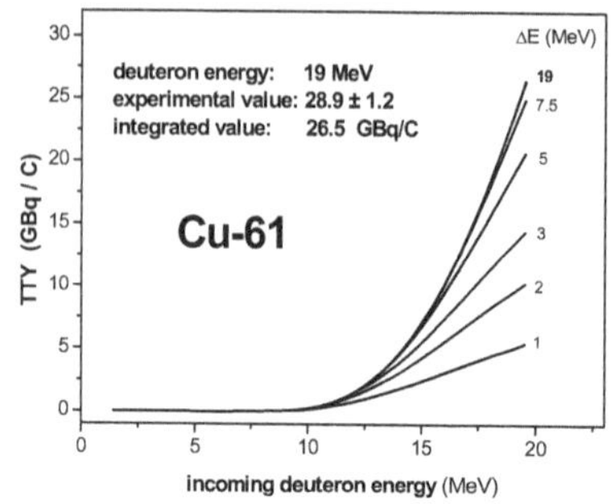

Figure 15b. Integrated thick-target yield for ^{61}Cu as a function of energy loss into the target. The data at 19 MeV deuteron energy are compared with the experimental value

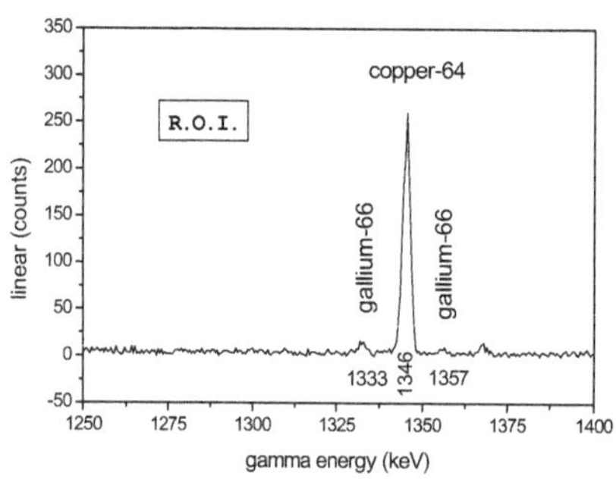

Figure 16. HPGe spectrum of pure Cu fraction after radiochemistry. The same zoom of region of interest of Figure 12b

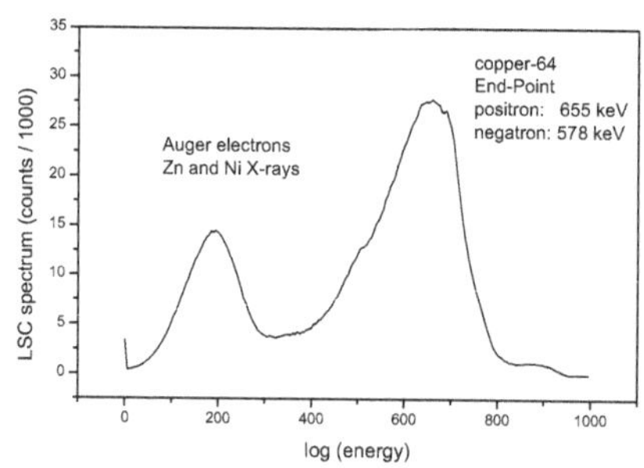

Figure 17. Log-energy scale LSC beta spectrum of radiochemical pure Cu fraction

4. Conclusion

Medium energy cyclotron and sometimes nuclear reactors, allow production of HSARNs for biomedical, environmental, toxicological and occupational studies about long-term exposure to low doses of ultra-trace elements present in the ecosystems. Specific activities of the order of GBq·µg^{-1} are achieved, so the experiments can be carried out in the sub-nanogram concentration range.

HPGe gamma spectrometry, allows both qualitative and quantitative analytical and radioanalytical investigations on activated targets and radiochemical fractions. Conversely, β spectrometry by LSC is intrinsically characterized by low-resolution capability, due to continuous spectra of beta emitters, but presents a very high counting sensitivity. After the End Of Chemical Processing of target itself, β spectrometry can be used effectively to follow either β or electron decay of pure RN. Elemental analysis techniques like AAS, ICP-OES, INAA, RNAA, ASV, allow the determination of A_S(NCA) and the chemical purity of the radiotracer itself. As expected, the real A_S is lower than the theoretical carrier-free one; nevertheless, in most cases, the NCA HSARNs obtained are suitable for purposes envisaged. The use of extra-pure chemicals, targets and equipment allows reaching IDF values in the range from some tens to a few hundreds in some selected cases. Much lower IDF values were obtained with very rare elements.

Acknowledgement: Our research work is funded in the mainframe of the experiments of the Commission V of INFN and the MIUR, Italy. The contributions of JRC-Ispra of EC and ENEA-Bologna is substantial.

References

1. Bonardi M et al. (2002) High specific activity radioactive tracers: a powerful tool for studying very low level and long term exposure to different chemical forms of both essential and toxic elements. Microchem J 73:153–166
2. Bonardi M L et al. (2004) Cyclotron production and quality control of "High Specific Activity" radionuclides in "No Carrier Added" form for radioanalytical applications in the life sciences. J Radioanal Nucl Chem 259:415–419
3. Bonardi M et al. (1988) Fundamental parameters for the optimization of yield and radionuclidic purity of accelerator produced radioisotopes. Part I: thin-target excitation functions. Phys Med 1:23–46
4. Bonardi M et al. (1988) Fundamental parameters for the optimization of yield and radionulidic purity of accelerator produced radioisotopes. Part II: beam control and monitoring. Phys Med 2:83–101
5. Bonardi M et al. (1995) Irradiation methods for production of high specific activity radionuclides in No Carrier Added Form. J Radioanal Nucl Chem 195:227–236
6. Birattari C et al. (2001) Review of cyclotron production and quality control of high specific activity radionuclides for biomedical, biological, industrial and environmental applications at INFN-LASA. Proceeding International Congress on Cyclotrons and Their Applications, Cyclotron2001, East Lansing, MI, May 2001. American Institute of Physics, Melville, NY
7. Bonardi M L et al. (2003) Thin-target excitation functions: a powerful tool for optimising yield, specific activity and radionuclidic purity of accelerator-produced radionuclides. Czech J Phys 53:A393–A403

8. Bonardi M (1988) The contribution to nuclear data for medical radioisotope production from the Milan Cyclotron Laboratory, IAEA Consultant's Meeting on "Nuclear Data Requirements for Medical Radioisotope Production", Tokyo, April 1987, IAEA Document, INDC(NDS)-195/GZ, IAEA, Vienna, Jan. 1988, pp. 98–112
9. IAEA-TECDOC-1211 Charged-particle cross-section database for medical radioisotope production. Co-ordinated Research Project (1995–1999). IAEA Vienna, Austria, May 2001 Available online at: http://www-nds.iaea.org/medical/
10. Menapace E et al. (2004) Experimental results and model calculation of excitation functions relevant to the production of specific radioisotopes for metabolic radiotherapy and for PET. Radiat Phys Chem 71:943–945
11. Groppi F et al. (2005) The use of liquid scintillation counting as a very sensitive radioanalytical tool for the determination of alpha, beta and electron emitting impurities in radiopharmaceutical compounds. J Radioanal Nucl Chem 263:521–525
12. Bonardi M L et al. (2005) Cross section studies on ^{64}Cu with zinc target in the proton energy range from 141 down to 31 MeV. J Radioanal Nucl Chem 264:101–105
13. Abbas K et al. (2006) Cyclotron production of ^{64}Cu by deuteron irradiation of ^{64}Zn. Appl Radiat Isotopes 64:1001–1005
14. Groppi F et al. (2005) Optimisation study of alpha-cyclotron production of At-211/Po-211g for high-LET metabolic radiotherapy purposes. Appl Radiat Isotopes 63:621–631
15. Groppi F et al. (2006) Results on accelerator production of innovative radionuclides for metabolic radiotherapy and PET and on related nuclear data. Nucl Instrum Meth A 562:1072–1075
16. Morzenti S et al. (2008) Alpha-cyclotron production of ^{211}At/^{211g}Po by ^{209}Bi(α, 2n) reaction. J Radioanal Nucl Chem 276:843–847
17. Zona C et al. (2008) Wet-chemistry method for the separation of no-carrier-added ^{211}At/^{211g}Po from alpha cyclotron irradiated ^{209}Bi target. J Radioanal Nucl Chem 276:819–824
18. Ridone S et al. (2004) Radioanalytical quality control on beta-emitting [^{186g}Re]- and [^{153}Sm]-radiopharmaceutical compounds for bone metastases pain palliation. World J Nucl Med 3:241
19. Persico E et al. (2006) Proton and deuteron cyclotron production studies of high specific activity ^{186g}Re for radiotherapy. Technetium, Rhenium and other metals in Chemistry and Nuclear Medicine, SGE Editoriali, Padova, Italia, 7:613–614
20. Groppi F et al. (2005) Accurate determination of radionuclidic purity and half-life of reactor produced Lu-177g for metabolic radioimmunotherapy, World SciCo Singapore, 710–714
21. Canella L et al. (2008) Accurate determination of radionuclidic purity of reactor produced ^{177g}Lu for metabolic radiotherapy. J Radioanal Nucl Chem 276:813–818

DETERMINATION OF RADIONUCLIDES IN ENVIRONMENTALS SAMPLES

PAVOL RAJEC*, ĽUBOMIR MÁTEL, OLGA ROSSKOPFOVÁ, SILVIA DULANSKÁ AND DUSAN GALANDA
Comenius University, Department of Nuclear Chemistry, SK-842 15, Bratislava, Slovakia

Abstract. The Radiochemical Analytical Laboratory (LARCHA) at the Department of Nuclear Chemistry, Comenius University was established to monitor environmental conditions in the country. Implementation of the ISO 17025:2005 standard is a very important step for the development of research work and accreditation of university laboratories is appreciated by the nuclear power plants (NPP) and other nuclear facilities, local governmental authorities and international agencies. The techniques used for the determination and separation of alpha and beta emitters in environmental soils, sediments, water, etc. are discussed in this paper.

Keywords: Extraction chromatography, natural radionuclides, alpha–beta detection

1. Introduction

Due to the existence of nuclear power plants (NPP) in Slovakia, monitoring of anthropogenic radionuclides have to be performed and concentration of alpha, beta and gamma radionuclides should be determined in surface water and groundwater in drilled holes, soil and sediments. The concentration of natural and artificial radionuclides emitting alpha and beta activity at low levels is determined in complex environmental matrices; therefore, separation chemistry plays an important role in their determination. The present paper describes some separation techniques used in radioecology, with a special emphasis on extraction chromatography. The very high necessary decontamination factors are pointed out together with the suggested techniques.

* To whom correspondence should be addressed. e-mail: rajec@fns.uniba.sk

There are two main groups of radionuclide which are usually monitored. Natural radionuclides uranium, thorium and their progeny occurring in soil, water and minerals belong to the first group. Some of the progeny are radiotoxic alpha emitters and there are limits for their intake from water. The limit for the uptake of radionuclides via drinking water has been established in the form of a "total indicative dose" of 0.1 mSv per year [1]. Radium, radon, polonium are the most important radionuclides and the radioactivity concentration in tap water is less than 0.1 Bq/L and the gross-beta activity cannot exceed 1.0 Bq/L [1]. The gross-alpha and beta activity are the first method for the activity determination in water. More and more laboratories started to measure individual radionuclides like ^{226}Ra, ^{210}Pb, ^{210}Po, ^{232}Th, and natural U.

The second group is formed by anthropogenic radionuclides produced during nuclear weapons tests, nuclear satellites burn-up in the atmosphere and nuclear power accidents that may release large amounts of activities. The nuclear power industry is permitted by health authorities to continually release small, controlled amounts of specified radionuclides into the atmosphere and into open waters [2]. The most important anthropogenic alpha radionuclides in environment are ^{241}Am, ^{237}Np and ^{239}Pu [3]. Plutonium can be found in environmental samples as fallout from nuclear weapons tests, nuclear reactor accidents, discharges from reprocessing plants and radioactive waste.

Activity of pure alpha and beta emitters is difficult to determine by direct measurement and separation processes prior to their determination is needed [3]. During the history, separation techniques have been under a continuous development and today modern, efficient and fast methods are used in many laboratories. Other factors that affect the separation processes are price, organic solvents and their toxicity, material quality namely in the labs working under the ISO 17025:2005. The main separation techniques used for chemical separations involve oxidation–reduction, complex-ion formation, distillation/volatilization, solvent extraction, precipitation, co-precipitation, electrochemistry, and chromatography. The success or failure of a radiochemical procedure often depends on the ability to separate extremely small quantities of radionuclides that might interfere with detection of the analyte. Oxidation and reduction processes play an important role in radioanalytical chemistry, particularly from the standpoint of the dissolution, separation, and detection of analytes, tracers, and carriers. Effectively separation of actinides depends on the stabilization of selected separation oxidation state [4]. There are standard radioanalytical procedures, which are in the radio analytical manuals at the present days [3, 5–17].

Solid Phase Extraction (SPE) is a widely used technique for the isolation and concentration of analytes from liquid samples to achieve increased sensitivity in analytical process. Empore solid phase extraction membranes offer an innovative solution to environmental and bioanalytical sample preparation problems. Empore extraction disks provide a sample preparation solution for large volume samples. The disk format provides a large surface area for sorbent/sample contact. Faster flow rates and higher through put are realized compared to liquid–liquid extraction or traditional packed column technology [17]. The Empore membrane technology incorporates solid phase sorbent particles within a network of polytetrafluoroethylene (PTFE) fibrils. The Empore SPE membranes applied as cartridges and plates offer fast sample flow rates (as the mass transfer kinetics of the tightly packed particles allow recoveries that are independent of sample flow rate); reduced solvent usage compared to liquid–liquid extraction and traditional packed particle SPE products; clean eluates (the PTFE fibrils minimize occurrence of fine particles, extending the column life) and potential elimination of solvent evaporation/ reconstitution steps [17].

3M Empore is material with Super Lig 644C bound into it. SuperLig is selective, ignoring competing ions. The Empore membrane is a mesh of polymer-based fibrous material that acts as a high-flow filter.

A variety of SPE materials called AnaLig have been developed using "molecular recognition" ligands on solid supports and commercialized by IBC Advanced Technologies [18]. These ligands are covalently bound to various polymer or silica-gel supports. These materials have been developed and investigated mainly for water, wastewater or effluent cleanup processes. AnaLig materials selective for Sr, Ra, and Tc were incorporated in Empore Rad disks marketed by 3M for use in analysis of water samples.

2. Experimental

2.1. PREPARATION OF SAMPLES

The samples of soil were grinded, homogenized and dried to a constant weight at 105°C for at least 8 h, then ashed at 550°C for decomposition of the organic matter. Tracers (^{242}Pu, ^{243}Am, ^{232}U) and carriers (Sr^{2+}, Y^{3+}) were added to the samples before digested with 8 M nitric acid and hydrogen peroxide. The schemes of sample preparation which were used in LARCHA Laboratory:

(a) Digestion in autoclave: 20 g samples were digested in a mixture 50 mL of 8 M HNO_3 + 2.0 mL of 30% H_2O_2 (for oxidation of a remaining organics in the sample) at 150°C for 8 h. Samples were cooled and centrifuged. The supernatant was transferred to a beaker and digestion in autoclave was repeated twice.

(b) Digestion using a hot-plate heating: 5 g sample was weighed into a beaker and digested (20 mL 8 M HNO_3 + 3 mL H_2O_2) during heating on a hot plate at 70°C for about 8 h. The solid and liquid phases were separated, cooled and centrifuged. Leaching was repeated three times. All the supernatants were collected.

(c) Microwave digestion procedure (Microwave laboratory system Milestone Ethos): 0.5 g dry samples, 4 mL conc. HNO_3 and 100 μl H_2O_2 were added into each microwave container. Digestion: 20 min ramp 20 min digestion to 190°C. Both phases were cooled and centrifuged.

2.2. SEPARATION TECHNIQUES

2.2.1. *Separation of plutonium*

Plutonium was separated by liquid–liquid extraction with 30% Aliquat-336 in toluene. Aliquat-336 (Tricapryl methyl ammonium chloride) is a quaternary ammonium salt extracting different species with an anion exchange mechanism [19, 20]. Aliquat-336 extracts tetravalent actinide nitrate complexes. Aliquat-336 was diluted in 30% toluene and converted to the nitrate form by equilibration with 4 and 8 M HNO_3. Before separation of plutonium 2–3 g of sodium nitrite were added to ensure conversion of Pu(III) to Pu(IV) [21]. The molarity of the solution has to be 8–8.4 M.

The aqueous phase contained ^{241}Am, ^{242}Cm, ^{90}Y, ^{90}Sr and ^{137}Cs were left aside for analysis of ^{241}Am and ^{90}Sr. Radionuclide impurities, such as uranium and thorium, were washed out from the organic phase containing plutonium by using 8 M HNO_3 (twice for uranium) and by using concentrated HCl (four times for thorium). Plutonium was eluted from Aliquat-336 using 0.15 M HCl–0.025 M $H_2C_2O_4.2H_2O$. The sample was evaporated and was ashed in a muffle furnace for 30 min at 550°C. After cooling, 4 mL of concentrated HNO_3 was added and evaporated to dryness. The sample was dissolved using 4 mL of concentrated HCl, evaporated and then again dissolved with 6 mL 1 M of HCl and transferred to a plastic tube. These were sample preparations for alpha spectrometry.

2.2.2. *Separation of Americium extraction method*

The solution was evaporated and consecutively dissolved at in 0.1 M HNO_3/4 M $NaNO_3$. Americium was separated by liquid–liquid extraction with 0.3 M TOPO/toluene [19]. Americium was then eluted from TOPO/toluene using 4 M HNO_3 and washed with toluene (twice). The solution was evaporated and dissolved with 6 mL of 1 M HCl. The final source for alpha spectrometry of americium was prepared by the micro co-precipitation with NdF_3 as plutonium [19, 23].

2.2.3. *Separation of uranium aqueous sample and soil samples [24]*

^{232}U tracer was added to 1 L aqueous sample and sample was evaporated to dryness. Solid was dissolved in 10 mL of 1 M $Al(NO_3)_3$ – 3 M HNO_3. *UTEVA Resin* – prepacked column was installed into vacuum box and was conditioned with 5 mL of 3 M HNO_3 (flow rate 1 mL min^{-1}). After introducing the sample, the sorbent was washed two times with 5 mL of 3 M HNO_3. The sorbent was converted to chloride form with 5 mL of 9 M HCl and washed with 20 mL 0.05 $C_2H_2O_4$ in 5 M HCl. This rinse removes plutonium, neptunium and thorium from the column. Uranium is eluted with 10 mL of 0.01 M HCl.

^{232}U tracer was added to soil and the soil was treated as it is described in preparation of sample method 1 (autoclave digestion). Eight molar nitric acid was evaporated to dryness and dissolved in 10 mL of 1 M $Al(NO_3)_3$ – 3 M HNO_3. The next steps were the same as it was mentioned above for aqueous sample.

2.2.4. *Separation of strontium extraction method*

A strontium fraction was evaporated to dryness and the residue was dissolved in concentrated HNO_3. Stronium-90 was determined by beta-counting of the daughter activity of yttrium-90. Yttrium was separated from the fraction, which also contains strontium and other components by liquid–liquid extraction with TBP (tributyl phosphate). The TBP extractant was pre-equilibrated with concentrated HNO_3 [21]. Samples were agitated for 5 min and phases were splited by a short standing. The organic phase contained yttrium and the aqueous phase contained the Am-Sr fraction. TBP phase was washed with concentrated HNO_3. Yttrium was eluted from TBP using 15 mL of deionized water and 15 mL of 2 M HNO_3. 30 mL of saturated ammonium oxalate solution was added to the beakers with ^{90}Y for precipitation of yttrium oxalate $Y_2(C_2O_4)_3.9H_2O$. Samples were heated on the hot plate (70°C) for 15 min with occasional stirring. After cooling the beakers to the room temperature, the precipitate of yttrium oxalate was filtered (Whatman No. 42 filter paper) and washed with 25 mL of water, followed by 25 mL of 95% ethyl alcohol. The count rate of the yttrium oxalate was determined using a low-level alpha–beta counter Tesla NRR 610 and Tesla NA 6201. The yield of yttrium was determined from the ratio of the weight of yttrium oxalate precipitated to the expected weight of yttrium oxalate (as determined from the yttrium carrier standardization).

2.2.5. *Procedure for AnaLig Sr-01 [18]*

The samples were traced with ^{85}Sr for gamma yield determination and acidified with concentrated nitric acid to 2 M HNO_3. The steps were as follows:

- Each AnaLig Sr-01 column was conditioned with 10 mL of 1 M HNO_3.
- Samples were transferred into the column and 5 mL of 1 M HNO_3 was added to rinse the beaker and each solution was transferred into the AnaLig.
- The time, when the last rinse completely drains through the each column, was recorded as the start of yttrium ingrowth.
- 15 mL 0.04 M Na_4EDTA was added to elute the strontium into the Cerenkov counting vial.
- For AnaLig column regeneration, 5 mL of 0.05 M Na_4EDTA and then 5 mL of 1 M HNO_3 were passed through the column.
- ^{85}Sr recoveries were measured using HPGE detector at 514 keV line.
- Samples were counted over a 2-week period to monitor the in-growth of ^{90}Y on TRI CARB LSC counter.

2.2.6. *Procedures for Empore^{TM} Strontium Rad disks*

The 3M Empore™ Sr Rad disks is commercially sold for the quantitative determination of radio-strontium in aqueous solutions. Model sample consisted of 150 μL of stock solution of ^{90}Sr in 2 M HNO_3 (activities between 57 Bq/L). The samples (30 mL of the model samples) were spiked with ^{85}Sr (for gamma yield determination).

- The Empore disk was preconditioned by 10 mL methanol and washed with 20 mL of 2 M HNO_3 at a flow rate of 10 mL/min.
- After entire sample has passed through the disk, the disk was rinsed with 20 mL of 2 M HNO_3. The end time of this rinse was recorded as the start of the ^{90}Y ingrowth.
- 15 mL of 0.04 M Na_4EDTA was added to elute the strontium into the Cerenkov counting vial.
- 10 mL of 0.05 M Na_4EDTA was passed through the disk for EmporeSr rad disk regeneration.

- Aliquots of the elution solutions were measured using HPGE detector for ^{85}Sr recoveries at 514 keV line and counted directly by Cerenkov counting over a 2-week period for the ingrowth of ^{90}Y using TriCarb LSC counter.

2.2.7. *Procedure for Sr resin*

Sr-Spec resin is a selective extraction chromatographic material produced by EIChroM Industries [22]. The Sr-resin is a cation exchange resin and consists of the crown ether 4,4′(5′)-bis(tert-butylcyclohexano)-18-crown-6 dissolved in 1-octanol and sorbent on an inert polymeric support. Procedures using Sr-resin are simple, short, high recovery and reproducible. Separation procedure was as follows:

- The Sr-resin column was washed with 10 mL of deionised water and pre-conditioned with 10 mL of 8 M HNO_3.
- The solution was loaded onto the column, and resin was washed with 5 mL of 8 M HNO_3.
- Sr was eluted with 15 mL of 0.05 M HNO_3 into a Cerenkov counting vial.
- The elution solutions were measured using HPGE detector for ^{85}Sr recoveries at 514 keV line and counted directly by Cerenkov counting over a 2-week period for the in growth of ^{90}Y using TriCarb LSC counter.

2.2.8. *Measurement of activity Pu and Am*

In this work, plutonium and americium were co-precipitated from 1 M HCl, with 40% HF as a precipitant and Nd as a carrier [19, 23]. Staying 30 min in the freezer (the precipitation was let to settle) the precipitation was homogenized using ultrasonic equipment and the neodymium fluoride suspension was filtered through a 25 mm, 0.1 μm. The filter was dried and the activity of actinides was determined using alpha spectrometry with 600 mm^2 active area Ultra™ ion implanted silicon detectors (EG&G ORTEC, Model 676A).

3. Results and Discussion

Separation technique of plutonium, americium, uranium, thorium and strontium-90 by a complex separation scheme using extraction with Aliquat 336 and purification of uranium and ^{241}Am with extraction chromatography was illustrated on an example of determination of radionuclides in soil samples. Soil samples were from the area of Podunajske Biskupice, a locality near Bratislava, Capital of Slovakia. The developed extraction separation process has proved to be very robust because it is suitable for 20 g mass of soil samples. Such large mass is necessary to measure low (plutonium and americium) specific activity for many samples from Podunajske Biskupice region, Table 1. ^{137}Cs was determined by gamma spectrometry and ^{90}Sr was determined by using TBP extraction method (Section 2.2.4).

TABLE 1. The interval of specific activities (Bq kg^{-1}) of the radionuclides in soil samples from the area of Podunajske Biskupice

Radionuclide	(Bq kg^{-1})
^{137}Cs	14.1–83.8
^{90}Sr	3.8–29.2
^{239,240}Pu	0.130–2.904
^{241}Am	0.074–0.580

The correlation of specific activity of ^{90}Sr with specific activity of ^{137}Cs for soils from different origins is linear, Figure 1. Therefore, the specific activity of ^{137}Cs in soil samples, easy measurable by gamma spectrometry, can be used for rough estimation of ^{90}Sr specific activity, which is not a simple procedure of analysis and detection of a radionuclide.

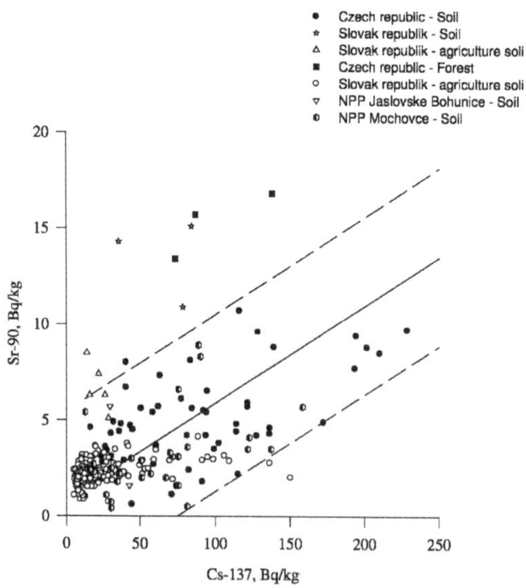

Figure 1. Dependence of ^{90}Sr specific activity on the specific activity of ^{137}Cs for different regions

Occurrence of natural ^{40}K as well as radionuclides from decay series of uranium and thorium in chosen soil profiles are in accordance with the average concentration of those radionuclides in dominant type of soils in the monitored areas of Slovak and Czech Republic, Figure 2. The average values of specific activity of potassium, uranium and thorium in soil from the locality Podunajske Biskupice are ^{40}K:481 ± 159; ^{238}U:27.3 ± 4.5; ^{232}Th: 29.2 ± 4.6 Bq kg^{-1}. The values are typical for soils activity measured in different areas of Slovak and Czech Republic.

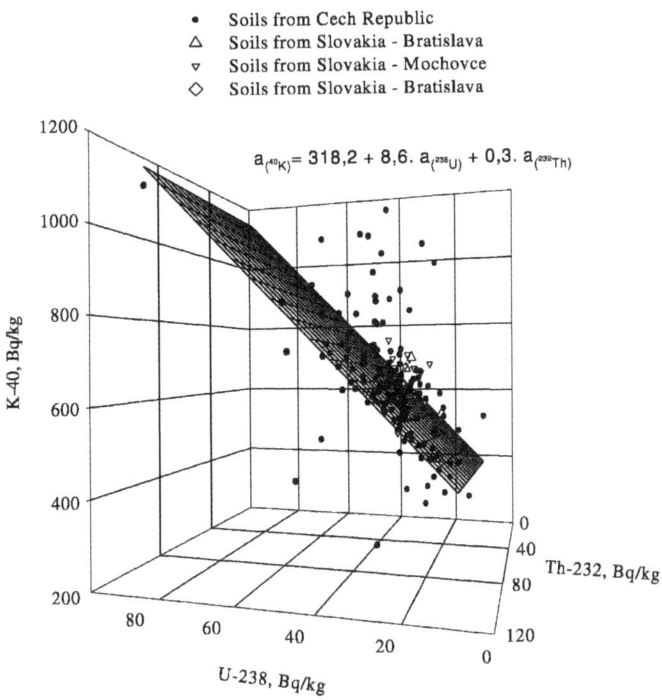

Figure 2. 3D graph of specific activity of uranium, thorium and ^{40}K in soil samples from different regions

Radiochemical yield of ^{232}U tracer was used for the determination of uranium. The alpha spectra measured with semiconductor detector shown in Figures 3 and 4 proved that the separation process is efficient because a pure alpha spectrum for natural uranium and thorium was collected. A correlation of ^{234}U volume activity with ^{238}U for the mine water samples collected in the areas Novoveska Huta of East Slovakia is shown in Figure 5. It was found that the activity ratio of $^{234}U/^{238}U > 1$ can be observed for some water samples [25].

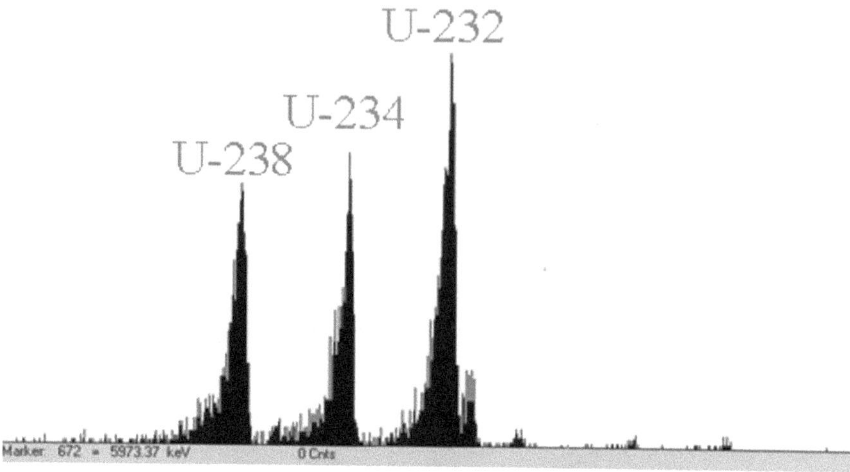

Figure 3. Alpha spectrum of U fraction after separation soil sample

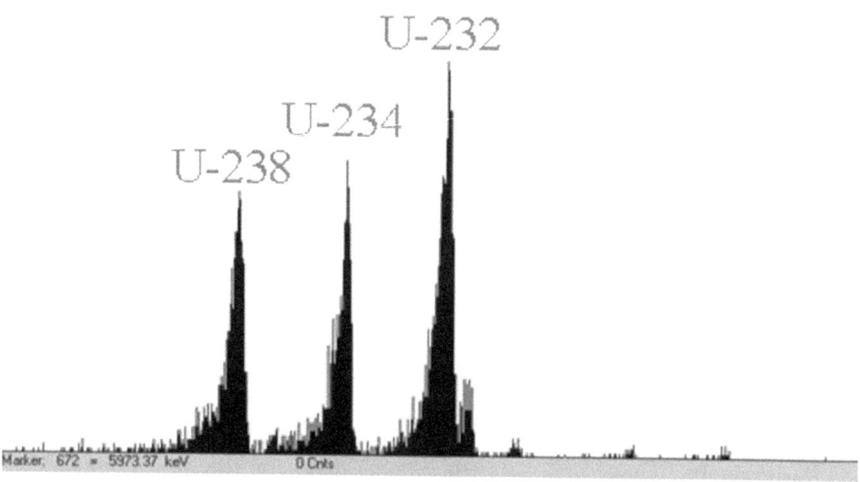

Figure 4. Correlation of ^{234}U volume activity versus ^{238}U

Method for determination of ^{90}Sr by extraction of ^{90}Y with TBP from concentrated HNO_3 is a reliable but tedious method and new development trend prefer extraction chromatography or solid phase extraction methods. The results of ^{90}Sr separation from model aqueous solutions are shown in Table 2, for 18 samples. Three commercial sorbents produced by three different manufacturers were tested. Empore extraction disks marketed by

3M, Sr-Spec resin; a selective extraction chromatographic material produced by EIChroM and AnaLig Sr-01 by IBC Advanced Technologies were compared. All materials are suitable for separation of ^{90}Sr; however, the radiochemical yield is the highest for AnaLig Sr-01 with average R = 89 ± 4.6% followed by Empore extraction disks R = 88 ± 5.2% and Sr-Spec resin R = 68 ± 22%. It seems that AnaLig Sr-01, as well as Empore extraction disks, are very suitable for quantitative and reliable strontium separation from aqueous solutions.

TABLE 2. Radiochemical yield and activity of model solution – 59 Bq/L ^{90}Sr

	Yield AnaLig Sr (%)	A ^{90}Sr (Bq dm^{-3})	Yield Empore Sr (%)	A ^{90}Sr (Bq dm^{-3})	Yield Sr resin EICHR OM (%)	A ^{90}Sr (Bq dm^{-3})
Average	89 ± 4	61.39 ± 2.16	88 ± 5	59.52 ± 1.45	68 ± 15	60.04 ± 2.7

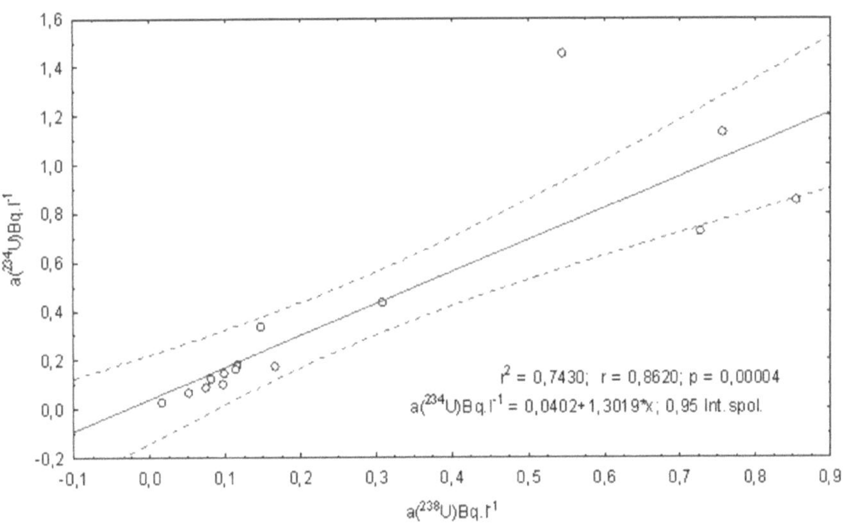

Figure 5. Correlation of ^{234}U volume activity versus ^{238}U

4. Conclusion

Separation methods are very important for hardly determined pure alpha and beta radionuclides. Development of separation processes during the history has been changing from co-precipitation, extraction, ion exchange chromatography to modern extraction chromatography methods. More and more laboratories are, due to ISO standards for testing, required to use commercial columns and standard operation procedures (SOP) especially designed for radionuclides determination. Of course, the fast, simple and reliable methods dominate in monitoring of radionuclides. For the determination of major alpha and beta radionuclides in soil samples complex schemes are used. Despite the differences in origin of samples, significant correlations were found for the activity concentrations between ^{90}Sr and ^{137}Cs. It was experimentally determined that AnaLig Sr-01 as well as Empore extraction disks are very suitable for quantitative and reliable strontium separation from aqueous solutions.

Acknowledgement: This work was supported by the Grant No APVV-20-007105 from the State Committee for Scientific Research, APVV Agency, Slovakia.

References

1. WHO (1996) Guidelines for Drinking-Water Quality, Volume 2, Health criteria and other supporting information. World Health Organization Geneva, Switzerland
2. Chopin GR, Rydberg J, Liljenzin JO (2002) Radiochemistry and Nuclear Chemistry, 3rd edition. Butterworth-Heinemann, Boston, MA
3. Multi-Agency Radiological Laboratory Analytical Protocols Manual (MARLAP), July 2004, United States Environmental Protection Agency
4. Seaborg GT, Loveland WD (1990) The Elements Beyond Uranium. Wiley, New York
5. Case FN (1964) ORNL Radioisotopes Procedures Manual ORNL-3633. Oak Ridge National Laboratory, Oak Ridge, TN
6. Manual of the Environmental Measurements Laboratory HASL-300 (1997) 28th Edition Environmental Measurements Laboratory, Department of Energy, US
7. Goheen SC, McCulloch M, Daniel JL (1990) Hanford Environmental Analytical Methods PNL-8534. Pacific Northwest Laboratory, Richland, WA
8. Goheen SC, McCulloch M (1997) DOE Methods for Evaluating Environmental and Waste Management Samples, Battelle Press ISBN 1-57477-021-7
9. Health and Environmental Chemistry Analytical Techiniques Data Management and Quality Assurance (1996) LA-10300-M Los Alamos National Laboratory
10. Wong KM, Jokela T, Noshkin VE (1994) Radiochemical Procedures for Analysis of Pu Am Cs and Sr in Water Soil Sediments and Biota Samples. Report UCRL-ID-116497. Lawrence Livermore National Laboratory, Livermore, CA
11. Kleinberg J, Smith HL (1975) Collected Radiochemical Procedures (Radiochemical Group CNC-11) UC-4 Chemistry Los Alamos National Laboratory
12. Lowestoft: Ministry of Agriculture Fisheries and Food Directorate of Fisheries Research Aquatic environmental protection Analytical methods (1990) ISSN 0953-4466
13. Chen QJ, Aarkrog A, Nielsen SP, Dahlgaard H, Lind B, Kolstad AK, Yixuan Yu (2001) Procedures for Determination of $^{239,240}Pu$, ^{241}Am, $^{234,238}U$ $^{228,230,232}Th$, ^{99}Tc and ^{210}Pb–^{210}Po in Environmental Materials. Riso National Laboratory ISBN 87-550-2871-3

14. Radiochemical Analytical Procedures Manual United States Transuranium and Uranium Registries Washington (2001) Washington State University Nuclear Radiation Center
15. Riekkinen I, Jaakkola T et al. (2002) Analytical Methods for Wide Area Environmental Sampling (WAES) for Air Filters STUK-YTO-TR 184 Helsinki: Radiation and Nuclear Safety Autority ISBN 0785-9325
16. Michael F, L'Annunziata (2003) Handbook of Radioactivity Analysis ISBN 0124366031 (1133)
17. http://solutions.3m.com/wps/portal/3M/en_US/Empore/extraction/
18. IBC Advanced Technologies AnaLig gel Data Sheet Strontium Column Series SR-01 IBC Advanced Technologies Inc
19. Sill CW (1987) Precipitation of actinides as fluorides or hydroxides for high resolution alpha spectrometry. Nucl ChemWaste Manag 7:201–215
20. Mátel L, Mikulaj V, Rajec P (1993) Determination of Pu-239,240 in environmental samples from surroundings of the Atomic Power Station Jaslovské Bohunice. J Radioanal Nucl Chem 175:41–46
21. Mikulaj V, Švec V (1993) Radiochemical analysis of strontium-90 in milk, soil and plants by solvent extraction. J Radioanal Nucl Chem 175:317–324
22. http://www.eichromcom/products/info/sr_resinc.fm
23. Hindman FD (1983) Neodymium fluoride mounting for alpha spectrometric determination of uranium, plutonium, and americium. Anal Chem 55:2460–2461
24. Analytical Procedures Rev (2005) ACS.07 Eichrom Technologies Inc
25. Osmond JK, Ivanovich M (1992) Uranium-Series Mobilization and Surface Hydrology in: Uranium-Series Disequilibrium. Applications to Earth Marine and Environmental Sciences, Ivanovich M, Harmon RS (Eds). Clarendon, Oxford

RADIOLOGICAL INVESTIGATION OF ISSY-KUL REGION OF KYRGYZ REPUBLIC

AZAMAT KALYEVICH TYNYBEKOV*
AND JEENBEK E. KULENBEKOV
Kyrgyz Russian Slavonic University, Kyrgyz Republic

Abstract. Lake Issyk-Kul located in north-east of Kyrgyz Republic and it is one of the biggest and deepest lakes on the world. The lake is tributed with 134 rivers. The lake has no flowing off, which accumulates mineral substances coming with only those rivers and rain. In order to define the radiological situation of southern-east part of Lake Issyk-Kul in 1998–2005 are done radiological investigations by International Science Center (Kyrgyz Republic) in the frame of INTAS project. In this article we tried to assess the radon situation on costal area of Lake Issyk-Kul, based on the studies done.

Keywords: Lake Issyk-Kul, radon, radiological monitoring

1. Introduction

Lake Issyk-Kul located in north-eastern part of Kyrgyz Republic on the bottom of tectonic cavity surrounded with Kungoi and Teskei Ala-Too ranges. Isyk-Kul is one of big and deep lakes in comparison with other lakes in the world. The originality of the geography, climatic, hydrogeological, hydrochemical characteristics make this region a unique peculiarity of the basin. The lake is supplied with 134 rivers. Many of them are spring from eternal glaciers which are 834 in the Issyk-Kul basin. The lake has no flowing off and mineral substances of those rivers and rains which accumulate in the lake. Major infrastructure recreational establishments and settlements formed on the lake coastal area [1, 2]. In order to define the radiological situation of south-eastern part of Lake Issyk-Kul radiological investigations

* To whom correspondence should be addressed. e-mail: azamattynybekov@mail.ru

by International Science Center of Kyrgyz Republic in the frame of INTAS (International Association) project were conducted in 1998–2005.

Natural radiation accompanies us during the whole life and it has been become known that radon is one of the most dangerous radionuclides for man health. Accounts conducted in different countries showed that its influence is about half of dose exposed to man from all source of radiation. Radon is gas and has no smell, no color, no taste and it is 7.5 times heavier then air as well as attract the most attention of scientists of the world in connection with issues of air quality in dwellings. In many countries as USA, European countries, Australia etc. it has been already conducted or being conducted the mapping of radon in territories for the purpose to define the areas with high radon concentration. Thousands of buildings are inspected to understand the flats and houses where the radon concentration

TABLE 1. Radiation levels of dwellings inside in different places of Issyk-Kul region

Coordinates		Locality	Sort of building	Radiation background	
Latitude	Longitude			Inside	Outside
42,73903	77,9764	Ton, northern part of Kuturgu village (v)., Reserve	Brick building without roof	18.3E–05	
42,15849	77,14992	State farm «Ton»	New breeze block house	2.75E–05	3.15E–05
42,15849	77,14992	State farm «Ton»	New pise building	3.48E–05	2.89E–05
42,206078	77,685500	Nearby Kuturga v.	Brick building	2.14E–05	1.83E–05
42,206816	77,683890	Ak-Terek v.	New house	2.40E–05	2.49E–05
42,152013	77,037504	Nearby Ak-Terek v.	Fisherman house	2.30E–05	1.806–05
		Bokombaev .	House (yard)	2.20E–05	2.25E–05
		Tort-Kul v.	New house	3.00E–05	2.10E–05
42,115	76,997	«	New pise building	3.50E–05	1.906–05
42,115	76,997	«	Pise building	3.10E–05	2.00E–05
		«	Brick building	3.20E–05	2.00E–05
		«	Two floor house	3.00E–05	
42,148	77,175	Kadji-Sai v.	Old school	1.60E–05	1.90E–05
42,5772	76,6386	"Shahter" Resort	Building	1.50E–05	
42,681	77,219	Balykchy town	Brick building	2.50E–05	1.80E–05
42,487	78,336	Kosh-Kel v	Pise building with cement base	2.10E–05	1.50E–05
42,488	78,3364	Nearby Karakol town	Pise building	2.10E–05	1,30E–05
42.487	78.336	Karakol town	University building	2.80E–05	1.80E–05
42.468	78.3364	Karakol town	Wooden house	6.00E–05	1.80E–05

exceeds the limited level. In Great Britain such houses consist of 10%, in Portuguese 8.6%, in Norway 10% of examined houses. In Kyrgyz Republic, the issue of radon is less investigated. In this article it is tried to conduct the works to assess the radon situation on costal area of Lake Issyk-Kul.

Radon 222 is the decay product of Radium 226, radioactive substance extended everywhere of soil in different composition and in different concentrations. In order to find the content of materials, which forms integral level of radiation, the radiological monitoring of costal area of Lake Issyk-Kul was performed. Gamma spectrometric analysis was done in selected range of probe from territory which comparatively higher radiation levels, Figures 1 and 2. It was ascertainment that high content of radioactive materials in soil probe consisted of thorium, potassium and uranium.

Radon often comes out from soil to percolate via base hole of construction materials, in consisting of in outside air, natural gas and water. According to the literature the relative contribution of every adopted source in forming "radon pressure" in dwellings reaches to the following values [3]:

- The soil under the building and construction materials 78%
- Outside air 13%
- Water use inside the house 5%
- Natural gas 4%

These average indices, however, allow precisely the assessment of above mentioned radon sources for dweller in concrete house. Average global equivalent dose of radiation is about 1.3 mSv, however in area with higher radon concentration this dose can be significantly higher. One of the main tasks at Kyrgyz Republic is to protect the man from radon and daughter product radiation (DPD). According to the assessment of United Nation of Organizations (UNO), radon with consisting of about ¾ annual equivalent radiation dose is received from earth source and half of this dose from all other radiation sources.

Main part of radon radiation dose created due to lung radiation via breezing in (DPD) in the dwelling, also due to radon consisting of in water. The drinking water with high content of radon is accumulated on the wall of stomach, inside of adipose tissue, spleen and adrenals. Some organs and tissues selectively concentrate secondary of radon. RaD (Pb-210) is accumulated into skeleton bones, in tooth and in fingers. RaE (Bi-210) is accumulated in liver and lungs. Polonium is detected in all organs and tissues but more in kidneys, liver and muscles. To make remove Po-210 from men organism is carried out through gastro-intestinal tract. According to the medical statistics in Kyrgyz Republic, the growth of death-rate because of cancerous is 61.7%. In Issyk-Kul province 100 patients with cancerous reaches to 81.6 in death, which are 1.5 times more than in republic general. Indices of population death rate on southern coast of Lake Issyk-Kul, in Djety-Oguz district is 78.5% and in Karakol town 86.2% exceeds the value for Issyk-Kul province 62.9%. This value is 61.7% for all around the Kyrgyz Republic. Among malignant edemas for Issyk-Kul province the first type with death rate indices is the stomach cancer, 10.0%; the second type is the lung cancer and the third type is malignant edemas lymph, 3.9%. These data require detailed investigation to study the reason of conditioned on this situation.

The investigations to ascertain radiological situation in Issyk-Kul region had been conducted for 2 years (1997–1998) in the project of "Radiological Monitoring of Issyk-Kul region" financed by SCOP (USA). The data of background radiation level of the territory was obtained and maps with indices of radiation level in the region were drawn up [3, 4].

2. Measurement Methods

The results of measurements represent simultaneous values of coordinates of the site by help of satellite device GPSR and measurement of the radiating background by exponometer-detector Eberline. During measurements the data of radiating background were saved in the memory of exponometer-detector. The satellite device automatically fixed the longitude and latitude of the sampling site with regular frequency and also kept these data in its memory.

The investigation area was selected at southern coastal area of Lake Issyk-Kul including seven settlements and five rivers in Djety-Oguz and Ton districts, Figures 2 and 3. It was found that the natural radiation background of the locality sourced by radioactive elements of thorium and radium. During the investigations, approximately 2,000 measurements were done in 400 probes selected.

Measuring point selection and sample preparation were performed by using the common methods: The probes were selected according to the envelope method by the expectation of every sample represent the soil itself. In case of the controlling of soil pollution in Kadji-Sai tailing dump; the probe areas were outlined along the vectors "wind rose". It was five point probes of one probe area. Common probe mass of one area is about 1 kg. All taken probe samples transported into a glass container and kept there to analyzing process. After delivery of the samples to the laboratory, the samples were separated into big clods, were cleaned up from the roots, insects, stones, glasses, etc. And then were sifted through the sieve with 1 mm of hole diameter.

Radiometric analysis of soil samples were carried out in the laboratories of Physics Institute, National Academy of Science at Kyrgyz Republic by using gamma spectrometer. This device gives the qualitative analysis of radioactive isotopes as well as the activity concentrations of radioactive isotopes which emit gamma radiation. The processing of obtained data was conducted by using special software. By evaluation of measurement results, the content of radioactive isotopes in soil has been investigated for this region.

Radiation level measurements were performed inside of the dwellings in different settlements of Issyk-Kul province. The selection of investigating houses was casual covering both though the relatively new and old buildings. Table 1 shows the radiation background levels of dwellings inside in fall-summer period. It was studied during this period because of air ventilation, the radon content decreases to its minimum.

Therefore, the radiological survey of radiation background inside of dwellings was conducted in coastal area of Lake Issyk-Kul, including coordinates, measuring date and times as well as sort of buildings. The detailed maps indicated the radiation levels at inside and outside were drawn up including coordinate characteristics.

3. Result and Conclusion

Radiation background measurements in selected water and soil probes were performed on southern coastal of Lake Issyk-Kul during these scientific-research studies. Measurement results evaluated by including the coordinate characteristics demonstrated that the most important areas, as radio-ecologically, situated at areas near the mouths of rivers flowing into the Lake, Figures 4 and 5.

The analysis results of gamma background measurements inside of dwellings in different settlements of Issyk-Kul province, showed the significant difference of exposure dose of inside of studded settlements, Figure 1.

It was defined that localities with high radioactivity values are settled on beach nearby Djenish village and in the territory of former producing lot No.7 nearby Ton village. The high radioactivity level in some places was interpreted with the higher content of radioactive element of thorium.

The results of fulfilled work showed that the outside radiation levels on investigated common territory are in limits of standard, but inside radiation levels in several places in different dates exceed the usual standard. According to our results, it is necessary to conduct more detailed further investigation to find out the reasons of high contents of radon in air of dwellings.

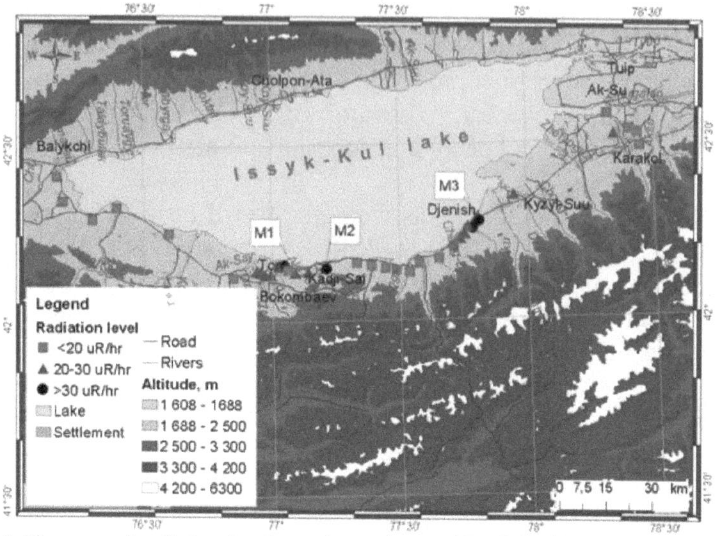

Figure 1. The map of radiation levels on the territory of Issyk-Kul Lake area, Parcel M1-94 uR/hr (11.8 uR/hr) an altitude-1 m; Parcel M2-118 uR/hr (12.3 uR/hr) an altitude-1 m, 150 uR/hr an altitude-1 m; Parcel M3-108 uR/hr (13.0 uR/hr) an altitude-1 m, 140 uR/hr an altitude-1 m

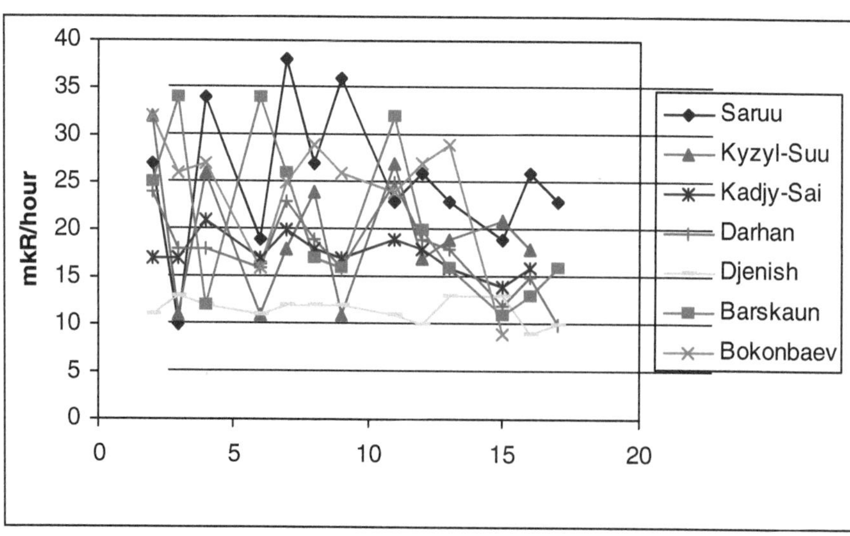

Figure 2. Exposured dose level of gamma radiation from soil selected from villages in southern coastal area of Lake Issyk-Kul

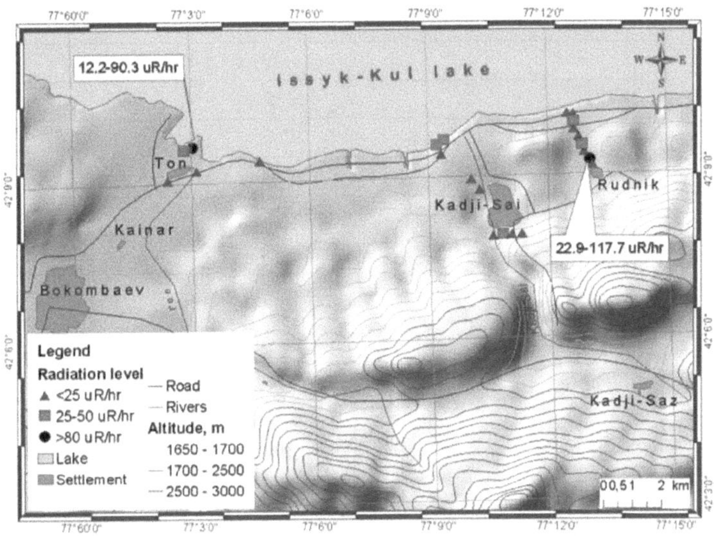

Figure 3. Indication of radiation level near the shop No. 7 and tailing dumps

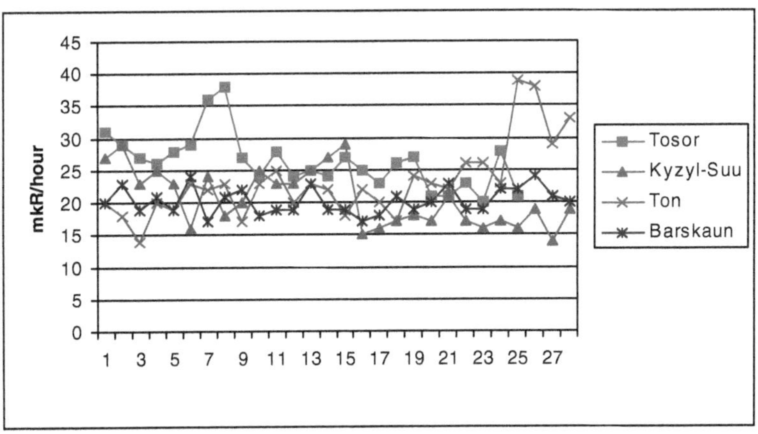

Figure 4. Power level of gamma radiation exposure dose of water probe are sampled in villages south coast of Lake Issyk-Kul

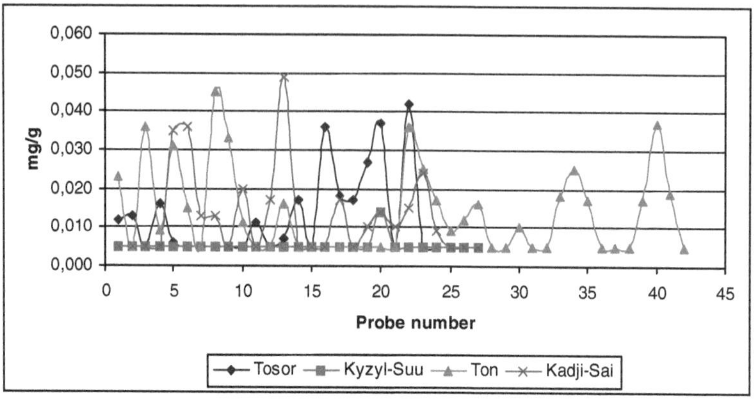

Figure 5. Content of thorium in river water in southern coastal area of Lake Issyk-Kul (mg/g)

References

1. Tynybekov AK (1999) Radiological characteristic of southern coastal area of Issyk-Kul Lake. Environment and Man Health Report. Collection of Scientific Works-T.VII.-Bishkek, Kyrgyz Republic
2. Tynybekov AK, Hamby DM (1999) Radiological characteristic of southern coastal area of Issyk-Kul Lake Report. Institute of Management, Business and Tourism, No.2-Bishkek, Kyrgyz Republic
3. Hamby D, Tynybekov AK (1999) Radiological monitoring of southern coast of Lake Issyk-Kul. Health Phys 77:427–430
4. Tynybekov AK (2000) Radiological investigation in Ton and Djety-Oguz districts of Issyk-Kul province. Surveying mountains and life in mountains International Conference, KSMI

WORKING TOGETHER FOR NUCLEAR SAFETY

OLEG UDOVYK*
National Institute for Strategic Studies, Kyiv, Ukraine

Abstract. This paper aims to provide an insight into the 15 year history of the TACIS Nuclear Safety Programme. It will present the nuclear safety issues that led to the programme's beginnings and portray the various aspects of its implementation, funding, challenges and achievements. Finally, the paper will outline the future direction of the programme.

Keywords: TACIS, nuclear safety

1. Introduction

On 26 April 1986, Unit 4 of the Chernobyl Nuclear Power Plant suffered the world's worst nuclear accident that spread radioactive particles across parts of Europe. The accident heightened public concern about nuclear power and in particular about the risks caused by design weaknesses and safety of ageing plants in central Europe and Soviet Union. The Chernobyl accident prompted unprecedented international cooperation to improve nuclear safety.

Since the early 1990s, the International Community has come together to make substantial financial and technical contributions to the upgrading of conditions both at the Chernobyl site and in the Commonwealth of Independent States (CIS). Safety measures have been implemented to prevent such a catastrophe from ever happening again.

The European Union (EU) set up TACIS (Technical Assistance to the Commonwealth of Independent States) after the break-up of the Soviet Union in 1991, and the following year created the TACIS Nuclear Safety Programme [1].

Since its inception, the programme has been contributing to an improvement of nuclear safety in the CIS by transferring technology and know-how and establishing fail-safe mechanisms that have collectively

* To whom correspondence should be addressed. e-mail: oleg_udovyk@hotmail.com

contributed to the improvement of nuclear power plants and the avoidance of accidents in the years since Chernobyl.

Nuclear safety considerations also extend to the mining of uranium, the production and safe transport of nuclear fuel, handling of fresh and spent fuel and the storage, treatment and safe disposal of nuclear waste products and emergency preparedness and management.

These have also been covered by the Programme together with a number of security related aspects such as assistance to safeguarding nuclear material (in order to prevent theft or diversion) and redirecting thousands of former weapons scientists, technicians and support staff to peaceful activities.

2. Origins of the Tacis Nuclear Safety Programme

The EU's strategy for improving nuclear safety was developed in accordance with the G7 strategy adopted in Munich in 1992 and reflected the International Atomic Energy Agency's classification of design and operating risks regarding nuclear reactors [2]. A 'Master Plan' identified 12 areas of investigation to ensure that all potential nuclear safety concerns were addressed. The areas covered include:

- Integrity of the primary circuit
- Upgrading of instrumentation and control systems
- Accident analysis
- Training of personnel (including the use of simulators and quality control)
- Fire protection
- Seismic analysis
- Containment/confinement
- Maintenance
- Emergency preparedness (including prevention and management)
- Nuclear fuel cycle and waste management
- Safety issues specific to RBMK reactors
- System analysis

Between 1991 and 1999 the Programme succeeded in making improvements in the above areas of concern; difficulties were encountered in implementation consequent upon its novelty and scale, differences in culture towards nuclear engineering and safety, operating in a closed environment, lack of common understanding, etc. However, the balance is generally considered as positive for the period.

In light of experience, the direction of TACIS was refined. An initial assessment of the energy programme led to the following recommendations being made:

- Sustainability: Projects must be selected which have a lasting impact in the countries concerned.
- Investment-oriented: Projects must increasingly pave the way for capital investment in the energy sector.
- Coordination: A high degree of coordination is essential to achieve the large-scale capital investment required.
- Local contribution: The potential must be tapped to ensure genuine transfer of know-how and skills.
- Local conditions: Economic, political, social and cultural factors should be fully taken into account.
- Reference projects: More references should be available so that the benefits of change can be demonstrated.

Nuclear safety took up a leading position in its own right. Although the TACIS funding to nuclear safety is significant, the costs for implementing appropriate safety upgrades at nuclear plants are immense and decommissioning costs are even more so. Careful targeting is therefore necessary to obtain the maximum benefit from the resources available. The nuclear safety strategy of the EC was subjected to a thorough assessment based on the policy agenda and the situation in the counties concerned. Nevertheless, the overriding common objective is to attain the highest level of nuclear safety in the partner countries.

In following years the TACIS Nuclear Safety Programme has concentrated on the following main areas:

- Design safety
- On-site assistance and operational safety
- Off-site emergency preparedness
- Waste management
- Regulatory authorities and their technical support organizations
- Control of nuclear materials
- Conversion of nuclear military scientists to civil jobs
- Chernobyl closure and the Shelter Implementation Plan

3. The Role of the EC

The EC manages the bulk of the EU's contribution to international nuclear safety assistance through the TACIS Nuclear Safety Programmers. In addition, several EU Member States have their own bilateral assistance programmers. The Directorate General (DG) External Relations and the Europe Aid Co-Operation Office (DG AIDCO) carry out strategic Planning and implementation of the Programme, respectively. The EC's Joint Research Centre (JRC) provides technical assistance to DG AIDCO addresses nuclear trafficking issues and runs training programmers for nuclear inspectors. The EC is assisted by a Committee of representatives of the EU Member States known as the PHARE-TACIS Expert Group.

The EC delegations are staffed by experts and play an important role in the implementation of the assistance programmers in the host countries as well as ensuring the coordination and involvement of the partner countries.

In addition, a Joint Management Unit (JMU) was set up in Moscow to facilitate project management and cooperation amongst all those participating in the Programme in Russia. The Joint Support Office in Kiev carries out a similar function in Ukraine. Finally, the European Court of Auditors verifies the sound financial management of the projects.

4. The Role of the International Atomic Energy Agency (IAEA)

The IAEA is a major player in international nuclear safety and safeguards [3]. It works with its Member States and the EC in coordinating joint programmers and is the watchdog on international nuclear issues. The IAEA's Safety Standards Series, which includes Safety Fundamentals, Safety Requirements and Safety Guides, have become a global benchmark.

In the early 1990s, the IAEA launched a major international programme to evaluate first generation Soviet-designed reactors and to provide safety assistance to plant operators and regulators. A significant part of the programme focused on developing Safety Issue Books for the two different types of reactors in the NPPs – VVER and RBMK. Safety Issue Books are lists of safety issues generic to all units of a plant type, ranking their associated safety significance and the corresponding recommendations for safety improvements.

The findings of the Safety Issue Books have been the basis for the development and implementation of the nuclear safety improvements provided by the TACIS Nuclear Safety Programme and the actions implemented by the Russian and Ukrainian NPPs through their own funding or with the assistance of other donor programmers. The adoption of the Convention on Nuclear Safety in Vienna on 17 June 1994 represents a major milestone, which was made possible by the cooperation between the IAEA, governments and national nuclear safety authorities. The Convention's objective was to legally commit participating States operating land-based nuclear power plants to maintain a high level of safety by setting international benchmarks to which they would subscribe.

5. How Safety Issues Have Been Addressed

The worst nuclear accidents have been caused by a combination of design shortcomings and operator error. Since the Chernobyl accident great emphasis has been placed on the behavior of nuclear plant management and staff and their attitude towards safety. The International Nuclear Safety Advisory Group (INSAG) introduced the term 'Safety Culture' in its Summary Report on the Post-Accident Review Meeting of the Chernobyl Accident. The concept is defined as: that assembly of characteristics and attitudes in organizations and individuals which establishes that, as an overriding priority, nuclear plant safety issues receive the attention warranted by their significance. This definition draws attention to the priority of safety, the importance of the staff and management attitude to reduce the risks of operational error by forward thinking and constantly assessing the safety significance of events and issues concerning safety, so that they receive the appropriate level of attention.

Much has been achieved during the TACIS Nuclear Safety Programmer's 15-year history [4]. The catalyst for this success has been close coordination with the national authorities and involvement of personnel in the partner countries as well as effective cooperation between the EU teams. This success story has led to effective solutions in areas vital to nuclear safety.

Nuclear safety is built on three independent pillars: prevention of accidents; mitigation of their consequences in terms of the release of radioactive material to the environment; and off-site emergency preparedness for use should prevention and mitigation proves insufficient.

6. Design Safety

The design, construction and operation of any nuclear plant should ensure safe operation and avoidance of accidents. Notwithstanding the very low chance of an accident with significant off-site consequences, all nuclear installations are required, as part of they are licensing, to develop plans to deal with any emergency. This is part of the "belt and braces" approach to nuclear safety that is endemic in this discipline. Plant operators, together with local and national authorities, must be prepared for any form of incident or accident. An integrated approach (emergency plan) and a high level of cooperation are necessary to control the situation if a release of radioactivity does happen. This has been developed over the years based on experience and the plans are frequently exercised to demonstrate their efficacy.

6.1. OFF-SITE EMERGENCY PREPAREDNESS

Off-Site Emergency Preparedness refers to measures to protect the population and the environment in a nuclear emergency following, or given a threat of, an accidental release of radioactive material.

Under the TACIS Nuclear Safety Programme, a step-by step coordinated approach was adopted, involving the partners in the NPPs in the CIS countries. Should an incident occur a special control facility takes over operations from the normal plant control room? This facility manages actions that need to be taken within the plant while keeping external agencies fully up-to-date with developments as they happen. Issues such as radiation monitoring, warning the general public, medical provisions and possible evacuation are focused on separately, so that the staff of the affected plant can concentrate on assessing what needs to be done to remedy the situation.

The backbone of off-site emergency preparedness is the Emergency Plan, which is compiled by the operator of the nuclear plant together with local authorities and regulators.

The plan outlines the procedures and responsibilities for alerting key personnel, assessing risks and mitigating actions, identifying incident response teams etc. Off-site, an emergency response centre must be in place and be equipped with all the necessary computer and communication links to deal with the situation. This off-site centre comes under the control of one person, who is high ranking and a suitably qualified member of the emergency services, with various experts assisting him/her.

Speed is a key in any emergency situation so there must be reliable communication systems to receive the best information from the staff at the affected plant. Off-site emergency preparedness staff must also be able to understand and collate large amounts of data from radiation monitoring points and meteorological data on the prevailing winds.

The process A “Needs Assessment” study was carried out in most of the CIS, which has enabled the EC to start a comprehensive programme of assistance. A vital element has been the provision of training for key personnel, including solid understanding of how to implement off-site emergency preparedness.

The objectives of the assistance for off-site emergency management were:

- To assist in improving local, regional and national offsite emergency response arrangements in Eastern Europe and in bringing them to an adequate level
- To improve the technical basis for the exchange of data and information in an emergency within Eastern Europe and with the Union, thus contributing to a more effective response to any future nuclear accident that may have widespread effects in Europe

The more urgent needs were met quickly, in particular the provision of equipment, drugs, etc., to enable the effective implementation of existing emergency plans should they ever be needed. This was followed by providing coherent support in four interlinked areas critical for effective emergency response: early warning, through monitoring systems, of radioactive material in the environment; the communication of monitoring and other data and information to local, regional and national emergency centers (and between national centers); decision support systems to process monitoring and other data to enable well informed and timely decisions on countermeasures; and training.

Some examples

- Supply of prophylactic iodine tablets and protective clothing, field communications, dosimeters and instrumentation
- Establishment of comprehensive training material on off-site emergency management and a regional training centre in Obninsk
- Early warning systems in national emergency centers in Belarus and Ukraine (GAMMA project)
- Ring of detectors around several NPP in Ukraine and Russian Federation to provide input to early warning systems
- Decision support system in national emergency centre in Ukraine and planned for Russian Federation
- Prototype data exchange system in national emergency centers in Belarus, Russian Federation and Ukraine for more effective communication of data and information, both within and between countries

6.2. ON-SITE ASSISTANCE AND OPERATIONAL SAFETY

On-Site Assistance falls into two categories: supply of hardware covering all activities relating to equipment supply, from tendering and following-up of the installation to final acceptance (Hard On-Site Assistance), and service activities not linked to the preparation and implementation of equipment supply projects (Soft On-Site Assistance). The second category includes actions such as training; site visits in the EU, twinning and management courses to improve safety procedures.

Since the start of operations in 1991 the objective of the TACIS Nuclear Safety Programme in the field of On-Site Assistance has been to provide general assistance to the Soviet designed NPPs, which has led to an enhancement of their operational safety. The assistance to the nuclear operators has been made possible by EU utilities specialized in the operation of nuclear power plants.

Operational safety has been increased through On-Site Assistance provided on a continuous basis by the EU operators at 14 sites in the NIS. The assistance has concentrated on the level of design safety, operating and surveillance conditions and the organization of operational safety. All assessments made of the nuclear safety programme to-date have confirmed the appropriateness of On-Site Assistance as a means of transferring safety culture and supporting the implementation of safety improvements at NPPs in the partner countries.

7. The Importance of Regulatory Authorities

Regulatory Authorities are essential in all aspects of nuclear use, whether it's electricity production, medical research or storage and disposal of radioactive material.

They have the authority to grant and withdraw licenses and regulate sitting, design, construction, commissioning, operation and decommissioning of nuclear installations.

To perform their function they must have adequate authority, financial and human resources and be independent from conflicting interests or political interference.

Since 1993 EU assistance has focused on:

- Technical assistance aimed at strengthening the capabilities of the nuclear regulatory and its Technical Support Organization (TSO), in the review of a number of safety relevant upgrading and modernization measures at NPPs and other types of nuclear sites (waste facilities) carried out jointly with EU TSO experts (“2 + 2” approach)
- Transfer of regulatory methodology including the formulation of legislation and regulatory documents
- Support to licensing assessments for specific plant improvement projects
- Assistance in overall safety assessments of specific installations

The independent regulatory authorities have been strengthened through EU technical and financial assistance and much of the necessary legal framework has been put in place in the CIS, although further work is still needed.

8. Nuclear Security

At Munich Airport in August 1994, 350 g of weapon grade plutonium were discovered aboard a flight from Moscow. Three couriers were arrested and 200 g of lithium confiscated. Both plutonium and lithium are required in the construction of thermo-nuclear weapons. The amounts found were significant. Although Russian authorities denied that any of their weapons-grade material had gone missing, the materials were believed to have originated in the former Soviet Union.

Recent evidence shows a significant worldwide rise in the smuggling of nuclear and other radioactive materials. In fact, according to the IAEA, there have been 300 confirmed cases since 1993, with 215 of them occurring since 1999.

Most of these concern materials being stolen or diverted from factories, hospitals or research laboratories. This is a serious development at a time of increased terrorist threats.

To counteract such a threat, the EC took a pioneering and leading role in developing, with the IAEA and the Dedicated International Technical Working Group, a Model Action Plan to respond to illicit trafficking cases. This Plan deals with the three lines of defense: prevention, detection and response.

Since the early TACIS projects dealing with nuclear safety, the EC addressed the first line of defense by supporting the enhancement of Nuclear Material Accountancy and Control in the Russian Federation. To-date, co-operation has been particularly successful in the areas of:

- Training on safeguards methodology for experts, operators and inspectors
- Improving Nuclear Material Accountancy and Control
- Developing instrumentation in co-operation with industry in the Russian Federation

At the same time, the EC has developed new methodologies that help in the fight against nuclear trafficking. It provides research facilities for rapid 1-day analysis and detailed nuclear forensic investigation with a response team available on a 24/7 basis. It also supplies joint analysis capabilities with the participation of local scientists to help national authorities and international organizations in their efforts to combat nuclear trafficking.

The detection of nuclear "fingerprints" is one such method that can help solve nuclear crimes. It involves identifying nuclear material that has been seized or accidentally released and assessing the immediate associated hazards. Most importantly, the scientists can identify the original source and therefore the possible smuggling routes of the material. The fine-tuning of these forensic procedures has been developed and applied at the JRC's Institute for Transuranium Elements in Karlsruhe, Germany.

Finally, a pilot project with Ukraine allowed to test the implementation of the Model Action Plan at a national level and was successfully duplicated in 12 candidate countries to provide an integrated and coordinated response to illicit trafficking of nuclear materials and to improve international collaboration in the field.

Overall, the measures set up by the EC in co-operation with the Russian Federation and the international community has all contributed to a greater culture of nuclear security in Russia and the CIS countries.

9. The Future of the Nuclear Safety

The safety of nuclear power is again at the forefront of international attention, 20 years after the Chernobyl accident. With global warming increasingly seen as a major threat to human life, coupled with concerns about security of supply for conventional energy, some governments are reconsidering the nuclear energy option. With the exception of renewable energy sources most other proven forms of power generation produce greenhouse gases that increase global warming. Furthermore, nuclear energy provides electricity less dependent on external sources of fuel. Already, nuclear power generates one third of the EU's and approximately half of Ukraine's electricity. For these reasons, many as essential to secure energy supply in future now consider the nuclear option. However, nuclear safety will remain of paramount importance.

Over the past 15 years the TACIS Nuclear Safety Programme made a very important contribution to the improvement of nuclear safety in the CIS. This is confirmed by the fact that the safety issues identified by the IAEA have mostly been resolved, safety culture has been improved, there have been no major accidents and the availability of the plants has, in general, been increased. Much has already been achieved but there is still further work to be done.

While it is not expected that there will be an immediate change to the current method of assistance, future programmers are likely to evolve towards 'soft assistance' (safety culture and regulatory support), as major safety related equipment has already been supplied. Future strategy is expected to favors increased international cooperation, drawing on the resources and expertise of other organizations. The collaboration with the IAEA will be developed to ensure a coherent approach towards the identification of problem areas, planning and joint projects.

References

1. Tacis Nuclear Safety Programme – EU/RUSSIA/UKRAINE: TACIS NUCLEAR SAFETY ACTION PROGRAMME 2001
2. Delegation of the European Commission to Belarus, Nuclear Safety in Central Europe & the New Independent Statesoperation Office: http://ec.europa.eu/europeaid/tender/index_en.htm
3. International Atomic Energy Agency (IAEA) – http://www.iaea.org
4. International Nuclear Safety Center of Ukraine – http://www.insc.gov.ua

ENVIRONMENTAL STUDIES IN UZBEKISTAN INSTITUTE OF NUCLEAR PHYSICS WITH THE USE OF NUCLEAR METHODS

BEKHZOD S. YULDASHEV, UMAR S. SALIKHABAEV, RAISA I. RADYUK, SERGEY V. ARTEMOV, GENNADIY A. RADYUK* AND ERKIN A. ZAPAROV
Institute of Nuclear Physics Academy of Science, Republic of Uzbekistan (AS RU), Tashkent, Uzbekistan

Abstract. Institute of nuclear physics is one of the biggest research centers in Central Asia that conducts a lot of studies in the field of environmental radioactivity. In this study, the evaluation of the data obtained by gamma spectrometry of natural radionuclides in river-bottom sediments and coastal soil in basins of Syrdarya and Amudarya rivers and their inflows on the territory of Uzbekistan are cited. Possible seasonal deviations of activity of radionuclides are also discussed.

Keywords: Radiation monitoring, gamma spectroscopy, radionuclide migration

1. Introduction

Radiation monitoring of environment in the basin of Syrdarya and Amudarya rivers was initiated in 2002 [1]. The first data of Syrdarya river analysis were performed in the work [2], in which water salinity, alpha–beta activity of water samples, river-bottom sediments and coastal soil samples were analyzed. It has been shown, that water salinity and alpha-beta activity of water is in accordance with each other depending on remoteness from river source. Alpha–beta activity of river-bottom sediments and coastal soil also is of regular and identical character, which is however completely distinct from that observed in water samples.

The data on distribution of natural radionuclides along the river showed that [3] distribution of radionuclides and thorium and uranium rows in soil samples and river-bottom sediments, correlate with behaviour of total alpha–beta activity, but have the prominent features, most probably

* To whom correspondence should be addressed. e-mail: radyuk@inp.uz

caused by anthropogenic activity. In general it became clear, that it is necessary to continue these researches and to increase the number of sampling points. In 2003, a 3-year project on radiation monitoring of trans-boundary rivers of Central Asia and Kazakhstan Syrdarya, Amu Darya and their inflows was approved.

In given research work, analysis of data on radionuclide composition of river-bottom sediment and coastal soil samples along the rivers on the territory of Uzbekistan, obtained in the first half of project activities is presented.

2. Experimental Techniques

Thirty sampling points have been selected across Uzbekistan. The first 15 sampling points are for monitoring purposes, they are permanent and the research results allow gathering extensive statistical data on radio-ecological situation in the basins of the researched rivers. The second 15 sampling points are for research purposes, and depending on obtained results, the variation of point selection is possible. Distribution of sampling points on the rivers as follows: on Syrdarya river – 25, 26, 27, 28, 04, 29, 30, 22, 05, 23, 24, 06; on Ahangaran river, together with Kuvasay inflow – 18, 19, 7 (inflow), 08, 20, 21, 09; on Chirchik river – 10, 16, 11, 12, 17. On southern rivers, six sampling points were selected: Zaravshan river – 13, 14, 15; Amudarya river – 03, 02, 01. Sampling of the water, river-bottom sediments and coastal soil in each sampling point were done, twice a year in spring and autumn, for research purposes. In the process of sampling, field researches of physical–chemical parameters of water with the help of "Hydrolab" device were carried out. Thus, the influence caused by human factor was reduced to a minimum. Gamma-spectrometer researches of volumetric samples were carried out by using ultra-pure germanium SILENA and CANBERRA detectors. Soil and river-bottom sediment samples were placed into Marinelli vessels with the following sizes: small vessel – D = 85 and 105 mm and H = 60 and 80 mm; big vessel – D = 83 and 120 mm and H = 60 and 85 mm. The weight of the samples in these vessels usually constituted about 0.3–0.6 kg for the small vessel, and 0.6–1.1 kg – for the big vessel. Depending on the weight of the samples, amendments for self-absorption of gamma-beams in the sample, through changing registration efficiency curve were introduced. On gamma-spectra nuclides of thorium (^{208}Tl, ^{212}Pb, ^{212}Bi, ^{228}Ac) and uranium (^{214}Pb, ^{214}Bi, ^{226}Ra, ^{238}U) radioactive rows and ^{40}K were defined. Exposition was 9–12 h. The weight of the samples, exposition time and background conditions of detectors have caused a range of statistical errors in identification of specific activity equal to 8–20% as 8% for ^{40}K, 15% for ^{226}Ra, 20% for ^{238}U and 10% for the other nuclides. The results of the analysis are expressed as activity per unit of dry sample (dried, crushed and sifted beforehand in Bq/kg.

3. Analysis of Results

The preliminary results of experiments have shown that activity of all nuclides can be divided into several groups based on the size of activity in the samples. Radionuclides of thorium radioactive row, have approximately identical activity both in samples of river-bottom sediment, and in samples of coastal soil, therefore, in further analysis we use average activity of thorium nuclides activities.

$$A_{cp}^{Th} = \frac{1}{4}(2,78 * A_{Tl208} + A_{Pb212} + A_{Bi212} + A_{Ac228})$$

The behaviour of all average activeness, in all 30 sampling points is represented by six curves: three for river-bottom sediments for three seasons and three for soil samples for three seasons. It was found that activities of thorium row nuclides do not represent seasonal dependence.

Activities of uranium row nuclides are more diverse. Here, first of all it is possible to allocate ^{214}Pb, ^{214}Bi nuclides into a separate group. We have attributed ^{226}Ra, ^{238}U nuclides to the second group, that possess higher specific activeness (1.5 times and more). However, in this group $2^{26}Ra$ nuclides slightly differ by specific activeness in comparison with ^{238}U nuclide.

All nuclides of both rows, in increasing order have identical activity and identical manner of distribution on sampling points. It is worth noting, that in certain points, activity of nuclides is three to four times higher than in majority of points, in which activity is close to each other and constitute the background.

In contrast to radioactive row radionuclides ^{40}K has higher specific activity and more uniformly dispersed on the area.

In those points, where nuclides of radioactive rows have significantly higher than background activity, activity concentration of ^{40}K is approximately two times higher than background.

In general, we can conclude that there is no seasonal effect on soil and river-bottom sediment samples. In certain points, there exist significant deviations from average level of activity. The activity levels of uranium along the inflow of Kuvasay is significantly higher and reaches 200 Bq/kg.

Thus, we have revealed one area with increased activity of radionuclides on the territory of Uzbekistan. The primary part of the basin of Syrdarya and Amudarya rivers, contains natural quantity of radionuclides [4].

It has already been stated that it is difficult to establish seasonal variations of activity concentration of measured samples. However, in Kuvasay where increased contents of all radionuclides are observed in spring soil samples, it can be noticeable the increasing in activity concentrations of uranium row nuclides. We investigated the seasonal changes of activity in water from Akhangaran river with its inflow – Kuvasay.

It was observed that Pb-214, Ac-228 radionuclides which represent uranium and thorium rows respectively, were more intensive than K-40. All measurement results of sample activities of water, soil, bottom sediments and Aqutic plants are presented in Table 1.

TABLE 1. Salinity and activity of water, soil, bottom and water plants in Sydarya

Sample site	Distance from source (km)	Salinity (g/L)	Activity water (Bq/L)		Activity soil (Bq/kg)		Activity bottom (Bq/kg)		Activity aqutic plant (Bq/kg)	
			β	α	β	α	β	α	β	α
Kg03	250	0.22	0.046	0.02	660	550				
Kg05	350	0.25	0.048	0.02	560	380				
Kg07	600	0.20	0.059	0.04	630	340				
Kg08	650	0.24	0.048	0.03	760	560				
Kg09	700	0.29	0.059	0.04	750	460				
Kg10	750	0.36	0.064	0.07	440	420				
Uz04	800	0.87	0.24	0.11	470	300	440	300	95	15
Tj13	1,000	0.96	0.26	0.16	500	310	450	270	120	10
Tj15	1,100	1.31	0.28	0.09	610	410	550	400	190	40
Tj14	1,150	0.93	0.29	0.20	600	370	540	430		
Uz05	1,200	1.13	0.34	0.25	470	220	460	410	100	50
Uz06	1,300	1.14	0.34	0.19	490	250	490	360	90	20
Kz01	1,350	1.39	0.35	0.29	650	440	690	560	165	45
Kz02	1,450	1.42	0.32	0.19	630	270	490	280		
Kz07	1,750	1.31	0.34	0.22	600	360	490	360	165	50
Kz08	1,850	1.26	0.34	0.15	570	350	590	280	180	20
Kz09	1,950	1.35	0.40	0.22	600	580	540	380	115	10
Kz10	2,050	1.09	0.26	0.16	570	430	520	320	145	30
Kz11	2,150	1.31	0.37	0.32	540	330	530	510	140	80
Kz12	2,250	1.35	0.34	0.29	520	370	570	420	155	35
Kz13	2,350	1.33	0.34	0.29	510	520	550	270	160	50
Kz14	2,500	1.40	0.38	0.24	480	290	570	300	100	30
Kz15	2,700	1.48	0.43	0.19	520	370	450	320	170	25
Average deviation		18%	22%	40%	12%	30%	10%	30%	35%	43%

Standard deviations in determination of gamma and beta activities are 10%. In determination of alpha activities standard deviations can reach 20% for low-active samples. Quick analysis of this material shows that the highest activity for samples in sampling locations 18, 19, and 07 for all seasons. And it was determined that the activity concentrations significantly changed by seasons. in these locations. The most significant change in activity concentrations were found to be in of uranium row nuclides and total alpha-activity. Activity distribution of Pb-214 in bottom sediment samples along the Akhangaran with Kuvasay inflow are shown on Figure 1.

In fall maximum activities are observed in location 19, and in spring in location 07. This tells us about the migration of nuclides of uranium row. Activities in all seasons decrease to 30–50 Bq/kg in 60 km and further stay at this level over 67 km until reservoir (location 21). After that activity goes down to 20–30 Bq/kg at the confluence of the Akhangaran and Syr-Darya.

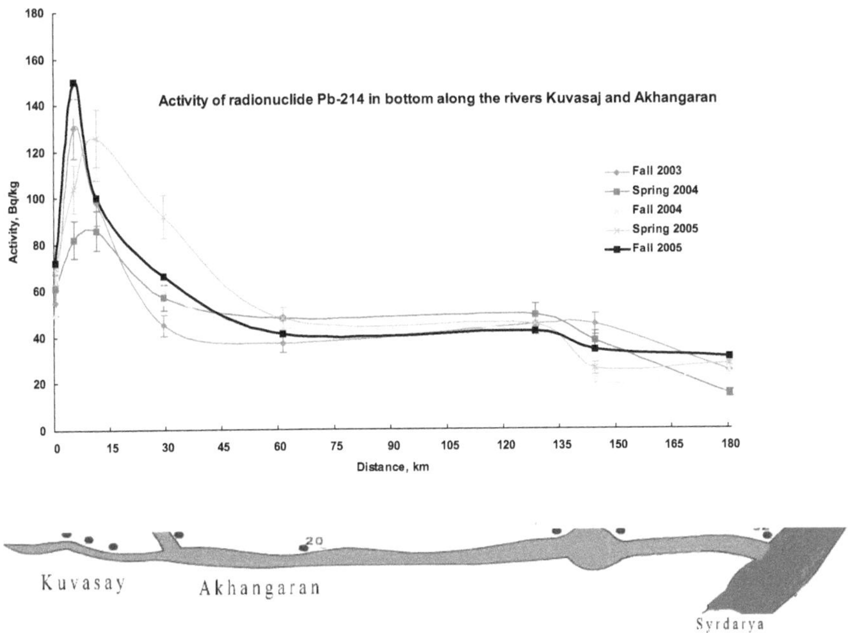

Figure 1. Distribution of activity Pb-214 along the river Akhangaran with inflow Kuvasay in coastal soil

We found that the amount of ^{214}Pb, ^{214}Bi, ^{226}Ra, ^{238}U radionuclides has exceeded in spring than the autumn level by three times. Activity concentrations of ^{238}U increases by two times in spring while activity concentrations of ^{214}Pb, ^{214}Bi, ^{226}Ra decrease. It is possible that this is due to various migration behaviours of the given nuclides. The amount of nuclides contained in thorium radioactive row does not change by season neither in soil samples nor in river-bottom sediment samples.

4. Conclusion

Based on the results of different seasonal measurements maps of distribution natural radionuclides in the basin of Syrdarya river and their inflows on the territory of Uzbekistan in the soil and river-bottom sediments were created. Everywhere except for Kuvasay (inflow of Ahangaran), for all radionuclides it had been observed that the average level of activity concentrations are almost same as the other parts of the earth. In one of the spots along the Kuvasay river, significant excess of average level by three and more times was revealed.

In majority of sampling points, seasonal changes in activity concentrations were not observed. In one spot along the Kuvasay river triple excess (compared to autumn) in the level of activity of nuclides of uranium row was seen in spring soil sample. For thorium nuclides row, such anomaly was not observed. Such effect is not presented as well in samples of river-bottom sediment. As a conclusion, the revealed possible seasonal effect of increasing in activity concentrations deserves special attention and longer period of observation in this region.

Acknowledgement: The authors acknowledge the organizing committee of NATO advanced training courses in Mugla, Turkey, May 2008 and personally Professor Gul Asiye Aycik for their support in our participation in the meeting and paper presentation.

References

1. Passel H D et al. (2002) The Navruz Project: Transboundary Monitoring for Radionuclides and Metals in Central Asia Rivers. Sampling and Analysis Plan and Operation Manual. Sandia National Laboratories, Uzbekistan
2. Barber D S, Yuldashev B S, Passel H D et al. (2003) Radiation monitoring of Syrdarya River. Eurasia Nucl Bull 2:82–87
3. Yuldashev B S, Passel H D et al. (2004) Radiation Monitoring of Syrdarya River (II), 3rd Eurasia Conference Nuclear Science and Its Application. October 5–8, 2004 Tashkent, Uzbekistan
4. Kozlov V F (1991) Handbook of Radiation Protection (in Russian) Energoatomizdat, Moscow

RADIOBIOLOGICAL EFFECTS OF ^{241}Am INCORPORATED IN CELLS OF ORGANISM AND METHODS OF PREVENTION OF THE MENACE OF COMBINED TOXICITY OF THE TRANSURANIUM ELEMENTS

NAMIK RASHYDOV* AND VALENTYNA BEREZHNA
Institute Cell Biology and Genetic Engineering NAS of Ukraine, 148 Zabolotnogo, Kyiv-03143, Ukraine

Abstract. After nuclear accident at the Chernobyl Nuclear Power Station (ChNPS) the radioactive isotopes ^{137}Cs, ^{90}Sr, transuranium and other elements transported vast amounts of the atmosphere, much of which was subsequently deposited not only in the immediate vicinity of power plant in Ukraine, Belarus and Russian Federation, but over the large parts of world. The contamination of wide territories not only in Ukraine with fission products of uranium and transuranium elements is an essential consequence of the accident ChNPS that is classified as a global ecological catastrophe. Radionuclide of this contamination is transferring through feed chains into drinking water, forage and foodstuffs and in the end these processes are responsible for accumulation of dose irradiation population of people.

Keywords: Nuclear accident, food chain, contamination, Chernobyl

1. Introduction

The contamination of wide territories not only in Ukraine with fission products of uranium and transuranium elements is an essential consequence of the accident at the Chernobyl Nuclear Power Station that is classified as a global ecological catastrophe. Radionuclide of this contamination is transferring through feed chains into drinking water, forage and foodstuffs and in the end these processes are responsible for accumulation of dose irradiation population of people. The highest Relative Biological Efficiency

* To whom correspondence should be addressed. e-mail: nrashydov@yahoo.com

(RBE) is intrinsic to the emitters of alpha radiation and Americium-241 (^{241}Am) belongs to this category of emitters. Radionuclide ^{241}Am is a daughter product of ^{241}Pu isotope. The activity of ^{241}Am is increasing with time owing to β-decay of this plutonium isotope. The amount activity of ^{241}Am that was ejected from the wrecked reactor in environment fall up to 0.14 PBq (1.1 kg) additionally. Maximum of ^{241}Am activity will be peaked at year 2059 when it should be in 40 times higher then it is today because estimated income about 184 PBq of ^{241}Pu was penetrated in environment. A fraction of dose resulted from radiation of ^{241}Am in internal dose from all radionuclide (^{137}Cs, ^{90}Sr) will account during this time approximate to 20%. Taking into account that the RBE of ^{241}Am irradiation is very high this radionuclide should exert the property of emitters of the most dangerous to health of people [1, 2]. The recovery process in cell with α-decay case practically was absent because in tissue cumulated capture dose the high RBE. As result the incorporated radionuclides in cell and tissue are more dangerous for organism. It is important to investigate exchange, migration, distribution and uptake and general oncology pathologies in case of incorporated isotope ^{241}Am appeared in the interval doses 0.25–8.3 Gy in lungs and in interval doses 0.07–27.4 Gy in skeleton of the mouse [3]. Compounds of ^{241}Am find their ways into tissue of respiratory organs or gastro-intestinal system and they are absorbed mainly in skeleton bones, liver and kidney [4]. The time of biological half time of renewal of ^{241}Am varies from 1 to 5 and more years according to age, composition of ration and life mode of people.

The isotope ^{241}Am environmental circulation deal with some problems below mentioned:

1. To fit methods ^{241}Am estimation to biological objects of different types
2. To estimate going over radionuclide from soil to plants and other biological materials
3. To elucidate the pathways of ^{241}Am migration along the food chains
4. To calculate capture doses from radionuclide in target tissues
5. To carry out forestall investigations of the early and late effects of ^{241}Am
6. To develop means for preventing absorption of ^{241}Am in tissue of different organisms and stimulation of recovery processes in damaged cells

Mechanisms of biological action of ^{241}Am on the levels of molecule, membrane, genome, cells, tissue and organ are suggested to be investigated when this radionuclide is accumulated by organisms through gastro-intestinal system or inhalation of as called “hot particles” [5, 6].

The study physico-chemical, transport, enzymatic peculiarities and peroxide modifications of membranes in their relation to deterioration of metabolic processes in damaged organism was investigated [7].

Moreover, the main indexes of blood and hemopoietic tissue directly concerned with radiation injury under the effect of ^{241}Am will be revealed. Investigations of the membrane bounding and complexing properties of ^{241}Am will provide a possibility to find substances capable of inhibition of the ^{241}Am absorption or increasing the rate of its resorption.

At the time, cytogenetic effects of the transuranium elements resulted from injuries of genetic apparatus of cell and their role in induction of genome instability and increasing of the rate of translocations connected with cancer genesis were studied [8–10].

The results of investigations formed the basis for elaboration of the proposals concerned the preventive and therapeutic means aimed at decreasing of negative effects of ^{241}Am on organisms which are inhabited on territories contaminated with ^{137}Cs, ^{90}Sr, ^{241}Am and ^{241}Pu or in the event of an nuclear emergency [11].

2. Experimental

The sample of plants and soil were collected around village Yaniv where the specific radioactivity for radionuclide ^{241}Am was 660 kBq/m^2. Mushrooms were find around villages Kopachi, Chistogalovka and Chernobyl district where were discovered radioactivity spot area contaminated with radionuclide ^{241}Am up to 30 kBq/m^2. All colleted biological samples from above mentioned contaminated sites dried and grinded for investigation to content isotopes ^{137}Cs, $^{239+240}$Pu and ^{241}Am [12, 13].

The activity of sample for isotope ^{241}Am in samples carry out with method measurement low level gamma-phone spectrometry with using semiconductor detector L0515R (Canberra) with high pure germanium (by iteration step 570 eV on the pick line 122 keV) and with portable spectrometry station InSpector MCA. The sample with mixed radionuclide ^{241}Am was extractives, precipitate and use for separation ion exchanges method with lanthanides as carrier.

To study autoradiographic investigation To study autoradiographic investigation seedlings *Arabidopsis thaliana* were aseptically grown in hard agar cultured medium containing ^{241}AmCl$_3$ with specific activity of $5 \cdot 10^4$ Bq/kg. After 25 days some leaves and parts of stems which did not directly contact with medium were carefully cut off to avoid contacting with medium. Selected parts of *Arabidopsis thaliana* were put on the microscopic glass slides and were dried for a few days. During this process the selected parts of *Arabidopsis thaliana* glued onto the slides. The slides with parts of plants were coated with photo emulsion gel LM-1 (Amersham – Biosciences UK) and were stayed at +4°C of temperature for 20 days. After development the sample slides, the α-particles track were observed by light microscope.

3. Results and Discussion

For calculation capture dose for target tissue of mans from contaminated radionuclide necessary data about the transfer coefficients it's from soil to plants. In Table 1 described data accumulation radionuclide and the transfer coefficients for plant samples collected around village Yanov.

As shown from Table 1 proportions of substance radionuclide $^{241}Am/^{239+240}Pu$ for plants scabish, white European birch and red oak were variation within from 3.2 to 8.3 whereas for soil this proportion amounted only 1.1. In order elucidate that radionuclide ^{241}Am uptake from soil to plants better than isotopes of the $^{239+240}Pu$. In Table 2 demonstrated data accumulation radionuclide and the transfer coefficients for mushroom samples collected around village Kopachi. As shown in Table 2, proportions of substance radionuclide $^{241}Am/^{239+240}Pu$ for mushrooms cep, toadstool, *Suillus luteus*, polish were variation within from 0.9 to 1.1 whereas for soil this proportion amounted only 1.2. In order elucidate that radionuclide ^{241}Am uptake from soil to mushrooms by help biomolecular carrier complex similarly as isotopes of the $^{239+240}Pu$.

TABLE 1. The accumulation radionuclide ^{137}Cs, $^{239+240}Pu$, ^{241}Am (Bq/kg) and the transfer coefficients (Tc, Bq/kg:Bq/m^2) for uptake from soil to plant samples collected around village Yanov

Plants	^{137}Cs	$^{239+240}Pu$	^{241}Am	Tc for ^{137}Cs	Tc for ^{241}Am	$^{241}Am/^{239+240}Pu$
Oenothera biennis L.	5.8 ± 1.5)·10^6	6.8 ± 1.2	47 ± 5	0.2	$7 \cdot 10^{-5}$	6.9
Betula alba L.	7.4 ± 1.1)·10^5	5.4 ± 0.5	45 ± 3	0.03	$7 \cdot 10^{-5}$	8.3
Quercus rubra L.	2.6 ± 0.2)·10^6	1.0 ± 0.2	3.2 ± 0.2	0.1	$5 \cdot 10^{-6}$	3.2
The specific activity of soil, Bq/m^2	$2.7 \cdot 10^7$	$6.0 \cdot 10^5$	$6.6 \cdot 10^5$	–	–	1.1

TABLE 2. The transfer coefficients of ^{137}Cs, $^{239+240}$Pu, ^{241}Am from soil to mushroom samples collected around village Kopachi, Bq/kg:Bq/m^2

Mushrooms	^{137}Cs	$^{239+240}$Pu	^{241}Am	Tc for ^{137}Cs	Tc for ^{241}Am	^{241}Am/$^{239+240}$Pu
Boletus edulis Bull. : Fr	(2.9 0.4)·10^5	14 ± 4	13 ± 3	0.1	0.5·10^{-3}	0.9
Amanita citrina (Schaeff.) S.F. Gray	(3.5 ± 1.1)·10^5	120 ± 10	130 ± 10	0.1	4.6·10^{-3}	1.1
Suillus luteus (L.: Fr) S.F. Gray	(1.9 ± 0.2)·10^7	16 ± 3	16 ± 2	4.0	0.5·10^{-3}	1.0
Boletus badius (L. : Fr) S.F. Gray	(3.5 ± 1.1)·10^6	5.2 ± 1.5	5 ± 1	1.0	0.2·10^{-3}	1.0
The specific activity of soil, kBq/m^2	4.7·10^3	24 ± 4	28 ± 4	–	–	1.2

We are also investigated coefficient uptake for vegetables cultures red beet, cultivated cabbage, potato, cucurbit and others which were grown around Chistogalovka and district Chernobyl for radionuclides ^{137}Cs and ^{241}Am. There are cultivated vegetables often used people as food. For this samples we found that the coefficient uptake for radionuclide ^{137}Cs amplitude variation was from 1.3·10^{-3} until 8.1·10^{-3} whereas for radionuclide ^{241}Am this coefficient was variety from 8.7·10^{-5} to 1.6·10^{-4}. Mentioned that radionuclide ^{241}Am cumulated in pulp of the vegetables of the times 3–3.9 less amount than in the rind part of plants.

Our data good concordant with results other authors [14–16] where mentioned that coefficients uptake for radionuclide ^{241}Am more than 2–10 times for isotopes ^{239,240}Pu for phytomass and for reproductive organs wheat, barley, lupine, lucerne, English ryegrass and et al plant. It is necessary mark that coefficient uptake had high variables. Two subject necessary note that the coefficient uptake very small for radionuclide ^{241}Am and this element maldistribution by organs and tissues. As result the capture dose also may be nonuniform distributed by tissues of plant. This fact proof of evidence our experiments where was investigated peculiarity distribution ^{241}Am in *Arabidopsis thaliana* plant used autoradiography method.

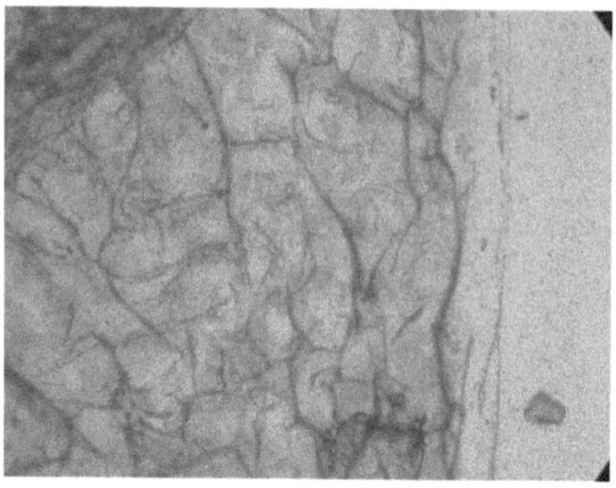

a – Control (without α-particles tracks in leaves)

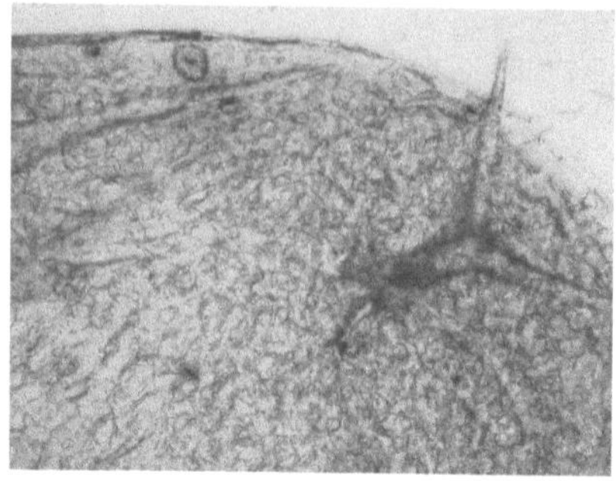

b – The α-particles tracks in leaves

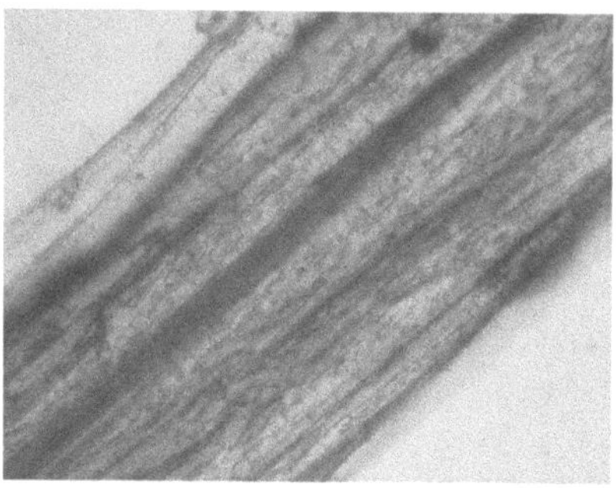

c – α-particles tracks in petiolule of low level layer leaves

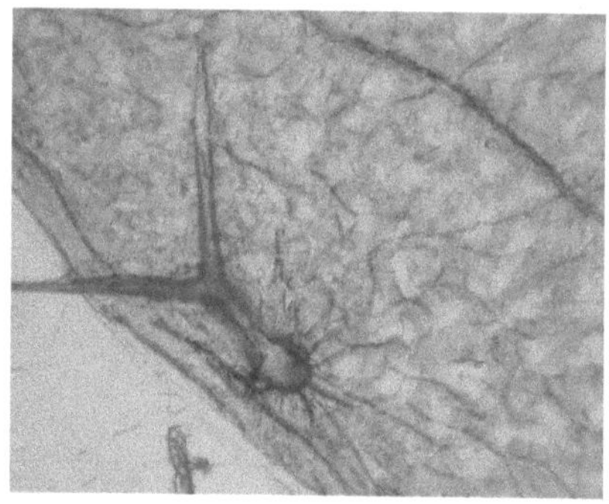

d – α-particles tracks from ^{241}Am localized around of the trichome

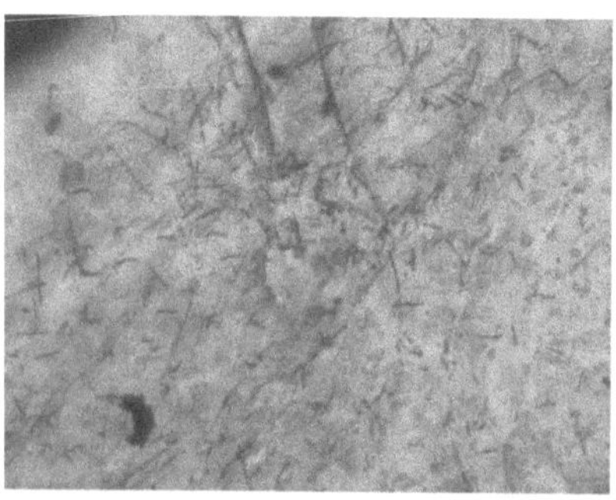

e – A lot of α-particle tracks in low level layer leaves

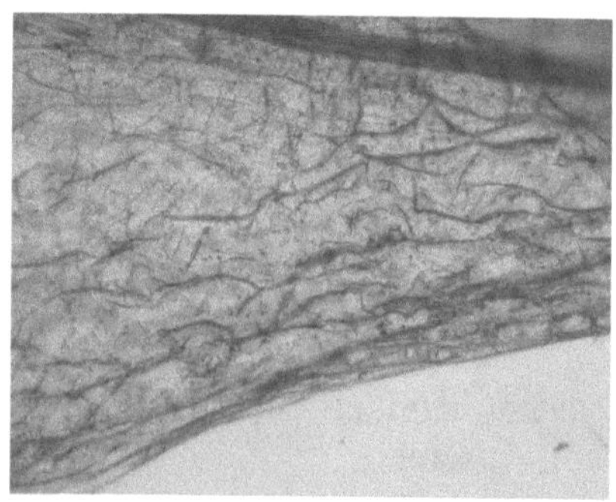

f – A lot of α-particle tracks localized in fascicles of the leaves

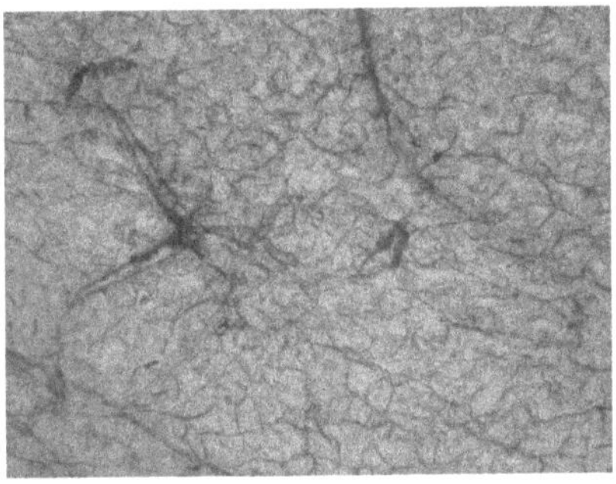

g – α-particle the tracks in high level layer of leaves occur very scarce

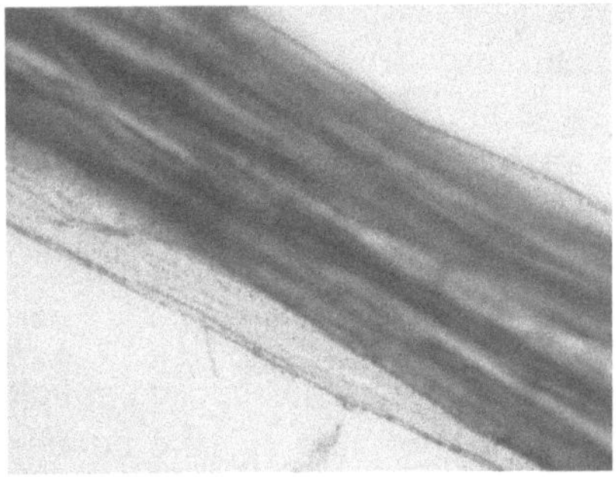

h – α-particle the tracks in petiolule high level layer of leaves occur very scarce

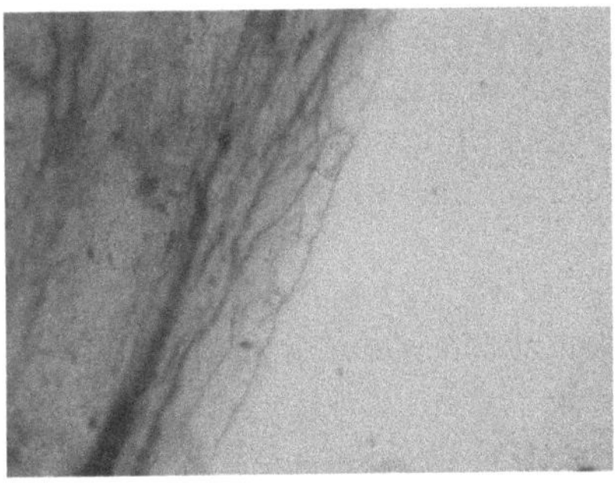

i – α-particle the tracks in apical meristem do not observed

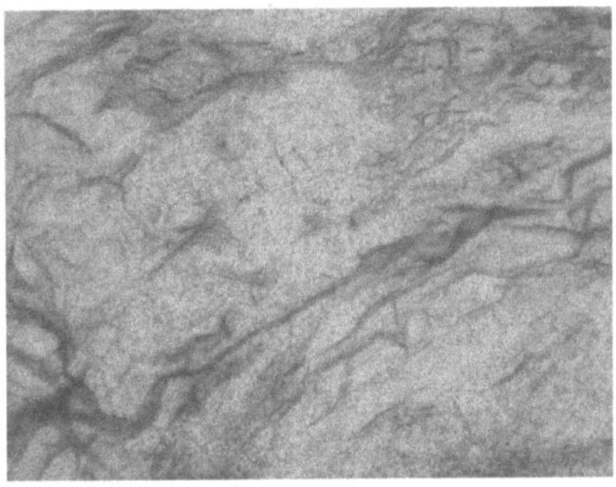

j – α-particle the tracks in petal flower do not observed

Figure 1. ^{241}Am distribution in organs and tissues of *Arabidopsis thaliana* plant

As result this investigations in slides for control variants which Arabidopsis thaliana seedlings were grown in hard agar cultured medium without $^{241}AmCl_3$ did not observed the track of α-particles. The density of spots on these slides was created only environmental radioactivity phone, Figure 1a.

When *Arabidopsis thaliana* seedlings grown in hard agar cultured medium containing $^{241}AmCl_3$ in concentration with specific activity $5{\cdot}10^4$ Bq/kg in the slides with plant parts which were cut off from low level layer observed a lot of spot of the tracks of α-particles from decayed radionuclide ^{241}Am. A lot of spot we were found around trichome, in stomatal cells, respiratory cells, in petiolule, in carry out fascicles for low level layer leaves of plant, Figure 1b–f. But in high level layer leaves the tracks α-particles occurred very scarce, Figure 1g and h. In the anther, petal flower, apical meristem and generative organs the tracks of the α-particles do not observed, Figure 1i and j.

We observed that accumulation the radionuclide of ^{241}Am depended of carry out fascicles system of the leaves and localization of the layer leaves not far from length root collar. The first layer leaves were taken up high level amount radionuclide ^{241}Am.

4. Conclusion

It is confirmed that the coefficient uptake is very small for ^{241}Am. As result the capture dose also may be nonuniform distributed by tissues of plant. The peculiarity distribution radionuclide ^{241}Am in Arabidopsis thaliana plant on high level layer leaves, in petiolule and in carry out fascicles of the leaves significatived that go into this isotope from root system to top of plant very slow and membrane of cells played as discrimination barrier in this processes as mentioned in our previously investigations. For calculated accumulated dose from radionuclides formed polulation of people may be presumably which used for foods vegetables cultures grown in cantaminated sites.

For rough estimated internal dose in the gastro-intestinal system necessary marked that only 30% radioisotope ^{137}Cs and approximately 1% radionuclide ^{241}Am accustom to it system. For this radionuclides the constant equilibrium dose had $\Delta = 135.8$ g·μGy/MBq·c for radionuclide ^{137}Cs and $\Delta = 886.4$ g·μGy/MBq·c for radionuclide ^{241}Am and taking into consideration this parameters internal dose formed its isotopes were 1.7 mGy and 1 μGy per year for hypothetical consumer. Our calculation shown that only 0.1% dose created from isotope ^{241}Am in total capture dose all radionuclide from vegetative foods in Chernobyl region for people in case used it for foods. Necessary remember that this calculated do not included meals foods and inhalation of as “hot particles” [17–20]. Our calculations evidencing that part of high density irradiation with high RBE in the total internal dose content low part, but this dose may be very strong radiobiological effect because they distribution in organism nonuniform and mainly may be localized in specific target tissues as result.

Acknowledgement: We express thanks to Drs. V. Tryshin and A. Berlizov for providing dosimeter measuring in this investigations.

References

1. Blincoe C, Bohman V R, Smith D D (1981) Uptake and release radionuclides. Health Phys 41:285–291
2. Ellender M, Harrison J D, Pottinger H E, Thomas J M (1996) Osteosarcoma induction in mice by the alpha-emitting nuclides, Pu-239, Am-241 and U-233. Proceedings of 9th

International Congress of the International Radiation Protection Association, Vienna, 4:67–69
3. van den Heuvel R, Gerber G B, Leppens H et al. (1995) Long term effects on tumour incidence and survival from Am-241 exposure of the BALB/c mouse in utero and during adulthood. Int J Radiat Biol 68:679–686
4. Labejof L, Berry J P, Duchambon P et al. (1998) Apoptosis of rat kidney cells after Am-241 administration. Anticancer Res 18:2409–2414
5. Bulman R A (1978) The movement of plutonium, americium, and curium through the food chain. Naturwissenschaften 65:137–143
6. Bulman R A, Johnson T E, Ham G J, Harrison J D, Clayton R F (1993) Speciation of plutonium in potato and the gastro-intestinal transfer of plutonium and americium from potato. Sci Total Environ 129:267–289
7. Grodzinsky D M (1995) Late effects of chronic irradiation in plants after the accident at the Chernobyl nuclear power station. Radiat Protect Dosim 67:41–43
8. Kutsokon N, Rashydov N, Berezhna V, Grodzinsky D M (2004) Biotesting of radiation pollutions genotoxicity with the plants bioassays: radiation safety problems in the Caspian region. Kluwer, Boston, MA
9. Schlenker R A, Thompson E G, Oltman B G, Toohey R E (1995) Bone surface concentrations and dose rates 11 years after massive accidental exposure to Am-241. Health Phys 69:324–328
10. Schmid E, Roos H (1996) Dose dependence of sister chromatid exchanges in human lymphocytes induced by in vitro alpha-particle irradiation. Radiat Environ Biophys 35:311–314
11. Fuhrmann M, Lasat M, Ebbs E, Cornish J, Kochian L (2003) Uptake and release of Cs-137 by five plant species as influenced by soil amendments in field experiments. J Environ Qual 32:2272–2279
12. Kabata-Pendias A (2000) Trace elements in soil and plants. Pulowy, Poland
13. Frissel M J (1998) FAO/IAEA/IUR Protocol for experimental studies on the uptake of radionuclides from soils by plants. Annual review "Soil–Plant-Relationship", Seibersdorf, Austria
14. Bunzl K, Kracke W (1987) Soil to plant transfer of Pu-239/240, Pu-238, Am-241, Cs-137 and Sr-90 from global fallout in flour and bran from wheat, rye, barley and oats, as obtained by field measurements. Sci Total Environ 63:111–124
15. Fresquez P R, Armstrong D R, Mullen M A, Naranjo L J (1998) The uptake of radionuclides by beans, squash, and corn growing in contaminated alluvial soils at Los Alamos National Laboratory. J Environ Sci Health B 33:99–121
16. Popplewell D S, Ham G J, Johnson T E et al. (1984) The uptake of plutonium-238, 239, 240, americium-241, strontium-90 and cesium-137 into potatoes. Sci Total Environ 38:173–181
17. Durbin P W (1973) Metabolism and biological effect of the transplutonium elements. In: Hodge H C (ed) Handbuch der experimentellen Pharmacologie, Springer, New York
18. Filipy R E, Kathren R L (1996) Changes in soft tissue concentrations of plutonium and americium with time after human occupational exposure. Health Phys 70:153–159
19. McInroy J F, Kathren R L, Toohey R E (1995) Postmortem tissue contents of Am-241 in a person with a massive acute exposure. Health Phys 69:318–323
20. Toohey R E, Kathren R L (1995) Overview and dosimetry of the Hanford americium accident case. Health Phys 69:310–315

TESTING AND PERFORMANCE EVALUATION OF ILLICIT TRAFFICKING RADIOACTIVITY DETECTORS

ANTON ŠVEC*
Slovak Institute of Metrology, Bratislava, Slovak Republic

Abstract. Installed radiation portal monitors are widely used screening devices for detection of illicit trafficking and potentionally harmful radioactive materials. However, they should be properly installed, maintained and tested. A prototype of a calibrating and testing source has been developed for the purpose. This remote and computer controlled device helps to check basic metrological parameters of radioactivity monitors *in situ*. Some possibilities to obtain more information from these instruments operated in activity measuring mode is presented and discussed. Recommendations for future progress are given.

Keywords: Portal monitor, radioactivity detector, calibration and testing, large uncertainty

1. Introduction

Experience in many parts of the world continues to prove that movements of radioactive materials outside of the regulatory and legal frameworks continue to occur. Such movements may either be deliberate or inadvertent. Deliberate, illegal movements of radioactive materials, including nuclear material, for terrorist, political or illegal profit are generally understood to be illicit trafficking. The more common movements outside of regulatory control are inadvertent in nature. An example of an inadvertent movement might be the transport of steel contaminated by a melted radioactive source that was lost from proper controls. Such a shipment may present health and safety threats to the personnel involved as well as to the general public [1].

So the interest in the detection of radioactive material movements is induced not only by law and international agreements but arises from practical considerations about possible consequences of leaving this issue

* To whom correspondence should be addressed. e-mail: svec@smu.gov.sk

abandoned. An important role is played by the fact that there is no alternative how to detect the presence of radioactive materials without measurements. On the other hand, an extremely large variety of possible sources of danger connected with radioactivity exists. There is no universal solution to this problem and taking some measures is (or should be) always preceded by a thorough cost/risk analysis.

Most instruments in use are based on the gamma radiation detection. This radiation is emitted at most nuclear transformations, however, with different yields and different physical characteristics. It is penetrating all known materials but with gradually decreasing intensity depending on the conditions characterized mainly by the radiation itself and by the nature and amount of the penetrated materials. The detection probability is also affected by the source/detector geometry arrangement and by the detector characteristics. An important parameter to be accounted for is the omnipresent background level and its variations.

The most efficient but expensive instruments for the radioactivity detection are installed portal-like monitors based on large plastic scintillation detectors. There are many types of these instruments produced by many companies and for different purposes: for monitoring pedestrian paths and luggage conveyors at airports, cars and trucks at transport corridors and border crossings, railroad carriages, cargo containers at seaports etc. There are instruments specialized to watching outputs from radionuclide production facilities and nuclear power plants and others assigned for monitoring inputs to scrap metal collection and separation sites and to steel works. Their common feature consists in their ability to check the investigated object for the presence of the lowest possible amount of radioactivity and to cause an alarm if detected. Another common feature in contrast to this positive function is that the limit for alarm settings is rather vague and the number of nuisance alarms may be eventually too high. This situation resembles notorious installed metal detectors intend for hidden weapon detections but often responding to a single coin. While a hidden weapon alarm can be easily checked by alternative ways and the situation can be fast solved the decision in the case of radioactivity is more difficult and the responsibility is much higher. Moreover, these detection systems are operated by non-expert personnel such as local and state law enforcement officials and customs and border patrol agents with low qualification in radiation physics. No wonder their decisions are usually exaggerated and inadequate the real situation.

Lately (in 2006) the US Department for Homeland Security (DHS) sponsored research, development, and testing activities that were designed to produce portal monitors that, in addition to detecting, would also identify the type of nuclear or radiological material. Portal monitors with this new identification technology currently cost about $377,000 or more per monitor. In these same tests, DHS also tested the performance of currently deployed portal monitors. Although this lead to a decision to spend some $1.2 billion for new technology it also raised a wave of criticism because of not sufficient cost/benefit analysis and because of low level of tests proving higher technical capabilities of these instruments.

Also recently a new international standard has been issued dealing with radiation monitors for the detection of radioactive materials [2]. Its final version contains a lot of useful information about the contemporary state of art in such instrument design and testing. Although it clearly states that it does not provide the data needed to determine the performance of a monitor in measuring the quantity of the radioactive material, the recommended acceptance test concerns using 0.6 MBq ^{137}Cs (and 17 MBq ^{241}Am and 0.15 MBq ^{60}Co respectively) radiation source for alarm triggered with 0.9 probability at confidence level of 95%. The test method assumes the source moved with appropriate speed through the detection zone or exposed to the monitor for about 1 s. This represents a good starting point for the real activity assessment derived from measured and other data.

2. Experimental

2.1. METHOD

It is self-evident that any testing of a radioactivity monitor must be performed by means of a radioactive material or source. The question is only how close the testing conditions should imitate the real situation when investigating an object for the presence of radioactivity.

There are several different approaches to this problem. One of the first studies described a method used for railroad monitor type tests including various radionuclide sources (^{137}Cs, ^{57}Co, ^{60}Co, ^{131}I and ^{241}Am) with activities from 0.4 MBq to 5 GBq placed in a wagon filled with scraps and moved along detection units [3]. The number of passing-by was more than 1,000 to obtain adequate statistics. This method provided useful information about such system behaviors and estimates of their responses which may be expected in real conditions. However, similar tests are too expensive and yet less representative because the real conditions may vary in wide ranges and can hardly be standardized.

Another extreme approach used by some producers for the system operability verification is using a test source fixed close to the individual detectors e.g. by screwing it directly on the detector front plate. This method enables to obtain fast response with low activity required but lacks for the reproducibility and traceability to any calibrations. It suits more frequent (daily) routine tests as prove of the system functionality mainly.

The already mentioned testing method recommended in standard [2] consists in statistical evaluation of the monitor response to a reference radionuclide source under reference conditions. Such a test requires using a suitable instrumentation and is rather time-consuming although the number of trials has been reduced from originally suggested 10,000 exposures down to 50. Similarly, the response is derived from the defined source activity rather than dose rate proposed in earlier versions. Unfortunately, the standard procedure ensures only that the tested monitor would be able to detect an unshielded radionuclide source of specified activity and does not deal with the actual minimum detectable activity and with this parameter stability under different weather conditions and disturbances. Yet it is understood that the incident radiations are converted into detector count rates which are recorded and tested for changes thus triggering alarm signals.

There are various strategies how to evaluate the count rate obtained. The simplest method consists in fixed alarm settings derived from long-term background observations. A more advanced method is presented as an adaptive algorithm with a moving alarm level according to short- to medium-term observations. There is also a possibility to account for background suppressions caused by the shielding effect of inactive objects in the detection zone. A careful analysis of these and similar criteria suggests that little improvement can be expected using rather sophisticated statistical data processing [4]. One promising method already mentioned consists in using multichannel analyzers instead of simple discriminating instruments. These so called Advanced Spectroscopic Portals, however, are much more complicated and expensive, yet their overall sensitivity is lower due to less sensitive detectors used.

Following method is proposed as a simple and transparent solution to outlined problems with the definition of the radioactivity monitor performance and its testing. In contrast to the standard [2], an installed radiation monitor can be regarded as a measuring instrument capable to be calibrated and tested under specific conditions. The most convenient conditions are represented by an unshielded point-like radionuclide source placed in the very center of the detector zone (or in a similar well-defined place). If the source is characterized by its activity A the monitor response n in counts per second can be described as

$$n = n_0 + A \cdot \sum_{i,j} p_i(E) \cdot \varepsilon_{i,j}(E) \qquad (1)$$

where n_0 is the background count rate, $p_i(E)$ is the emission probability per decay for a photon with energy E and $\varepsilon_{i,j}(E)$ is the given photon counting efficiency obtained at calibrations in given conditions. The index i denominates i-th emission component of the radionuclide radiations and j concerns the detection response by some of physical interactions (photoelectric, Compton, pair production) as they all should be considered in a discriminating (wide window) detector. A similar model can be constructed for each channel of a monitor possessing spectroscopic properties. If we limit ourselves to one or several "typical" radionuclides, say ^{137}Cs, the instrument response can be determined for any of these radionuclides by means of radionuclide calibration coefficients C_n.

$$n = n_0 + A \cdot C_n \tag{2}$$

Now consider the activity measurement of an unknown source providing it contains an *identified* radionuclide. Its activity can be simply calculated

$$A = \frac{n - n_0}{C_n} \tag{3}$$

This formula can be used for testing the system stability using a defined radionuclide source under conditions identical with those used for calibrations. Whatever simple this formula is, it comprises several sources of uncertainty at least: the statistical variations in count rate determinations and the uncertainty of the calibration coefficient value including the uncertainty of the radionuclide identification. Up to this point the method corresponds with considerations given in a very useful publication [5]. Two important parameters of a radioactivity monitor can be derived this way: the sensitivity to gamma radiation and a minimum detectable activity. Both parameters are radionuclide-specific and the latter also depends on conditions of measurement namely on the background level and its variations and on the length of time available for measurements. The sensitivity (also named efficiency) is represented by the radionuclide calibration coefficient C_n and must be determined experimentally using one or several sources:

$$C_n = \frac{n - n_0}{A} \quad \text{or} \quad C_n = \frac{\Delta n}{\Delta A} \tag{4}$$

The minimum detectable activity MDA is usually understood as the alarm criterion deduced from the long-term background level and its variations and calculated for some alarm probability if an excess in the expected number of counts has been detected during the measuring time interval. Not regarding the way the criterion is obtained it can always be expressed as a multiple of the long-term background level $\overline{n_0}$

$$MDA = \frac{k \cdot \overline{n_0}}{C_n} \tag{5}$$

where the multiplication coefficient k may depend on the length of the measuring time interval, the background level and other factors which may be affect the system.

2.2. INSTRUMENTATION

Several types of calibrated radionuclide sources were used for real calibrations and verifications of installed radioactivity monitors. The monitors themselves need not be identified as they are owned by our customers but can be described as both railroad and truck portal monitors based on large plastic scintillation detectors. Only ^{137}Cs was selected as the most typical radionuclide for testing. The sources used first were simple small sealed sources with activities of ~3–1,200 MBq calibrated in our standard ionization chamber and placed on a tripod in the geometrical center of the detection zone. Later a special adjustable radionuclide source was constructed consisting from a 2 mL calibrated pipette placed in 25 mm thick lead shielding. The pipette has been filled with ^{137}Cs water solution of 5 MBq/mL nominal specific activity and sealed. Moving the pipette partly out of the shielding enables to change the amount of radioactive solution exposed to the irradiated object.

The idea of using an adjustable radionuclide source in field conditions was further developed in a technical proposal of a computer controlled source working on the principle known from digital-to-analog converters (DAC). The source consists of seven individual pneumatically activated sources in a common shielding which activities are graded in a geometrical series with quotient 2. Their combinations enable to set $2^7 = 128$ different intensities of the resulting radiation field and so the irradiation of the object of interest can be adjusted in less than 1% steps. A prototype of this irradiator has been constructed under EC support as its JRC 2005 Innovation Project which is now operated and tested by the IRMM. First results were already presented [6] and showed that this device was fast, safe and flexible in operation. It enables to perform various independent pre-programmed tests which may simulate various static and dynamic conditions and can be reliable reproduced many times.

3. Results and Discussion

Up to now, only limited results are available from regular testing of radiation portals which number in Slovakia is rather low. Yet, the experience with those few performed since 2002 is instructive at least. Six monitors operated by a steel company were tested in 2 year intervals using ^{137}Cs sources of various constructions.

As an example, results of one monitor testing with the adjustable source based on the pipette filled with 5.66 MBq/ml solution of ^{137}Cs is displayed in the Figure 1.

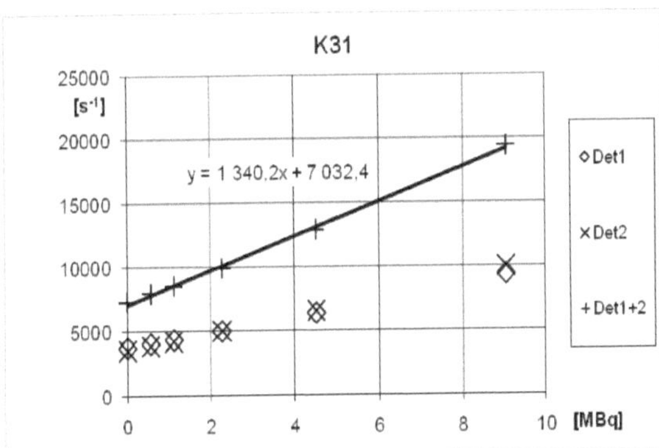

Figure 1. A two-side portal monitor calibration curve

The slope of the curve C_n = 1,340 $s^{-1}.MBq^{-1}$ represents the instrument sensitivity. The other parameter 7,032 s^{-1} represents the background reading increased in the presence of not perfectly shielded testing source. A real background value was about 4,600 s^{-1} which is equivalent to about 3.4 MBq of ^{137}Cs in testing position. The alarm threshold was set to 1.1 multiple of background reading which represented 10% of the actual value so the MDA was estimated as 0.34 MBq of ^{137}Cs in the testing point. It is essentially less than 0.6 MBq recommended in the standard [2] and this instrument would pass the test. However, it must be stressed that this value depends on the preset time of measurement and for 1 s intervals it equals 6.8× standard deviation. There are recommended values based on statistics which enable to estimate expected number of false alarms and also the probability that a radionuclide source with MDA would not be detected.

4. Application

4.1. EXTENSION FROM DETECTION TO MEASUREMENT

Presented method represents a simple means for testing and performance evaluation of illicit trafficking radioactivity detectors, mainly installed radiation portal monitors. Proposed parameters of sensitivity and minimum detectable activity are generally acceptable, well defined and can be easily verified. Also the conditions in which they are obtained are convenient and partly simulate real activity measurements. However, in real conditions the situation is much more complicated and one needs to account for additional influences: the difference in geometry, shielding effects to both background and source radiations, temperature and weather conditions, electromagnetic noise etc. These influences can be described by a set of corrective factors namely

$$A = \frac{n - n_0}{C_n} \cdot C_1 \cdot C_2 \cdot C_3 \; .... \qquad (6)$$

To find and determine correct (most probable) values of these factors is an extremely difficult task. There are several ways how to accomplish it: by a qualified estimation, experimental modeling [3], Monte Carlo simulation, a statistical evaluation of data from real alarm situations etc. It is evident that the individual corrective factors may achieve values not only considerably different from 1 but the uncertainty of these values may be extremely large and incomparable with other fields of measurement. This is probably the reason that the radiation monitors are not considered as measuring instruments: up to now there is no recommendation how to express and deal with extremely large uncertainties. Yet, similar situations occur in various sciences [7] and this justifies using logarithmic-normal distribution of results around the mean proposed for the activity estimation under poorly defined conditions of measurement. The literature [7] contains also valuable reasoning in favor of this approach even in cases when normal distribution is commonly used: a multiplication of several independent effects leads to a non-symmetric distribution of results which the log-normal distribution law fits the best.

The general problem with the determination of uncertainty in measurement is even more enhanced in fields connected with radiations and radioactivity. Lately an international standard [8] was issued dealing with the uncertainty in radiation protection instrumentation based on a generalized model of activity measurements (Equation 6). Unfortunately, some persistence in traditional way of thinking resulted in using the conventional distribution models only and the conventional uncertainty notation as well. In contrast, the paper [9] presents an attempt how to deal with extremely large uncertainties occurring in special cases like radioactivity detectors. It may come handy to bring some notes in favor of the log-normal distribution which seems to be the right solution.

Because of the common preference of normal or Gaussian distribution usually described as $\mu \pm \sigma$ (or $\bar{x} \pm s$ for the quantity estimations) some common characteristics with log-normal distribution has been searched for enabling to switch between these two. Using logarithmic transformation for individual observations of a positive variable x with assumed log-normal distribution we obtain

$$\bar{x}^{\otimes} = \exp\left(\frac{1}{n}\sum_{i=1}^{n}\log(x_i)\right) = \sqrt[n]{\prod_{i=1}^{n} x_i} \quad \text{and}$$

$$s^{\otimes} = \exp\left[\sqrt{\frac{1}{n-1}\sum_{i=1}^{n}\left(\log\frac{x_i}{\bar{x}}\right)^2}\,\right] \tag{7}$$

If the same data would be processed for normal distribution the estimation $\bar{x} \pm s$ of its parameters would be obtained and the mutual relationship between these two sets is

$$\bar{x}^{\otimes} = \frac{\bar{x}}{\sqrt{1+\left(\frac{s}{\bar{x}}\right)^2}} \quad \text{and} \quad s^{\otimes} = \exp\left(\sqrt{\log\left(1+\left(\frac{s}{\bar{x}}\right)^2\right)}\right) \tag{8}$$

When the coefficient of variations (or the relative uncertainty) $s/\bar{x}$ is small then $\bar{x}^{\otimes} \doteq \bar{x}$ and $s^{\otimes} \doteq 1 + s/\bar{x}$. This case corresponds to the usual situation of a symmetric normal distribution with relative small variations and the formulae enable to transform them into an equivalent log-normal distribution described in short as $\bar{x}^{\otimes} \times/\, s^{\otimes}$ where the symbol $\times/$ denotes an operation of multiplication or division analogical to the $\pm$ symbol. This also explains why both distributions fits data with small variations equally well. Normal or Gaussian distribution can be replaced by log-normal distribution and used without any dramatic consequences. Although the back transformation is possible too it makes less sense as the log-normal distribution is appropriate non-symetric in a linear scale especially for larger variations and therefore should be better noted as $\bar{x}^{\otimes}{}^{+s_u}_{-s_l}$ where s_u and s_l are above and bottom parameters replacing the common symmetric standard deviation, respectively. This operation seems therefore unpractical and the $\bar{x}^{\otimes} \times/ s^{\otimes}$ notation offers a simple representation of data distribution with $s^{\otimes} \geq 1$.

Concerning the considered model of measurements there are no obvious difficulties using it provided the log-normal distributions of concerned quantities and parameters. Resulting mean can be calculated directly from the formula (4) providing the mean (most probable) values are introduced. If all concerned uncertainties are expressed by means of their equivalent log-normal standard deviation estimates $s_i^{\otimes}$ they need to be combined unto resulting uncertainty by means of the formula

$$s^{\otimes} = \exp\sqrt{\sum_i \left(\log s_i^{\otimes}\right)^2} \tag{9}$$

which is only a little bit more complicated than that commonly used for small uncertainty propagations.

For illustration, let us consider a multiple of four correction coefficients with equal nominal values 1 and large (but not extremely large) uncertainties of ±50%. A combination of these coefficients and their uncertainties by common way yields the result 1 ± 100% which is evidently nonsense. However, converting them into equivalent parameters of a log-normal distribution provides $\bar{x}_i^{\otimes} = 0{,}894$ and $\bar{s}_i^{\otimes} = 1.604$ or 0.894 ×/ 1.604 in short which covers approximately the same range as the original. Combining four of them yields the result 0.64 ×/ 2.572 which is much more realistic for 68% probability or 0.64 ×/ 6.62 for 95% probability of result distributions (coverage coefficient 2).

4.2. FURTHER DEVELOPMENT

Correction factors need not necessary be treated like passive numbers representing most probable mean values with large uncertainties when an increased count rate has been registered. If more physical parameters are measured as influencing parameters these factors can be related to them either functionally or statistically and their uncertainty narrowed. A good starting point is created by introducing the spectroscopic facility in the system which enables to identify the radionuclide(s) and so bring calibration and correction coefficients closer to reality. Another possibility is weighing the vehicles utilizing the correlation between the shielding mass and its effect. Also a different response of two or more detectors can be utilized for a rough geometry assessment.

Finally, low cost electronics enables to use more sophisticated functions including vehicle self-identifications and testing, continuous self-monitoring, route recording etc. Although any system can be fooled some way, more advanced measurements may bring a progress in the radioactivity investigation.

5. Conclusion

The outlined procedure may help to deal with the radiation portal monitors in a similar way like with other, more precise instruments. They can be calibrated and tested as usual and their results can be treated in the familiar way however, accounting for unusually large uncertainties of their results. Further improvements can be expected.

References

1. Detection of radioactive materials at borders (2002) IAEA-TECDOC-1312, IAEA, Vienna
2. Radiation protection instrumentation – Installed radiation monitors for the detection of radioactive and special nuclear materials at national borders. Standard IEC 62244:2006
3. Dryák P, Šuráň J, Kovář P (1997) Type tests of systems for monitoring of radioactive sources in freights. Metrologie 3:2–9 (in Czech)
4. Burr T, Gattiker J R, Myers K, Tompkins G (2007) Alarm criteria in radiation portal monitoring. Appl Radiat Isotopes 65:569–580
5. Technical and functional specifications for border monitoring equipment (2006): Technical Guidance. Reference Manual IAEA, Vienna
6. Paepen J, Švec A, Camps J, Van Ammel R, Pommé S, Wätjen U (2007) Prototype of a radiation source for calibration of installed radiation monitors. IRPA Regional Congress Brasov, Romania
7. Limpert E, Stahel WA, Abbt M (2001) Log-normal distributions across the sciences: keys and clues. BioScience 51:341–352
8. IEC/TR 62461 (2006) Determination of uncertainty in measurements radiation protection instrumentation, 2006-12
9. Švec A (2004) Measuring instrument assessments and evaluation of measurement results with extremely large uncertainties. Metrol Test 2:4–11 (in Slovak)

Other relevant literature (not cited):

ANSI N42.35-2006 American National Standard Evaluation and Performance of Radiation Detection Portal Monitors used for Homeland Security.

Pibida L, Unterweger M, Karam L (2006) Development of gamma-ray emitting sources for portal monitors. Appl Radiat Isotopes 64:1271–1272

Krnáč Š, Povinec P (1996) Semiconductor gamma-ray spectrometry with whole spectrum processing. J Radioanal Nucl Chem 204:57–64

DETERMINATION OF LEAD-210 AND POLONIUM-210 IN MARINE ENVIRONMENT

AYSUN UGUR* AND GUNGOR YENER
Ege University, Institute of Nuclear Sciences, 35100, Bornova, İzmir, Turkey

Abstract. Naturally occurring ^{210}Pb and ^{210}Po are important tracers used in marine environment studies. In this study, ^{210}Po and ^{210}Pb measurements in marine samples (sediment, sea water and biota) were performed using alpha spectrometry. Chemical recoveries are obtained using ^{209}Po tracer. Two different approaches for ^{210}Pb determinations are also compared. First, to keep the original samples for about 1 year to eliminate the excess ^{210}Po and to reach ^{210}Po–^{210}Pb equilibrium before radiochemical separation of ^{210}Po. The second is to extract the excess ^{210}Po in the samples, then wait for attaining ^{210}Pb–^{210}Po equilibrium in the solution to obtain ^{210}Pb activities. In samples, time corrections for excess ^{210}Po originated from radioactive decay of grand parent are discussed.

Keywords: Sediment, sea water, mussel, radioactive equilibrium, ^{210}Pb, ^{210}Po

1. Introduction

Lead-210 (^{210}Pb) and its daughter ^{210}Po are naturally occurring radionuclides often used as oceanographic tracers. The measurement of low level ^{210}Pb is a useful tool to study sedimentary dynamics, geochronology, accumulation rates and sediment transport. ^{210}Po is a high energy alpha particle emitter in the uranium decay chain and is considered as an important source of internal radiation dose to marine organisms. In the most part of marine environment ^{210}Pb and ^{210}Po is mainly produced from the radioactive decay of ^{226}Ra dissolved in seawater and from the atmospheric deposition of ^{222}Rn daughters. Moreover, the concentration can be locally enhanced by the discharge of radium rich phosphogypsum

* To whom correspondence should be addressed. e-mail: aysun.ugur@ege.edu.tr

which is a by-product of some industrial processes, uranium and lead ore processing industries and burning of fossil fuels, [1–5].

In coastal waters, $^{210}Po/^{210}Pb$ activity ratio is mostly smaller than unity. In biological samples, spoiling the equilibrium, it is higher due to preferential uptake of ^{210}Po by marine organisms, especially algae. Masque et al. [6] remark that common values of the $^{210}Po/^{210}Pb$ ratio are 3 for phytoplankton and 12 for zooplankton. Boisson et al. point out that the ratio is generally close to 1 in sediments, approximately 2 in fecal pellets, 7 in phytoplankton and 30 in zooplankton detrial material. The behavior of ^{210}Po in the ocean differs from that of ^{210}Pb, especially because of the higher affinity of ^{210}Po for organic matter [7].

If the radioactive equilibrium exists between ^{210}Pb and its decay product, the activity of any member of the decay chain can be determined by the 5.3 MeV alpha-particles of ^{210}Po. The present study was aimed to discuss radioactive equilibrium and time corrections for ^{210}Po and ^{210}Pb determinations in organic and inorganic samples in marine environment.

2. Experimental

Fifteen mussel, sea water and sediment samples from different stations at Aegean Sea Coastal region were studied. For mussel (*Mytilus galloprovincialis*) and shell (*Chlomys sp.*) samples, in order to minimize the size effect, specimens in standardized groups by size were selected. Immediately after collection, the shells were cleaned with a nylon brush. The soft tissues and shells were submitted and rinsed carefully with abundant distilled water in order to eliminate the impurities. The soft parts, including interstitial fluid, were extracted from each sample. Then the composite samples were weighed and oven dried at 80°C. After oven dried to the constant weight, they were grounded, passed through a 2 mm mesh, homogenized and mixed thoroughly.

Sediment samples from different locations were recovered using a Van-Veen grab (5 L) near the shore (10–40 m) of the sampling stations. They were oven-dried to the constant weight and were sieved before analysis.

Coastal sea water samples (20 L) were collected using Nansen sampling bottles (2 L). Two liters of sub samples of sea water were stored in a pre-cleaned polyethylene container and acidified to pH 2.

The accurate and precise determinations of radionuclide concentrations in marine samples are essential in marine radioactivity assessments and in the use of radionuclides in studies of oceanographic processes. Regarding, our laboratory has participated in the intercomparison exercise IAEA-437 for Mediterranean Mussel of IAEA. The results are given in Table 1. This intercomparison exercise was organized to give the participating laboratories the possibility of testing the performance of their analytical methods on marine biota samples with low radionuclide levels such as mussel. The intercomparison samples were collected from the Mediterranean Sea in the framework of the CIESM Mediterranean Mussel Watch Program.

2.1. DETERMINATION OF ^{210}PB AND ^{210}PO

The classical approach to decay series studies was formulated by Bateman [8]. ^{210}Pb decays into a daughter radioactive nuclide ^{210}Po,

$$^{210}Pb\left(t_{1/2} \cong 22y\right) \longrightarrow {}^{210}Bi\left(t_{1/2} = 5.02d\right) \longrightarrow {}^{210}Po\left(t_{1/2} = 138.3d\right)$$

where λ_1 and λ_2 are the decay constants of ^{210}Pb and ^{210}Po, respectively. Since the decreasing is proportional to the amount of radioelement present, the rate of change in number of daughter atoms is given by the difference between its production rate from its parent and its decreasing rate to the daughter.

$$\frac{dN_2}{dt} = N_1\lambda_1 - N_2\lambda_2 \tag{1}$$

where N_1 and N_2 represent the number of atoms of each species present at any time, t. The solution of the linear differential equation gives ^{210}Po as

$$N_2 = N_{10}\left[\lambda_1/(\lambda_2 - \lambda_1)\right]\left[e^{-\lambda_1 t} - e^{-\lambda_2 t}\right] + N_{20}e^{-\lambda_2 t} \tag{2}$$

reduces to

$$N_2 = \frac{\lambda_1}{\lambda_2} N_{10}\left(1 - e^{-\lambda_2 t}\right) \tag{3}$$

Because $\lambda_2 - \lambda_1 \cong \lambda_2$ and $e^{-\lambda_1 t} \cong 1$.

If the mean life of daughter atom is much shorter than that of parent atom then the daughter atom is said to be in "permanent or secular equilibrium" with its parent [A1]. Thus, the activities are accepted to be equal, namely

$$N_1\lambda_1 = N_2\lambda_2 \tag{4}$$

at any time.

2.2. RADIOCHEMICAL PROCEDURE

After addition of polonium tracer as standard, each sample was completely dissolved in HCl and HNO_3. For sediment samples HF is also used in dissolving process. Polonium was spontaneously plated onto a silver disks in 0.5 N HCl in the presence of ascorbic acid to reduce Fe^{+3} to Fe^{+2} [9]. Recovery was found to vary between 70% and 90% for mussel and between 70% and 81% for sediment samples. ^{210}Po were measured through its 5.30 MeV alpha particle emission, using ^{209}Po (4.88 MeV alpha emission, $t_{1/2}$ = 109 years) as the internal tracer.

The unfiltered water was spiked with ^{209}Po tracer. The radionuclides were collected from sea water samples by co-precipitation with manganese dioxide then the precipitate was dissolved in a small volume of HCl/H_2O_2. ^{210}Po was plated onto copper and counted immediately.

2.3. ALPHA-RAY SPECTROMETRY

Concentrations of alpha activities of polonium were measured by PIPS detector, Ortec 450 mm^2 with 20 μm of depletion depth. The efficiency is 38% and 44% using ^{233}U, ^{244}Cm sources, respectively. Recovery corrections of ^{210}Po activity concentrations were performed by comparing the measured activity of the ^{209}Po yield tracer and calculated activity decayed by sampling time. Concentrations of ^{210}Po in samples were observed to be much higher than the detection limits (0.0003 Bq). Counting period was adjusted to obtain relative standard error of approximately 5%. Final activity calculations were attained to include the appropriate corrections for blanks and also for collection date.

Following the 5 h (mussel) and 6 h (sediment and sea water) of initial plating, copper disks were suspended in the plating solutions which were stirred overnight to remove remaining traces of ^{210}Po and ^{209}Po. The solution was then re-spiked with a known activity of ^{209}Po and kept for about 1 year to allow the ingrowth of ^{210}Po from ^{210}Pb and attain radioactive equilibrium. The sample was re-plated and the ^{210}Po activity concentration was determined. So, the second deposition provided the information of the ^{210}Pb content of the samples and hence on the extent to which the initial ^{210}Po was supported by its grand parent.

In the samples waited more than 3 months, before applying any chemical procedure, the activity measured in the first deposition comes from excess ^{210}Po and also the ingrowth ^{210}Po from ^{210}Pb exist in the sample during this period.

$$A_{total}\left({}^{210}Po\right) = A\left({}^{210}Po\left[{}^{210}Pb\right]\right) + A\left({}^{210}Po\right) \tag{5}$$

To get excess ^{210}Po we need ^{210}Pb content in the samples. The ^{210}Po activities obtained after the second deposition is equal to the ^{210}Pb activity in the sample. Activity concentration of ^{210}Po (A) is derived from Equation (5) and corrected for time using Equation (6).

$$A = A_0 e^{-\lambda t} \tag{6}$$

3. Results and Discussion

3.1. ANALYTICAL QUALITY CONTROL

The certified reference material of IAEA-437, has been analyzed to check the sensitivity of alpha spectrometry. The results are given in Table 1. It is seen that ^{210}Po values obtained in the samples are in good agreement with the recommended values. This fact legitimates that the ^{210}Po determination in our laboratory gives the reliable results.

TABLE 1. ^{210}Po concentrations (Bq kg^{-1} dry weight) in IAEA-437 reference materials

Sample code	^{210}Po Recommended value (confidence interval α = 0.05)	W_{Po}[1] (g)	Chemical recovery (%)	^{210}Po experimental value (Bq kg^{-1} total error of two σ[2])
IAEA-437 Mediterranean Mussel	7.5 (4.9–9.0)	1.0004	69	9.38 ± 0.66
		1.0005	59	8.02 ± 0.59
		1.0007	69	9.38 ± 0.66
		1.0011	62	9.33 ± 0.65
		1.0010	69	6.50 ± 0.47

[1]Sample weight for ^{210}Po analysis
[2]All possible sources of error (counting, blank, calibration)

3.2. EXPERIMENTAL DATA

3.2.1. *Sea water*

Marine samples collected in the Aegean Sea in 2004 have been analyzed. Figure 1 shows the activity concentrations of ^{210}Po and ^{210}Pb (mBq L^{-1}) in sea water samples were found to vary between 2 ± 1 and 12 ± 3 mBq L^{-1} and 3 ± 1–12 ± 3 mBq L^{-1} for ^{210}Po and ^{210}Pb, respectively. The difference between the high activity concentrations in surface water in present work and other studies in the literature might be due to the fact that samplings were realized near the shore [6].

The mean $^{210}Po/^{210}Pb$ ratio was estimated to be 0.85 (range = 0.50–1.40). In sea water main sources of ^{210}Po and ^{210}Pb are atmospheric deposition and ^{226}Ra in sediment. Sea is a very dynamic media as far as ^{210}Po and ^{210}Pb are concerned, because there are several factors that continuously change their concentrations, such as scavenging, feeding of biological materials. So, as Masqué indicates, although the half-life of ^{210}Po is much less than that of ^{210}Pb, disequilibrium between both radionuclides is often observed, leading to a typical total $^{210}Po/^{210}Pb$ ratio of 0.5 [6]. If sea water samples are not analyzed immediately after collection, a time correction will not be much meaningful, because one will not be able to take all the factors mentioned above into consideration in the calculations.

3.2.2. *Organic samples*

Figure 2 shows the concentrations of ^{210}Po and ^{210}Pb (Bq kg^{-1}) in organic samples. Activity concentrations in organic samples were found to vary between 242 ± 22 and 1,072 ± 35 Bq kg^{-1} dry weight and 12 ± 3–46 ± 5 Bq kg^{-1} for ^{210}Po and ^{210}Pb, respectively. The $^{210}Po/^{210}Pb$ ratio is calculated as 6.10–51.33. Thus, ^{210}Po data found for organic samples is little affected by the decay of ^{210}Pb.

In organic samples ^{210}Po concentrations are much higher than ^{210}Pb concentrations due to their feeding habits. In general, organic samples contain low ^{210}Pb activity therefore if the polonium measurement is realized at time t after sample collection, the ingrowth of this radionuclide from ^{210}Pb is not remarkable. So, it would be enough to wait about 6 months after the first deposition for the second deposition.

3.2.3. *Inorganic samples*

Activity concentrations of ^{210}Po and ^{210}Pb (Bq kg^{-1}) in inorganic samples are given in Figure 3. The activity concentrations of ^{210}Po and ^{210}Pb in inorganic samples were found to vary between 17 ± 4–83 ± 5 Bq kg^{-1} and 15 ± 4–86 ± 9 Bq kg^{-1}, respectively. The $^{210}Po/^{210}Pb$ ratio is calculated to be 0.62–1.56. The main source of ^{210}Po in inorganic sample is the ^{210}Pb content. As shown in Figure 3, $^{210}Po/^{210}Pb$ ratio is close to unity in inorganic samples. So, the ingrowth of ^{210}Po from ^{210}Pb contributes fairly important amount to total ^{210}Po in the samples. In this case to obtain the real ^{210}Pb content in the sample, namely the equilibrium state between

^{210}Po and ^{210}Pb, it is necessary to wait at least 12 months before second deposition to attain the radioactive equilibrium. As it is seen in Figure 4, only 54% of equilibrium attained in 6 months whereas 85% is attained in 12 months, theoretically.

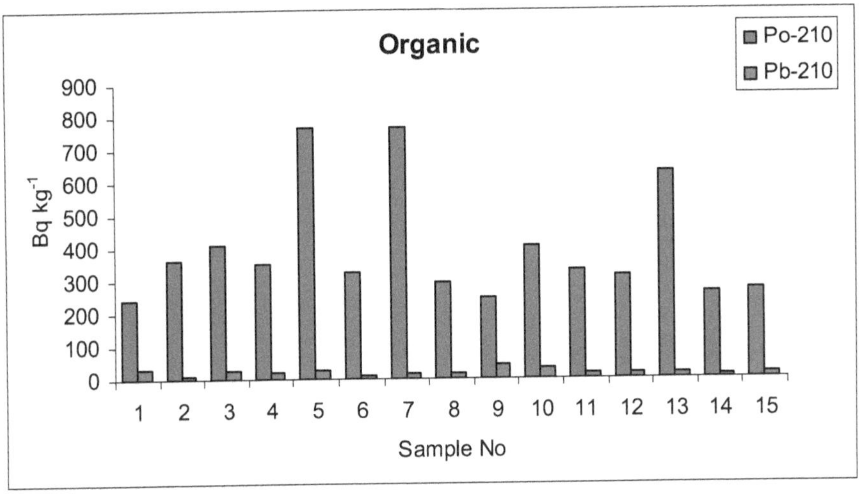

Figure 1. Activity concentrations of ^{210}Po and ^{210}Pb in unfiltered sea water

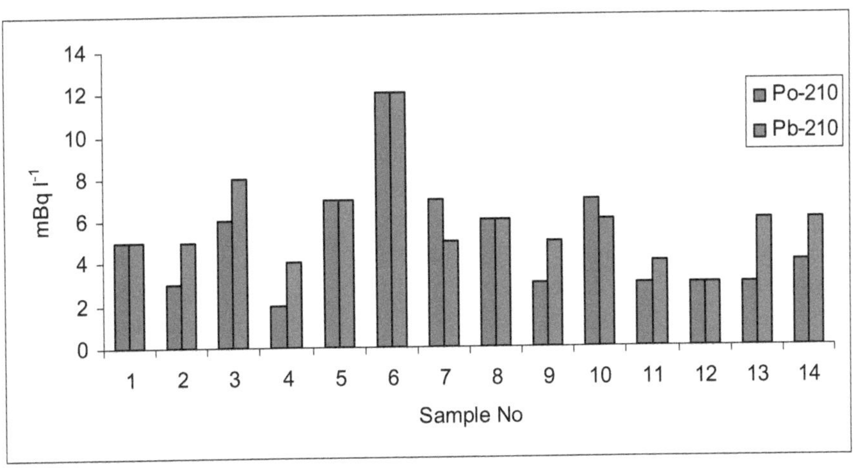

Figure 2. Activity concentrations of ^{210}Po and ^{210}Pb (Bq kg^{-1}) in organic samples

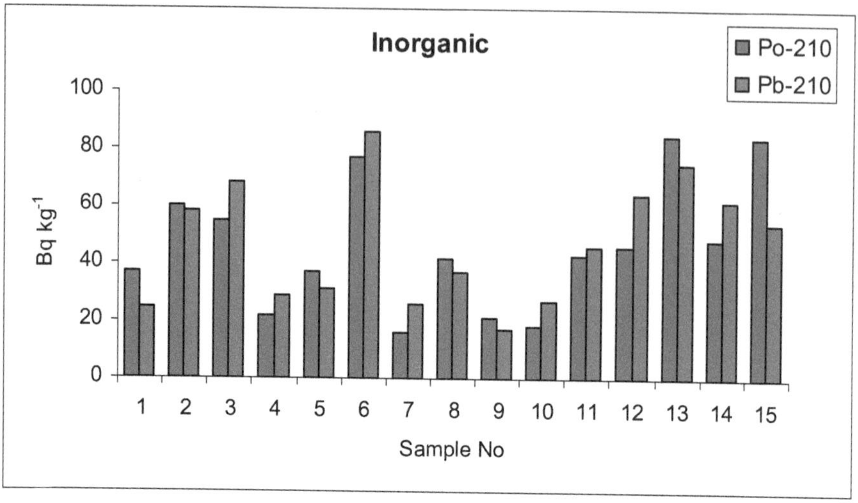

Figure 3. Activity concentrations of ^{210}Po and ^{210}Pb (Bq kg^{-1}) in inorganic samples

If it is interested in ^{210}Po activities at sampling dates such as pollution studies, biologic cycling and so on, it is preferred to measure ^{210}Po immediately after the sample collection. But due to several reasons it wouldn't be possible to do so. In that case it is necessary to get ingrowths ^{210}Po from ^{210}Pb within the period between collection and deposition dates. ^{210}Po measurements in the solution used for ^{210}Pb determinations should ideally be realized after waiting about 1 year.

But in case it is needed to get experimental results without waiting that long time, the second deposition could be done in a shorter period and then the real ^{210}Pb could be obtained by extrapolation as it is seen in the Figure 4. So, in the present study a series of experiments is planned to examine ^{210}Pb activity without waiting for 1 year period. Figure 4 also shows the ingrowth ^{210}Po data taken 1–6 months after the first deposition. The experiments will go on till the equilibrium state is attained. Then the extrapolated ^{210}Pb data will be compared with the measured value to see if this technique could be used properly when needed.

Figure 4 represents the theoretical and experimental activity trend of ^{210}Po in inorganic sample. It is clearly seen that the experimental data obtained so far is in good agreement with the theoretical plot. So, although the 12 month data is not available yet, it could be said that extrapolation made using earlier data to get equilibrium ^{210}Pb value, would not bring any serious mistake Table 2.

In case one is interested only in ^{210}Pb contents, the dry samples might preferably be kept for about 1 year to make excess ^{210}Po disappear. So, the ^{210}Po activities measured in the samples would be equal to ^{210}Pb activity.

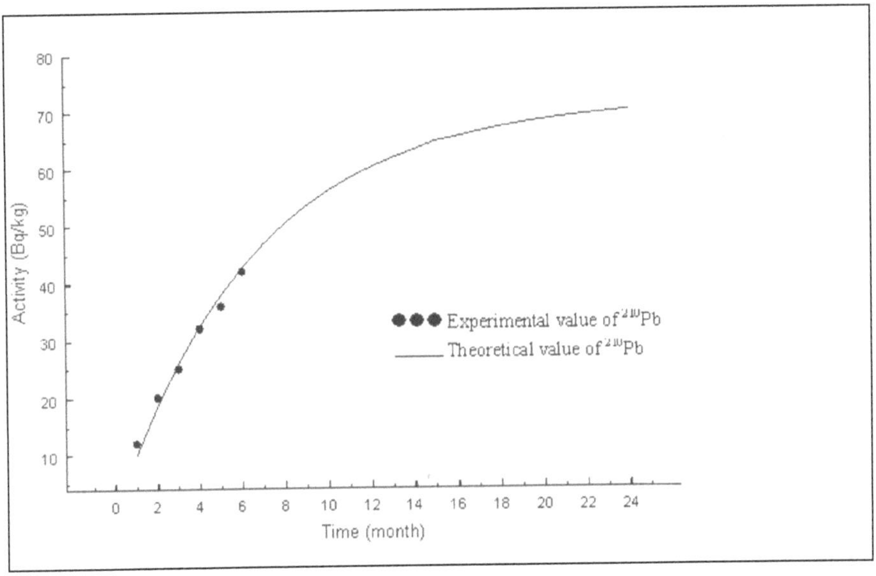

Figure 4. Trend of experimental activity of ^{210}Po in inorganic samples

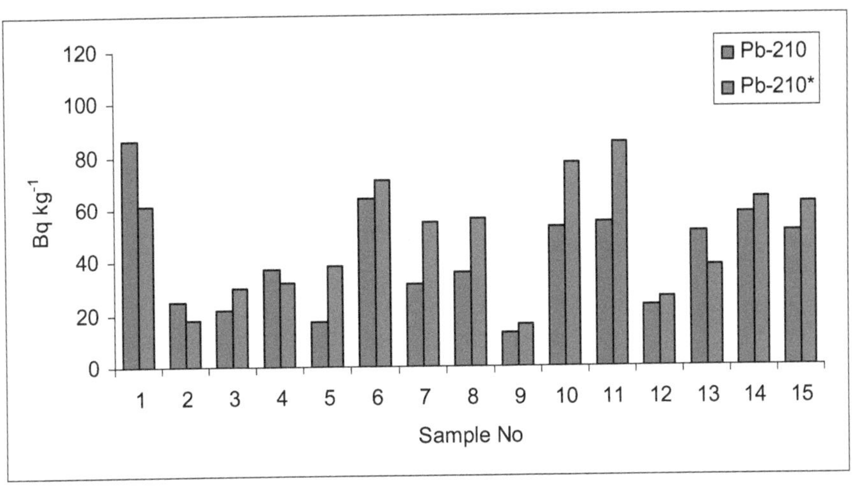

^{210}Pb*, refers to results obtained from the second deposition.

Figure 5. Concentrations of ^{210}Pb (Bq kg^{-1}) in inorganic samples which prepared by two different method

TABLE 2. The activity ratio of $^{210}Po/^{210}Pb$ in organic, inorganic and sea water sample

Sample No.	$^{210}Po/^{210}Pb$		
	Organic sample	Inorganic sample	Sea water sample
1	7.56	1.48	1.00
2	30.25	1.03	0.60
3	17.21	0.81	0.75
4	18.47	0.76	0.50
5	29.46	1.19	1.00
6	29.45	0.90	1.00
7	51.33	0.62	1.40
8	19.60	1.14	1.00
9	6.10	1.24	0.60
10	12.33	0.67	1.17
11	18.33	0.93	0.75
12	17.50	0.72	1.00
13	37.00	1.13	0.50
14	20.23	0.77	0.67
15	19.50	1.56	0.86

Figure 5 shows the comparison of two different approaches for ^{210}Pb determinations. The first is to keep the original samples for about a year to eliminate excess ^{210}Po and to reach ^{210}Po–^{210}Pb equilibrium before radiochemical separation of ^{210}Po. The second is to extract excess ^{210}Po in the samples, then wait for attaining ^{210}Pb–^{210}Po equilibrium in the solution to obtain ^{210}Pb activities. Concentrations in original samples which kept for about 1 year were found to vary between 16 ± 4 and 85 ± 9 Bq kg^{-1}. The activity concentrations of ^{210}Pb in the other samples that performed rural second depositions were found to vary between 17 ± 4–86 ± 9 Bq kg^{-1}.

References

1. Arya AP (2001) Fundamentals of Nuclear Physics, Allyn & Bacon, Boston, MA
2. Carvalho FP et al. (1993) An experimental study on the bioaccumulation and turnover of ^{210}Po and ^{210}Pb in Marine Shrimp. Mar Ecol Prog Ser 102:125–133
3. Carvalho FP (1997) Distribution, cycling and mean residence time of ^{226}Ra, and in the Tagus Estuary. Sci Total Environ 196:151–161
4. Ryan TP et al. (1999) ^{210}Po in mytilus edulis in the Irish Marine Environment. J Environ Radioact 43:325–342
5. Ugur A et al. (2002) Trace metals and ^{210}Po (^{210}Pb) concentrations in mussels (mytilus galloprovincialis) consumed at Western Anatolia. Appl Radiat Isotopes 57:565–571
6. Masque P et al. (2002) Balance and residence times of ^{210}Pb and ^{210}Po in surface waters of the Northwestern Mediterranean Sea. Cont Shelf Res 22:2127–2146
7. Boisson F et al. (2001) ^{210}Po and ^{210}Pb cycling in a hydrothermal vent zone in the coastal Aegean Sea. Sci Total Environ 281:111–119
8. Bateman H (1910) Solution of a system of differential equations occurring in the theory of radioactive transformations. Proc Cambridge Philos Soc 15:423–427
9. Flynn WW (1968) The determination of low levels of ^{210}Po in environmental materials. Anal Chim Acta 43:221–227

SUBJECT INDEX

FSC
www.fsc.org
MIX
Papier | Fördert gute Waldnutzung
FSC® C083411

Kolibri 360 GmbH
Ferdinand-Jühlke-Straße 7
99095 Erfurt, Deutschland
produktsicherheit@kolibri360.de